Robustness in Statistical Pattern Recognition

Mathematics and Its Applications

Volume 380

Robustness in Statistical Pattern Recognition

by

Yurij Kharin

Department of Mathematical Modelling and Data Analysis,
School of Applied Mathematics and Informatics,
Belarussian State University,
Minsk, Republic of Belarus

KLUWER ACADEMIC PUBLISHERS

DORDRECHT / BOSTON / LONDON

A C.I.P. Catalogue record for this book is available from the Library of Congress

ISBN 978-90-481-4760-1

Published by Kluwer Academic Publishers,
P.O. Box 17, 3300 AA Dordrecht, The Netherlands.

Kluwer Academic Publishers incorporates
the publishing programmes of
D. Reidel, Martinus Nijhoff, Dr W. Junk and MTP Press.

Sold and distributed in the U.S.A. and Canada
by Kluwer Academic Publishers,
101 Philip Drive, Norwell, MA 02061, U.S.A.

In all other countries, sold and distributed
by Kluwer Academic Publishers Group,
P.O. Box 322, 3300 AH Dordrecht, The Netherlands.

This is an expanded and updated translation from the
original Russian work of the same title.
Minsk, Universitetskoj, 1992 © Yu. S. Kharin

Printed on acid-free paper

To my Mother

and

to the Memory of my Father (1921-1968)

Contents

Preface

This book is concerned with important problems of robust (stable) statistical pattern recognition when hypothetical model assumptions about experimental data are violated (disturbed).

Pattern recognition theory is the field of applied mathematics in which principles and methods are constructed for classification and identification of objects, phenomena, processes, situations, and signals, i.e., of objects that can be specified by a finite set of features, or properties characterizing the objects (Mathematical Encyclopedia (1984)).

Two stages in development of the mathematical theory of pattern recognition may be observed. At the first stage, until the middle of the 1970s, pattern recognition theory was replenished mainly from adjacent mathematical disciplines: mathematical statistics, functional analysis, discrete mathematics, and information theory. This development stage is characterized by successful solution of pattern recognition problems of different physical nature, but of the simplest form in the sense of used mathematical models.

One of the main approaches to solve pattern recognition problems is the statistical approach, which uses stochastic models of feature variables. Under the statistical approach, the first stage of pattern recognition theory development is characterized by the assumption that the probability data model is known exactly or it is estimated from a representative sample of large size with negligible estimation errors (Das Gupta, 1973, 1977), (Rey, 1978), (Vasiljev, 1983)). Another characteristic of the first stage is the assumed simplest Gaussian model of data (Das Gupta, 1973, 1977). Let us list the main problems which have been solved at the first development stage of statistical pattern recognition theory:

- the problem of synthesis of an optimal decision rule that minimizes the risk functional (expected losses of decision making) (Sebestyen, 1962), (Fukunaga, 1972), (Zypkin, 1970);
- the problem of optimal decision rule search in a given restricted family of decision rules (e.g., in the family of linear or piecewise linear decision rules) (Sebestyen, 1962), (Duda and Hart, 1973), (Vapnik, 1974), (Fomin, 1976), (Gorelick and Skripkin, 1984);
- the problem of optimal decision rule construction when rejections of decisions are allowable (Anderson, 1958), (Patric, 1972);

- the problem of construction of convenient lower and upper bounds for error probability (Chernoff, 1962), (Fukunaga, 1972), (Kailath, 1967);

- the problem of accounting for feature measurement cost and sequential decision rule construction (Fu, 1971);

- the problem of decision rule construction using the maximum likelihood criterion for recognition of some simplest random processes (Fukunaga, 1972).

By the middle of the 1970s it became evident that the methods from adjacent mathematical disciplines and their simplest modifications are insufficient to solve many applied problems efficiently (Lachenbruch, 1975), (Chen, 1978), (Gorelick and Skripkin, 1984). The gap between the theory and practical applications began to grow. This was the beginning of the second development stage of the statistical pattern recognition theory. Demands for new research directions, based on new adequate models of data and corresponding mathematical techniques, have emerged (Das Gupta, 1977), (Verhagen, 1980). One of such directions, developed in this book, is Robust Statistical Pattern Recognition.

In the statistical pattern recognition theory and its applications optimal decision rules are often used. They minimize the risk functional only for definite loss functions under an exactly fixed hypothetical (classical) model of data. Traditionally, such model is based on certain assumptions, such as sample homogeneity for each class, absence of missing feature values, exact parametric form of conditional probability densities for observations (often the densities are assumed to be multivariate normal ones) and independence of observations.

These assumptions are often violated for real observed data (Launer and Wilkinson, 1979), (Smoljak and Titarenko, 1980), (Hampel *et al.*, 1986), (McLachlan, 1992): there are outliers, missing values, nonhomogeneity and dependence of sample observations, noncoincidence of probability densities with prescribed hypothetical densities, etc. For example, it was detected (Orlov, 1991) by testing of 2,500 samples in real statistical data archives that the hypothesis about normal (Gaussian) distribution is invalid for 92% of samples. Model distortions result in non-optimality of classical decision rules and in noncontrollable increase of classification risk. That is why the following problems investigated in this book are especially urgent in practice:

- robustness (stability) evaluation for classical decision rules for situations with distortions in data and in the hypothetical model;

- estimation of critical values of distortions for a given level of robustness factor;

- construction of robust (low sensitive to definite types of distortions) decision rules.

Investigation of these problems is stimulated by development of the new direction of scientific research, namely, robustness analysis in statistics, which is influenced by the investigations of J.Tukey (1960), P.Huber (1981), F.Hampel (1986), R.Berger

(1979), J.Kadane (1978), M.Tiku (1986), Ya.Zypkin (1984), H.Rieder (1994) and other researchers.

At present, there are no monographs and textbooks devoted to robust statistical pattern recognition. This monograph is an attempt to make up the deficiency. The book is primarily intended for mathematicians, statisticians and engineers in applied mathematics, computer science and cybernetics. It can also serve as a basis for a one-semester course for advanced undergraduate and graduate students training in these fields.

Let us briefly present the scope of the book.

CHAPTER 1 contains descriptions of main probability models of observed data encountered in pattern recognition problems. We define optimal (Bayesian) decision rules minimizing the risk of classification; these rules are defined in discrete and continuous spaces of feature variables; the formulae for risk computation are given.

In CHAPTER 2 we investigate the adequacy of classical models for observed data. We give mathematical descriptions for distortions of classical hypothetical models. We define basic notions for decision rule robustness analysis: guaranteed (upper) risk; robustness factor; critical distortion level; robust decision rule, etc.

CHAPTER 3 is devoted to risk stability analysis for common in practice plug-in rules, which are derived from the Bayesian decision rule by substitution of parametric probability distribution estimates for the unknown true multivariate distributions. These parametric probability distribution estimates are constructed using the family of minimum contrast estimators (including ML-estimators, LS-estimators, etc.). For regular families of probability distributions we for the first time give general asymptotic expansions and approximations of risk and robustness factors. High accuracy of these approximations is illustrated by computer experiments.

CHAPTER 4 is devoted to the same problems as in CHAPTER 3, but for the situations where no parametric model of probability distributions is known and nonparametric decision rules (Rosenblatt - Parzen, k-Nearest Neighbors) are used for classification. We find optimal values for smoothness parameters that optimize the robustness factor.

In CHAPTER 5 we investigate new problems of robust pattern recognition for common in practice types of multivariate data distortions: Tukey–Huber type contaminations; additive distortions of observations (including round-off errors); distortions defined by mixtures of probability distributions; distortions defined in L_2-metric, χ^2-metric, variation metric; and random distortions of distributions. For the first time we find (by the method of asymptotic expansions) estimates of the robustness factor and of critical distortion levels (breakdown points) for classical decision rules. We construct robust decision rules that minimize the guaranteed (upper) risk.

In CHAPTER 6 we investigate pattern recognition problems in which hypothetical assumptions about training samples are violated in various ways: there are contaminations of "i-th class training sample" by observations from alien classes, there are outliers in samples, or elements of the training sample are stochastically dependent. We give estimates for the robustness factor and investigate its dependence on sample sizes, distortion levels and other factors. We construct new decision rules with higher robustness order and illustrate their stability by computer results.

CHAPTER 7 is devoted to new urgent problems of robustness evaluation for traditional cluster analysis procedures and to construction of robust cluster-procedures in situations with distorted models. These problems are solved for the following sources of distortion: small size of samples; outliers; Markov dependence of random true class indices; runs (series) in random sequences of class indices. The constructed cluster-procedures can be used for recognition of disorder in stochastic dynamical systems.

This book is featured by bringing theoretical results to simple formulae for evaluation of robustness characteristics and to practical algorithms for robust decision rule construction, with multiple illustrations by Monte-Carlo modeling and real applied experimental data processing. Also, the book gives an opportunity for western researchers to become acquainted with statistical pattern recognition investigations carried out by researchers from the New Independent States.

The author thanks the USA Council for International Exchange of Scholars for Fullbright Research Grant in 1993, Belarussian Scientific Foundation for Mathematical Support Grant F40-263, and Belarussian Research Institute of Oncology and Medical Radiology for real experimental data, which were used for illustration of theoretical results. Special thanks to Professor William S. Mallios from California State University, Fresno, for reading the manuscript and editing some sections and to Professor Sergei Aivazyan from the Moscow State University for reviewing the manuscript in Russian language. The author thanks all colleagues from the Department of Mathematical Modeling & Data Analysis and all colleagues from the Laboratory of Statistical Analysis and Modeling of the Belarussian State University for many fruitful discussions and help. Great thanks to S. Agievich, R. Fursa, A. Kostevich, and E. Zhuk for their help in TEX-setting of the manuscript. The author is greatly indebted to Dr. N. Korneenko for editing and processing the English version of the manuscript.

Finally, the author is profoundly grateful to Kluwer Academic Publishers, especially Ms. Angelique Hempel, and Dr. Paul Roos, for their kind and helpful cooperation.

Yu. Kharin
Minsk, 1996

Chapter 1

Probability Models of Data and Optimal Decision Rules

This chapter describes main probability models of observed data in pattern recognition: random variables, random vectors, random processes, random fields, and random sets. Optimal (Bayesian) decision rules minimizing the classification risk are specified. These decision rules are defined in discrete and continuous spaces of feature variables. The computational formulae for risk are given.

1.1 Probability Models of Observation Data

Pattern recognition consists in finding optimal (in some sense) decision rules for classification of an observed object (phenomenon, situation) to one of L fixed classes (patterns, types). In practice, physical nature of the objects to be classified, as well as the experiments on observing them, are usually featured by uncertainty, variability, lack of determinism, and stochasticity. Adequate mathematical means for handling pattern recognition problems are provided by probability theory and mathematical statistics.

Let Ω denote the set (or space) of objects to be classified, i.e., an element $\omega \in \Omega$ is an *elementary object* subject to classification. Let us define the σ-algebra $\mathcal{F}$ of the subsets from Ω and a probability measure $\mathbf{P} = \mathbf{P}(A), A \in \mathcal{F}$ on the measurable space $(\Omega, \mathcal{F})$. The resulting probability space $(\Omega, \mathcal{F}, \mathbf{P})$ is a general mathematical model of a random experiment on appearance and registration of an object to be classified.

A *class (pattern)* is a set of elementary objects possessing fixed common attributes. Let L denote the number of classes; any pattern recognition problem makes sense only for $L \geq 2$. The classes $\Omega_1, \ldots, \Omega_L \subset \Omega$ are subsets of Ω and they constitute a $\mathcal{F}$-measurable partition of Ω :

$$\Omega = \bigcup_{i=1}^{L} \Omega_i, \ \Omega_i \in \mathcal{F}, \ \Omega_i \bigcap \Omega_j = \emptyset \ (i \neq j).$$

Let us construct the discrete random variable

$$\nu = \nu(\omega) = \sum_{l=1}^{L} l\, \mathbf{I}_{\Omega_l}(\omega),\ \omega \in \Omega, \tag{1.1}$$

where

$$\mathbf{I}_{\Omega_l}(\omega) = \begin{cases} 1, \omega \in \Omega_l, \\ 0, \omega \notin \Omega_l \end{cases}$$

is the *indicator function* for the class Ω_l. As follows from (1.1), $\nu \in S = \{1, 2, \ldots, L\}$ is the ("true") class number of the class to which a randomly observed object ω belongs. The random variable ν is characterized by the discrete probability distribution:

$$\pi_l = \mathbf{P}\{\nu(\omega) = l\},\ l \in S,\ \sum_{l \in S} \pi_l = 1, \tag{1.2}$$

where the value π_l is called the *prior probability* of the class Ω_l.

An observation (registration) of an object $\omega \in \Omega$ to be classified is carried out by means of measuring its attributes. The *result* of this observation process is described by a random element $x = x(\omega)$. The *observation space* X is the set of all possible observation results. A random element is defined by the $\mathcal{F}$-measurable mapping $\Omega \to X$:

$$x = x(\omega), \omega \in \Omega,\ x \in X. \tag{1.3}$$

Depending on the type of a scale used in the measurements of attributes (1.3), observation results can be described by numerical (quantitative) or nonnumerical data. Three types of measuring scales are distinguished in the theory of measurements (Pfanzagl, 1976): nominal scale, ordinal (qualitative) scale and quantitative scale.

In the *nominal scale* the scale values (gradations) are the names of equivalence classes for the object attribute. For example, in medical diagnostics such patient's attributes as sex, surname, name, patronymic name, place of birth, and nationality are evaluated by nominal scales.

The *ordinal scale* differs from the nominal scale in that a relation of linear order is introduced, and this allows us to estimate qualitatively (i.e., "better—worse") the degree of manifestation of the attribute for an object. For example, such variables as "patient's education level", "patient's profession", "stage of hypertonic disease" are evaluated by ordinal scales.

In general case, the data evaluated by nominal and ordinal scales are nonnumerical. Numerical data are derived from *quantitative scales.* Here the results of observations are evaluated by real numbers. Thus, in (1.3), X is derived from the real axis $R^1 = (-\infty, +\infty)$. For example, patient's temperature, height, and blood pressure are biomedical variables evaluated by quantitative scale.

We shall investigate mainly situations where the attributes (1.3) are measured by quantitative scale. Note that vast amount of literature concerns with pattern recognition in spaces of variables of mixed scale types (see, e.g., (Lbov, 1981)). In addition, special methods, algorithms, and computer software for transformation of data to quantitative scale (so-called *quantification methods* for nominal and ordinal variables) are developed.

Depending on the complexity of observation space X, the following types of data models (1.3) are known: random variable, random vector, random process (time series), random field, and random set (Andrews *et al.*, 1985). Let us consider them briefly.

Observation as Random Variable

In this case, $X \subseteq R^1$, where X is the observation space, and the observation result (1.3) is described by a random variable $x = x(\omega) \in R^1$ with distribution function

$$F_x(z) = \mathbf{P}\{x < z\}, z \in R^1.$$

If the observation space is discrete:

$$X = \{u_0, u_1, \ldots, u_M\}, 1 \leq M \leq \infty,$$

and different values $\{u_i\}$ of the attribute x are observed with elementary probabilities

$$p_i = \mathbf{P}\{x(\omega) = u_i\} > 0, i = 0, \ldots, M, \sum_{i=0}^{M} p_i = 1, \tag{1.4}$$

then the random variable x has discrete probability distribution. Otherwise (singular distribution model is not considered here) the random variable $x = x(\omega)$ has absolutely continuous probability distribution with probability density function

$$p_x(z) = \frac{dF_x(z)}{dz}, z \in X \subseteq R^1. \tag{1.5}$$

The complete definition of the mathematical model of observation requires the specification of the discrete distribution (1.4) or the probability density (1.5). At present, at least six main discrete models (1.4) and 13 main continuous models (1.5) are used in practice (Aivazyan *et al.*, 1983), (Kendall *et al.*, 1958), (Kharin *et al.*, 1987)) of statistical pattern recognition. Probability characteristics for these main data models are given in parametric form. As an example of a discrete data model, let us take the binomial probability distribution with parameter $\theta \in [0, 1]$:

$$x \in X = \{0, 1, \ldots, M\}, p_i = \mathbf{P}\{x = i\} = C_M^i p^i (1-p)^{M-i}, i \in X.$$

Plots of the binomial probability distribution for some values of θ and M are shown at Figure 1.1.

As an example of a continuous data model, we may consider the normal (Gaussian) probability distribution $N_1(\theta_1, \theta_2)$ with two parameters $\theta_1 \in R^1, \theta_2 > 0$:

$$p_x(z) = n_1(z \mid \theta_1, \theta_2) = \frac{1}{\sqrt{2\pi\theta_2}} e^{\frac{-(z-\theta_1)^2}{2\theta_2}}, \ z \in X = R^1. \quad (1.6)$$

Plots of normal probability density functions for some values of θ_1, θ_2 are shown at Figure 1.2. The parameter θ_1 specifies the mean (or mathematical expectation), and θ_2 specifies the variance (or dispersion):

$$\theta_1 = \mathbf{E}\{x\}, \ \theta_2 = \mathbf{D}\{x\} = \mathbf{E}\{(x - \theta_1)^2\}.$$

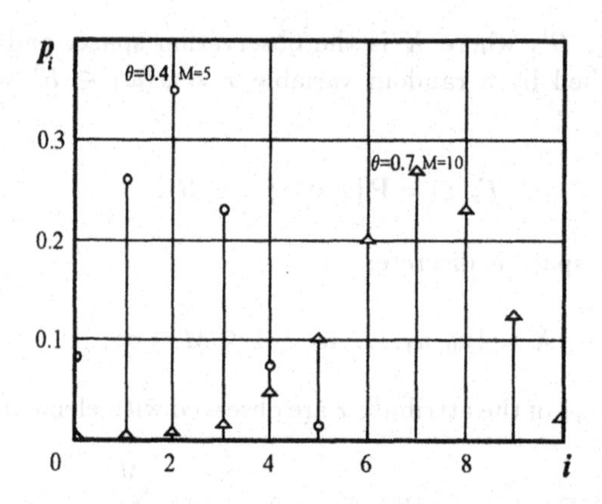

Figure 1.1: Binomial probability distributions

Observation as Random Vector

In this case, the observation space X is in the N-dimensional Euclidean space, $X \subseteq R^N$, i.e., an experiment records a set of N attributes specified by N random variables ($N \geq 1$)

$$x_1 = x_1(\omega), \ldots, x_N = x_N(\omega)$$

constituting a random vector:

$$x = \begin{pmatrix} x_1 \\ x_2 \\ \vdots \\ x_N \end{pmatrix} = \begin{pmatrix} x_1(\omega) \\ x_2(\omega) \\ \vdots \\ x_N(\omega) \end{pmatrix}.$$

Like in the previous scalar case, both discrete and continuous multivariate data models are used (Gnanadesikan, 1977). Among discrete multivariate data models, the most investigated one is the model described by the polynomial probability distribution (Kendall *et al.*, 1958) with the parameter vector

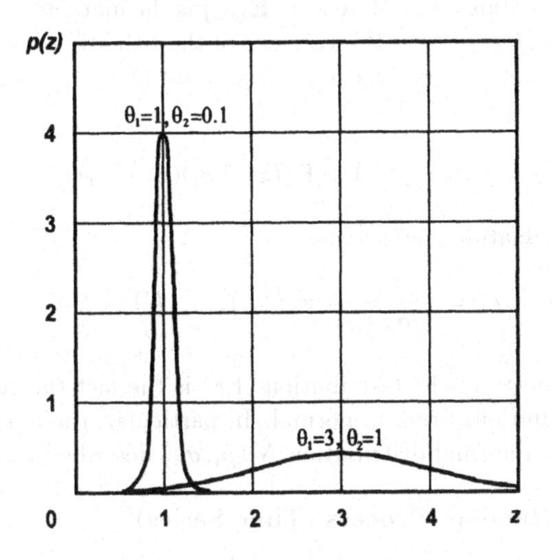

Figure 1.2: Gaussian probability densities

$$\theta = (\theta_1, \ldots, \theta_N), 0 < \theta_i < 1 \; (i = 1, \ldots, N, \sum_{i=1}^{N} \theta_i = 1):$$

$$\mathbf{P}\{x_1 = z_1, \ldots, x_N = z_N\} = \frac{M!}{z_1! \ldots z_N!} \theta_1^{z_1} \cdot \ldots \cdot \theta_N^{z_N}, \tag{1.7}$$

$$z_i \in \{0, 1, \ldots, M\} \; (i = 1, \ldots, N), \; \sum_{i=1}^{N} z_i = M.$$

This model generalizes the binomial model considered above: each component x_i has binomial distribution with parameters θ_i and $M, i = 1, \ldots, N$. The polynomial model is useful when the measured attributes (1.3) of the observed object are the frequencies of some events (outcomes).

Among continuous multivariate data models in pattern recognition, the most popular one is the model described by the N-variate normal (Gaussian) probability distribution $N_N(\mu, \Sigma)$ with mathematical expectation vector $\mu = (\mu_i) \in R^N$ and nonsingular covariance matrix $\Sigma = (\sigma_{ij}) \; (i, j = 1, \ldots, N)$ (Anderson, 1958), (Tong, 1990):

$$p_x(z) = n_N(z|\mu, \Sigma) = (2\pi)^{-\frac{1}{2}N} |\Sigma|^{-\frac{1}{2}} \exp(-\frac{1}{2}(z - \mu)^T \Sigma^{-1} (z - \mu)), \tag{1.8}$$

$$z \in R^N,$$

where T denotes matrix transpose. Here $\mu_i = \mathbf{E}\{x_i\}$ is the mathematical expectation and $\sigma_{ii} = \mathbf{D}\{x_i\} = \mathbf{E}\{(x_i - \mu_i)^2\}$ is the variance of the i-th attribute ($i = 1, \ldots, N$). Non-diagonal elements of the covariance matrix Σ are the cross-covariances of different attributes:

$$\sigma_{ij} = \mathbf{Cov}\{x_i, x_j\} = \mathbf{E}\{(x_i - \mu_i)(x_j - \mu_j)\}$$

and determine the correlation coefficients:

$$\rho_{ij} = \frac{\sigma_{ij}}{\sqrt{\sigma_{ii}\sigma_{jj}}} (i \neq j = 1, \ldots, N).$$

An important property of the distribution (1.8) is the fact that all its marginal and conditional distributions are also normal. In particular, the i-th component x_i has univariate marginal normal distribution $N_1(\mu_i, \sigma_{ii})$ described by (1.6).

Observation as Random Process (Time Series)

In this case the observation space X is the functional space, i.e., measuring of the attributes (1.3) of an object ω consists in recording of a real function in variable $t \in [0, T]$, which is usually interpreted as time variable:

$$x = x(\omega) = \{\xi(t, \omega) : 0 \leq t \leq T\}, \omega \in \Omega, \xi \in R^1. \tag{1.9}$$

Observations of such type are encountered in recognition of voice signals, electrocardiograms, radar signals, seismograms (Vasiljev, 1983), (Verhagen, 1980), (Vintsjuk, 1968). The mathematical model of observations (1.9) is a random process defined over the probability space $(\Omega, \mathcal{F}, \mathbf{P})$ within time interval $[0, T]$.

On input into computer the data for observation (1.9) are discretized: the values of the process are recorded over a finite set of M time moments:

$$0 \leq t_1 < t_2 < \ldots < t_M = T,$$

rather than over the whole interval $[0, T]$. This results in a time-ordered sequence of samples of the random process:

$$x_1 = x_1(\omega) = \xi(t_1, \omega), \ldots, x_M = x_M(\omega) = \xi(t_M, \omega), \tag{1.10}$$

which is usually called *time series* (Anderson, 1971). If ω is fixed, this sequence is called a *realization of the time series.*

The following basic mathematical models for time series $\{x_\tau\}$ (Anderson, 1971) are useful in pattern recognition:

1. **Stationary Time Series** with mean $\mu = \mathbf{E}\{x_\tau\}$, covariance function

$$\sigma(i) = \mathbf{Cov}\{x_\tau, x_{\tau+i}\} = \mathbf{E}\{(x_\tau - \mu)(x_{\tau+i} - \mu)\}, \ i = 0, \pm 1, \pm 2, \ldots,$$

and spectral density

$$g(\lambda) = \frac{1}{2\pi} \sum_{i=-\infty}^{+\infty} \sigma(i)\cos(i\lambda), \ -\pi \le \lambda \le +\pi.$$

2. Trend Model of Time Series

$$x_\tau = f(\tau) + u_\tau,$$

where $f(\tau)$ is a deterministic function called *trend*; $\{u_\tau\}$ is a sequence of independent identically distributed random variables:

$$\mathbf{E}\{u_\tau\} = 0, \mathbf{D}\{u_\tau\} = \sigma^2.$$

3. Autoregressive / Moving-average Time Series of order (p, q) with parameters

$$\alpha = (\alpha_0, \dots, \alpha_q), \beta = (\beta_0, \dots, \beta_p).$$

It is a sequence $\{x_\tau\}$, satisfying the difference equation with random right side

$$\sum_{j=0}^{p} \beta_j x_{\tau-j} = \sum_{i=0}^{q} \alpha_i u_{\tau-i} (\tau = p+1, p+2, \dots)$$

under given initial conditions (Broemeling, 1987).

Note that these models may be generalized to observations described by multivariate (vector) random processes and by vector time series $\xi(t, \omega), x_\tau \in R^N (N \ge 1)$.

Observations as Random Field

As in the previous case, here X is a functional space, but in contrast to (1.9), the variable t is a vector:

$$t = (t_i) \in T \subseteq R^m, i = 1, \dots, m; m \ge 2.$$

Here the mathematical model of observation $x(\omega)$ is a random field defined on the probability space $(\Omega, \mathcal{F}, \mathbf{P})$ and the region $\mathcal{T}$.

The observation model of such type is adequate for recognition of two-dimensional images (Verhagen, 1980), (Dubes, 1989): $m = 2$, $\mathcal{T}$ is the screen plane onto which the image ω is cast. Like in case (1.10), the discretization of the random field must be done (different types of scanning are possible). The mathematical models defined for time series may be generalized to random fields, including vector random fields.

Observations as Random Set

A *random set* $x = x(\omega) \in X$ defined on the probability space $(\Omega, \mathcal{F}, \mathbf{P})$ arises as a generalization of the term "random variable" for situations where the observations

x are sets. This model is relatively new in the probability theory (Matheron, 1975), (Orlov, 1979) and is promising in pattern recognition for patterns of complex nature in such fields as metallography, biology, medicine, etc.

Note that any realization of time series (1.10) (or of random field after discretization) may formally be represented as an M-dimensional vector $x = (x_1, \ldots, x_M)^T$. That is why when discussing general results concerning the synthesis of pattern recognition algorithms we shall use this universal form of representation of an observation $\omega \in \Omega$, i.e., we shall represent it as a vector of sufficiently large dimension N:

$$x = x(\omega) = (x_1, \ldots, x_N)^T \in X \subseteq R^N.$$

The list of attributes x to be recorded is usually chosen bearing in mind some general ideas about the classified objects. Therefore the dimension N of the observation space X may occur to be too large and render the algorithms unfeasible because of limited computer capabilities. In such situations it is advisable to pass to the so-called *feature space* $Y = R^n$ of smaller dimension $n < N$ by means of a measurable mapping $X \to Y$:

$$y = (y_1, \ldots, y_n)^T = T(x) = (T_1(x), \ldots, T_n(x))^T, x \in X, y \in Y. \tag{1.11}$$

Here the vector y is called the *vector of feature variables,* or *feature vector* In the cases when mapping (1.11) is not advisable, the feature space becomes equal to the observation space: $Y = X$.

Finally let us note that in analysis and synthesis of pattern recognition systems, computer simulation is one of powerful techniques; computer algorithms for simulation of observations to be classified are described in the existing literature, e.g., in (Kharin *et al.*, 1987).

1.2 Elements of Statistical Decision Theory

Let us formalize the problem of pattern recognition in terms of classical statistical decision theory.

Given the feature space $X = R^N$, assume that objects from classes $\Omega_1, \ldots, \Omega_L$, $L \geq 2$, are observed with prior probabilities $\pi_1, \ldots, \pi_L$. An object to be classified is formally specified by a composed $N+1-$component random vector $(X^T \vdots \nu)^T$ defined on probability space $(\Omega, \mathcal{F}, \mathbf{P})$. Here $x = (x_k) = x(\omega)$ is the observed random vector of N features $x_1, \ldots, x_N$, and $\nu = \nu(\omega)$ is an unobserved discrete random variable, with given probability distribution

$$\mathbf{P}\{\nu = i\} = \pi_i, \quad i = 1, \ldots, L,$$

indicating the true number of the class to which an observed object $\omega \in \Omega_\nu$, $\nu \in S = \{1, 2, \ldots, L\}$ belongs. The conditional probability distribution of the random vector $x = x(\omega)$ subject to the condition $\nu(\omega) = i$ is given in the form of conditional distribution function for class Ω_i:

$$F_i(z) = \mathbf{P}\{x_1(\omega) < z_1, \dots, x_N(\omega) < z_N | \nu(\omega) = i\}, \ z = (z_k) \in R^N. \qquad (1.12)$$

The pattern recognition problem consists in constructing a decision rule to forecast the unknown class number ν given an observation x.

There are L possible decisions about the unknown value ν. We shall call the set $S = \{1, 2, \dots, L\}$ the *decision space.* A decision $d \in S$ means that the observed object with the vector of feature variables $x \in X$ is assigned to the class Ω_d, i.e., it is recognized to be an object from the class Ω_d. Then the pattern recognition problem amounts to finding an $\mathcal{F}$-measurable mapping $X \to S$:

$$d = d(x), x \in X, d \in S, \qquad (1.13)$$

called *nonrandomized decision rule.*

We shall consider a generalization of decision rule (1.13) called *randomized decision rule.* Under this rule, a decision $d \in S$ determined (by coin tossing) at each observation point $x \in X$ is a discrete random variable with the following probability distribution:

$$\mathbf{P}\{d = i\} = \chi_i(x), i = 1, \dots, L. \qquad (1.14)$$

The functions $\{\chi_i(\cdot)\}$ satisfy the following restrictions:

$$0 \le \chi_i(x) \le 1, \sum_{i=1}^{L} \chi_i(x) = 1, x \in X; \qquad (1.15)$$

they are called the *critical functions.* In particular, if $\chi_i(x) \in \{0, 1\}$, we have a nonrandomized decision rule (1.14) that can be presented in the following form:

$$d = d(x) = \sum_{i=1}^{L} i\mathbf{I}_{V_i}(x), \qquad (1.16)$$

where

$$V_i = \{x : x \in X, \chi_i(x) = 1\}$$

is the region of the observation space that corresponds to the decision $d = i$ made. It follows from (1.14)–(1.16) that

$$\bigcup_{i=1}^{L} V_i = X, V_i \bigcap V_j = \emptyset, i \ne j = 1, \dots, L.$$

Thus, construction of a decision rule amounts to finding the critical functions $\{\chi_i(\cdot)\}$, and for the case of nonrandomized decision rule this amounts to finding the partition $\{V_i\}$ of observation space X.

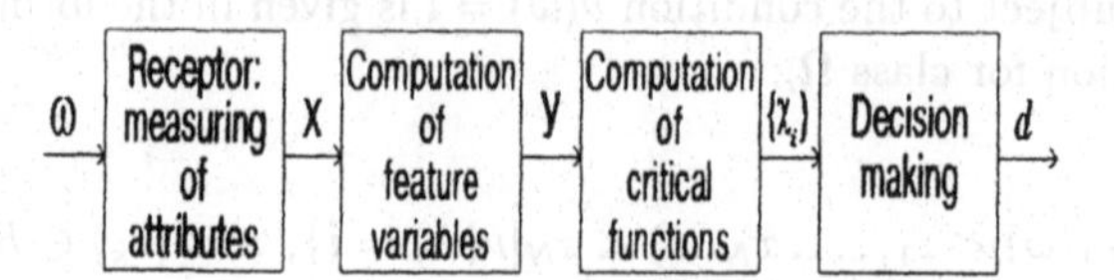

Figure 1.3: Flow-chart of a pattern recognition system

In the situation where mapping (1.11) is performed, the decision rule is constructed similarly to (1.14)–(1.16), but in this case we use feature space Y. A general flow-chart of the pattern recognition algorithm is shown at Figure 1.3.

Let us formulate now the problem of synthesis of a pattern recognition algorithm as an optimization problem. According to (1.14), (1.15), any collection of critical functions $\{\chi_i(x)\}$ defines some decision rule. One may construct infinitely many sets of critical functions for the same pattern recognition problem, and hence, there exists an infinite set of decision rules for the same problem. Not all of them are of equal value in practice. They may be compared by many characteristics, e.g., classification accuracy (percentage of wrong decisions), computational complexity of the algorithm. Usually it is convenient to quantify the degree to which the decision rule satisfies the demands. This may be done by a functional

$$r = r(\chi_1(\cdot), \dots, \chi_L(\cdot))$$

that gives a measure of optimality of the decision rule specified by critical functions $\{\chi_i(\cdot)\}$.

The value of r allows to compare any pair of decision rules $\{\chi_i(\cdot)\}$ and $\{\psi_i(\cdot)\}$. If $r(\{\chi_i\}) < r(\{\psi_i\})$, then we tend to believe that the rule $\{\chi_i(\cdot)\}$ is more effective than the rule $\{\psi_i(\cdot)\}$. The decision rule with critical functions $\{\chi_i^\circ(x)\}$ for which the functional r is minimal with value $r_o = r(\{\chi_i^\circ(\cdot)\})$, is called *optimal decision rule*, and the functions $\{\chi_i^\circ(\cdot)\}$ are called *optimal critical functions*. Hence the problem of construction of an optimal pattern recognition algorithm becomes the problem of minimization of functional r:

$$r(\{\chi_i(\cdot)\}) \to \min_{\{\chi_i(\cdot)\}}, \tag{1.17}$$

$$0 \le \chi_i(x) \le 1, \sum_{i=1}^{L} \chi_i(x) = 1, \; x \in X.$$

Methods for the solution of problem (1.17) substantially depend on the amount of prior information about classes $\Omega_1, \dots, \Omega_L$ and on the kind of the functional. When the prior information about classes is sufficiently complete, optimal decision rules are found by methods of statistical decision theory developed by A. Wald (1947). Unfortunately, in applied problems prior information is usually so scant that direct usage of results of statistical decision theory is impossible. But before we proceed to the description of this situation, let us present the results of the construction of optimal pattern recognition algorithms by means of classical statistical decision theory.

Assume that an $(L \times L)$ loss matrix $W = (w_{ij})$ is given, whose elements in general depend on the vector of features

$$w_{ij} = w_{ij}(x) \geq 0, x \in R^N, (i, j = 1, \ldots, L);$$

w_{ij} is the loss value for the recognition act when an observation x of an object from the class Ω_i is assigned to the class Ω_j (i.e., in fact $\nu = i$, but the decision $d = j$ was made). The diagonal element w_{ii} is the loss value at correct decision $(\nu = d = i)$, and the nondiagonal element $w_{ij} (i \neq j)$ is the loss value at wrong decision $(\nu = i \neq d = j)$; therefore $w_{ij} \geq w_{ii}$. In practice the so-called $(0-1)$*-loss matrix* W^o with elements

$$w_{ij}^o = 1 - \delta_{ij}, i, j = 1, \ldots, L, \tag{1.18}$$

is often used, where δ_{ij} is the Kronecker symbol. In this case, the loss for each correct decision w_{ii} is 0 and for each wrong one $w_{ij}, i \neq j$ the loss is 1.

The *loss function* for decision rule (1.14), (1.15) determined by critical functions $\{\chi_j(\cdot)\}$ is the function

$$w = w(i, z; \{\chi_j\}) = \mathbf{E}\{w_{id}(z)\} = \sum_{j=1}^{L} w_{ij}(z)\chi_j(z) \geq 0, i \in S, z \in X, \tag{1.19}$$

where $d \in S$ is a random variable with distribution (1.14). One can see the meaning of the loss function from (1.19); its value w specifies the expected loss in the situation where the observation $x = z$ corresponds to an object from the class $\Omega_i (\nu = i)$ and the decision rule (1.14), (1.15) with critical functions $\{\chi_j(\cdot)\}$ is used for classification.

The loss function (1.19) assumes random value, if a random observation vector x from randomly chosen class Ω_ν is being classified. The *classification risk* for the decision rule determined by critical functions $\{\chi_j\}$ is the functional $r = r(\{\chi_j\})$ defined as an expected value of a loss function:

$$r = r(\{\chi_j\}) = \mathbf{E}\{w(\nu, x; \{\chi_j\})\} \geq 0, \tag{1.20}$$

where $\mathbf{E}\{\cdot\}$ denotes mathematical expectation with respect to the distribution of ν, x. Substituting (1.19) into (1.20) and using joint probability distribution of ν, x (1.12), (1.2), we obtain an explicit expression for risk:

$$r = r(\{\chi_j\}) = \sum_{i,j=1}^{L} \pi_i \int_{R^N} \chi_j(z) w_{ij}(z) dF_i(z) \geq 0. \tag{1.21}$$

Let us consider the case of (0-1)-loss matrix (1.18). Substituting (1.18) into (1.21) and using (1.2), (1.15), we obtain

$$r = r(\{\chi_j\}) = \sum_{i=1}^{L} \pi_i r_i = \mathbf{P}\{d \neq \nu\}, \tag{1.22}$$

where

$$r_i = \mathbf{P}\{d \neq \nu | \nu = i\} \tag{1.23}$$

is the conditional error probability for the classification of class-Ω_i observations. Therefore, in the case of (0-1)-loss matrix the classification risk is the unconditional error probability.

The risk functional (1.21) is used as a measure of quality of decision rules. The less the classification risk, the better the decision rule. This leads to the problem of construction of an optimal pattern recognition algorithm, and the latter amounts to the risk minimization problem (1.17), (1.21). In the following sections we give the solutions of this problem for the cases of continuous and discrete observation models.

1.3 Optimal Decision Rules in Space of Continuous Feature Variables

Let us assume the continuous model of observations x in feature space $X = R^N$, i.e., the conditional distribution functions (1.12) are absolutely continuous, so that the conditional probability density of the random vector x of feature variables for class Ω_i (i.e., conditional on $\nu = i$) exists:

$$p_i(z) = \frac{\partial^N F_i(z)}{\partial z_1 \ldots \partial z_N}, z = (z_k) \in R^N, i \in S. \tag{1.24}$$

Let us denote:

$$f_k(x) = \sum_{l=1}^{L} \pi_l p_l(x) w_{lk}(x), x \in R^N, k \in S. \tag{1.25}$$

Assume that for arbitrary $K, L, 2 \leq K \leq L$, and arbitrary set of indices $1 \leq j_1 < j_2 < \ldots < j_K \leq L$ the manifold in R^N

$$\Gamma = \{x : f_{j_1}(x) = f_{j_2}(x) = \ldots = f_{j_K}(x)\}$$

is an $(N-K)$-dimensional hypersurface of zero Lebesgue measure.

Theorem 1.1 *If random observations to be classified are absolutely continuous, the loss matrix $W = (w_{ij})$, the prior probabilities $\{\pi_i\}$ and the conditional probability distribution densities $\{p_i(\cdot)\}$ for L classes are given, then up to a set of zero Lebesgue measure in the feature space $X = R^N$ the optimal under minimum risk criterion decision rule is unique, nonrandomized, and of the form*

$$d = d_0(x) = \arg\min_{k \in S} f_k(x), x \in R^N, d \in S. \tag{1.26}$$

Moreover, the minimal classification risk, attainable under decision rule (1.26), equals to

$$r_0 = \int_{R^N} \min_{k \in S} f_k(x) dx. \tag{1.27}$$

Proof. Taking (1.24), (1.25) into account, let us transform the risk functional (1.21) to the equivalent form

$$r(\{\chi_j\}) = \int_{R^N} \rho(x) dx, \tag{1.28}$$

where

$$\rho = \rho(x) = \sum_{j=1}^{L} \chi_j(x) f_j(x) \geq 0 \tag{1.29}$$

is a function whose meaning is *risk density*. The set of optimal critical functions is the solution of the optimization problem (1.17), (1.28). Since the restrictions in (1.17) are local (pointwise), i.e., they are imposed for each point $x \in R^N$, (1.28) implies

$$\min_{\{\chi_j\}} r(\{\chi_j\}) = \int_{R^N} \min_{\{\chi_j\}} \rho(x) dx. \tag{1.30}$$

The problem of minimization of risk density $\rho(x)$ at each point $x \in R^N$ as stated in (1.30), taking (1.17), (1.29) into account, is the simplest problem of linear programming:

$$\rho(x) = \sum_{j=1}^{L} \chi_j(x) f_j(x) \to \min_{\{\chi_j\}}, \tag{1.31}$$

$$\chi_j(x) \geq 0, \sum_{j=1}^{L} \chi_j(x) = 1.$$

For a fixed point $x \in R^N$ let us define the set of index values

$$J = \text{Arg} \min_{k \in S} f_k(x) = \{j : j \in S, f_j(x) = \min_{k \in S} f_k(x)\} = \{j_1, \ldots, j_Q\},$$

where $1 \leq Q \leq L$. The definition of J implies

$$f_{j_1}(x) = \ldots = f_{j_Q}(x) = \min_{k \in S} f_k(x) < f_l(x), l \notin J. \tag{1.32}$$

The case $Q > 1$ is possible only for the observations $x \in R^N$ that lie on the $(N - Q + 1)$-dimensional hypersurface determined by $Q - 1$ equalities from (1.32). The

Lebesgue measure of the set of points of this hypersurface in R^N is zero. Therefore, excluding such observations, we have $Q = 1$, i.e., $J = \{j_1\}$ is a single-point set. Hence the solution of (1.31) is unique:

$$\chi^*_{j_1}(x) = 1, \chi_k(x) = 0, k \neq j_1,$$

$$\min_{\{\chi_j\}} \rho(x) = f_{j_1}(x) = \min_{k \in S} f_k(x).$$

This, together with (1.16), (1.30), implies (1.26), (1.27).

■

The optimal decision rule (1.26) that minimizes the classification risk (actually, the Bayesian optimality principle amounts to this), is called the *Bayesian decision rule* (BDR). The minimal risk value r_o (1.27) characterizing the potential classification accuracy is called the *Bayesian classification risk.*

Let us clarify the probabilistic sense of the functions $\{f_k(x)\}$ that define the Bayesian decision rule. For this purpose let us denote the unconditional probability density of the random observation vector x by

$$p(x) = \sum_{i=1}^{L} \pi_i p_i(x).$$

By the Bayes formula, the posterior probability of the class Ω_l at observation $x = z$ equals to:

$$\mathbf{P}\{\nu = l | x = z\} = \pi_l p_l(z)/p(z), l \in S.$$

Then (1.25) implies

$$\frac{f_k(z)}{p(z)} = \sum_{l=1}^{L} w_{lk}(z)\mathbf{P}\{\nu = l | x = z\} = \mathbf{E}\{w_{\nu k}(x) | x = z\}.$$

Therefore, to within the factor $p(z)$ (negligible for the Bayesian decision rule (1.26)) $f_k(z)$ is the conditional mathematical expectation of the classification loss provided the condition that the observation $x = z$ occurred and it was assigned to the class Ω_k.

Let us define $L(L-1)/2$ the so-called *discriminant functions:*

$$f_{ij}(x) = f_i(x) - f_j(x), 1 \leq i < j \leq L. \tag{1.33}$$

Corollary 1.1 *To within a set of zero Lebesgue measure, the Bayesian decision rule may be presented as follows:*

$$d = d_0(x) = \sum_{i=1}^{L} i \prod_{\substack{k=1 \\ k \neq i}}^{L} \mathbf{1}(f_{ki}(x)) = \sum_{i=1}^{L} i \mathbf{I}_{V_i}(x), x \in R^N, \tag{1.34}$$

where

$$V_i = \{x : f_{1i}(x) \geq 0, \ldots, f_{i-1,i}(x) \geq 0, f_{i,i+1}(x) < 0, \ldots, f_{iL}(x) < 0\}$$

is the region of the decision $d = i$ *making.*

We shall pay special attention to the situation where there are two classes, i.e., $L = 2$, for the following reasons. First, this case frequently occurs in practice. Second, analytical results for $L = 2$ are visually intuitive and may be easily generalized to the case of many classes ($L > 2$).

Corollary 1.2 *If* $L = 2$, *then the Bayesian decision rule assumes the form*

$$d = d_0(x) = \mathbf{1}(f_{12}(x)) + 1, x \in R^N, \tag{1.35}$$

where

$$f_{12}(x) = \pi_2 p_2(x)(w_{21}(x) - w_{22}(x)) - \pi_1 p_1(x)(w_{12}(x) - w_{11}(x))$$

is the discriminant function.

The geometrical sense of such Bayesian decision rule is as follows. There exists a *discriminant hypersurface*

$$\Gamma_0 = \{x : x \in R^N, f_{12}(x) = 0\} \tag{1.36}$$

that partitions the feature space R^N into two subspaces. The subspace V_1^0 with $f_{12}(x) < 0$ is the subspace of decision $d = 1$ in favor of class Ω_1, and the subspace V_2^0 with $f_{12}(x) \geq 0$ is the subspace of decision $d = 2$ in favor of class Ω_2.

Let us consider some particular cases of the Bayesian decision rule (1.35) that are often used in practice.

1. Classical Bayesian Decision Rule

$$d = d_1(x) = \begin{cases} 1, \text{ if } l(x) \leq h_1 \\ 2, \text{ if } l(x) > h_1 \end{cases} = \mathbf{1}(l(x) - h_1) + 1, \ x \in R^N, \tag{1.37}$$

where

$$l(x) = \frac{p_2(x)}{p_1(x)} \geq 0$$

is the *likelihood ratio statistic* and

$$h_1 = \frac{\pi_1(w_{12} - w_{11})}{\pi_2(w_{21} - w_{22})} \geq 0$$

is the *decision rule threshold*. Here it is assumed that the loss matrix $W = (w_{ij})$ does not depend on observations $x \in R^N$:

$$w_{ij}(x) = w_{ij}(i, j = 1, 2). \tag{1.38}$$

Substituting (1.38) into (1.35), we obtain the expression for the discriminant function:

$$f_{12}(x) = c_2 p_2(x) - c_1 p_1(x), \tag{1.39}$$

$$c_1 = \pi_1(w_{12} - w_{11}) > 0, \ c_2 = \pi_2(w_{21} - w_{22}) > 0.$$

Then from (1.39) follows that decision rule (1.37) is a particular case of Bayesian decision rule (1.35):

$$d = d_1(x) = \mathbf{1}(f_{12}(x)) + 1, x \in R^N.$$

2. Siegert—Kotelnikov Decision Rule (Decision Rule of Ideal Observer)

$$d = d_2(x) = \mathbf{1}(l(x) - h_2) + 1, h_2 = \pi_1/\pi_2, x \in R^N. \tag{1.40}$$

Comparing with (1.37), we see that this decision rule is the special case of the Bayesian decision rule for the case of $(0 - 1)$-loss matrix W^0 (1.18). The decision rule (1.40) minimizes the unconditional error probability (1.22) $r = \pi_1 r_1 + \pi_2 r_2$.

3. Maximum Likelihood Decision Rule

$$d = d_3(x) = \mathbf{1}(l(x) - 1) + 1, x \in R^N. \tag{1.41}$$

This decision rule differs from the Bayesian decision rule (1.27) and the Siegert-Kotelnikov decision rule (1.40) only by the threshold: $h_3 = 1$. Therefore the decision rule (1.41) minimizes the risk when the prior probabilities and the loss matrix satisfy the equality:

$$\pi_1(w_{12} - w_{11}) = \pi_2(w_{21} - w_{22}).$$

In particular, this decision rule minimizes the classification error probability if classes Ω_1, Ω_2 are equiprobable: $\pi_1 = \pi_2 = 1/2$.

4. Maximum Posterior Probability Decision Rule

$$d = d_4(z) = \arg\max_{i \in S} \mathbf{P}\{\nu = i | x = z\}, z \in R^N, \tag{1.42}$$

$$\mathbf{P}\{\nu = i | x = z\} = \frac{\pi_i p_i(z)}{p(z)}, \ p(z) = \sum_{j=1}^{L} \pi_j p_j(z).$$

Let us rewrite (1.42) to an equivalent form:

$$d = d_4(z) = \begin{cases} 1, \text{ if } l(z)/h_2 \le 1 \\ 2, \text{ if } l(x)/h_2 > 1 \end{cases} = d_2(z), z \in R^N.$$

Thus, the maximum posterior probability decision rule coincides with the Siegert-Kotelnikov decision rule (1.40).

5. Minimax Decision Rule

In applied pattern recognition problems, situations with unknown prior probabilities π_1, π_2 are encountered. The risk functional (1.21) is undefined in such cases, because it depends on π_1, π_2. In these situations it is advisable to construct a decision rule that minimizes the maximum of the classification risk over π_1, π_2:

$$\max_{\pi_1,\pi_2} r(\{\chi_j\}) \to \min_{\{\chi_j\}}. \tag{1.43}$$

This decision rule is called *minimax decision rule.*

Under the conditions (1.38), (Wald, 1947) demonstrated that the minimax decision rule is the Bayesian decision rule with respect to the least favorable prior probabilities π_1^*, π_2^*, for which the Bayesian risk (1.27) is maximal ($r_0 \to \max_{\pi_1,\pi_2}$):

$$d = d_5(x) = \mathbf{1}(l(x) - h_5) + 1, h_5 = \frac{\pi_1^*(w_{12} - w_{11})}{\pi_2^*(w_{21} - w_{22})}. \tag{1.44}$$

Considering π_1, π_2 as variables ($\pi_2 = 1 - \pi_1$ by the normalization condition) and using the obvious identity

$$\min\{f_1(x), f_2(x)\} \equiv f_1(x) - (f_1(x) - f_2(x))\mathbf{1}(f_1(x) - f_2(x)), x \in R^N,$$

we are able to evaluate the Bayesian risk according to (1.27), (1.33), (1.35):

$$r_0 = r_0(\pi_1) = \pi_1 w_{11} + (1 - \pi_1)w_{21} - \int_{R^N} f_{12}(x)\mathbf{1}(f_{12}(x))dx,$$

$$f_{12}(x) = (1 - \pi_1)(w_{21} - w_{22})p_2(x) - \pi_1(w_{12} - w_{11})p_1(x).$$

Now let us find the least favorable prior probabilities π_1^*, $\pi_2^* = 1 - \pi_1^*$ from the condition

$$r_o(\pi_1) \to \max_{0 \le \pi_1 \le 1}.$$

The solution π_1^* of this extremum problem is defined from the equation

$$\frac{dr_0(\pi_1)}{d\pi_1} = w_{11} - w_{21} + \tag{1.45}$$

$$\int_{R^N} ((w_{21} - w_{22})p_2(x) + (w_{12} - w_{11})p_1(x))\mathbf{1}(f_{12}(x))dx +$$

$$\int_{R^N} ((w_{21} - w_{22})p_2(x) + (w_{12} - w_{11})p_1(x))f_{12}(x)\delta(f_{12}(x))dx = 0,$$

where $\delta(z) = d\mathbf{1}(z)/dz$ is the generalized Dirac delta-function (Gelfand *et al.*, 1959). By the well-known property $z\delta(z) \equiv 0$, the second integral in (1.45) equals to zero. Let us use the expressions of conditional error probabilities (1.23) for the Bayesian decision rule:

$$r_1 = r_1(\pi_1) = \mathbf{P}\{d \neq \nu | \nu = 1\} = \int_{R^N} p_1(x)\mathbf{1}(f_{12}(x))dx,$$

$$r_2 = r_2(\pi_1) = \mathbf{P}\{d \neq \nu | \nu = 2\} = 1 - \int_{R^N} p_2(x)\mathbf{1}(f_{12}(x))dx.$$

Then the equation (1.45) assumes the form

$$(1 - r_1)w_{11} + r_1 w_{12} = (1 - r_2)w_{22} + r_2 w_{21}. \tag{1.46}$$

It means that for the least favorable values of prior probabilities the conditional classification losses for Ω_1 and Ω_2 coincide. In particular, for the (0-1)-loss matrix (1.18) the equation (1.46) results in the equality of conditional error probabilities: $r_1 = r_2$. Note that (1.46) is a transcendental equation, and numerical methods are used for finding its solution π_1^*.

6. Neyman—Pearson Decision Rule

Sometimes, not only prior probabilities π_1, π_2 are unknown, as in the previous case, but the loss matrix W is unknown as well. For such problems the Neyman-Pearson principle is used instead of the Bayes principle. According to the Neyman-Pearson principle, the critical functions

$$\chi_1(x), \ \ \chi_2(x) = 1 - \chi_1(x)$$

must be chosen in such a way that the conditional classification error probability is a known fixed value ϵ for observations from class Ω_1, i.e.,

$$r_1 = \int_{R^N} (1 - \chi_1(x))p_1(x)dx = \epsilon, \ 0 < \epsilon < 1,$$

and it must be minimal for the class Ω_2:

$$r_2 = \int_{R^N} \chi_1(x)p_2(x)dx \rightarrow \min_{0 \leq \chi_1(x) \leq 1}.$$

Constructed in such a way, the Neyman-Pearson decision rule has the form

$$d = d_6(x) = \mathbf{1}(l(x) - h_6) + 1, x \in R^N, \tag{1.47}$$

where the threshold h_6 is the root of the equation

$$\int_{l(x) < h_6} p_1(x)dx = 1 - \epsilon.$$

Note in conclusion that all six considered types of decision rules used in practical pattern recognition systems are of common form

$$d = d_0(x) = \mathbf{1}(l(x) - h) + 1, x \in R^N$$

and differ by the threshold h only.

1.4 Optimal Classification of Multivariate Gaussian Observations

Until now in optimal decision rule construction we did not make any assumptions about the concrete form of multivariate probability density of the random observation vector. Now we shall consider the *Gaussian model* (1.8) of observations to be classified. This model is commonly used in practice. Let objects from $L \geq 2$ classes $\Omega_1, \ldots, \Omega_L$ be observed in space R^N with prior probabilities $\pi_1, \ldots, \pi_L$. An observed object is described by a Gaussian random vector $x \in R^N$. It means that the conditional probability distribution density of the random vector of feature variables x for an object from class Ω_i is the N-variate normal density:

$$p_i(x) = n_N(x|\mu_i, \Sigma_i), x \in R^N, \tag{1.48}$$

where the vector of mathematical expectation $\mu_i = (\mu_{ik}) \in R^N$ and $(N \times N)$-symmetric nonsingular covariance matrix $\Sigma_i = (\sigma_{ikl}), k, l = 1, \ldots, N$ are parameters of the density. Let us take the unconditional error probability $r = \mathbf{P}\{d \neq \nu\}$ to be the performance measure of classification. According to (1.22), this is the particular form of risk functional for the case of (0-1)-loss matrix (1.18). For the construction of optimal decision rule let us use Theorem 1.1. From (1.18), (1.25) we have

$$f_k(x) = p(x) - \pi_k p_k(x), \; p(x) = \sum_{l=1}^{L} \pi_l p_l(x),$$

therefore, by monotonicity of the logarithmic function,

$$\arg\min_{k \in S} f_k(x) = \arg\max_{k \in S}(\pi_k p_k(x)) = \arg\max_{k \in S} \ln(\pi_k p_k(x)).$$

From (1.26), (1.27), (1.48) it follows that

$$d = d_0(x) = \arg\min_{k \in S}((x - \mu_k)^T \Sigma_k^{-1}(x - \mu_k) + \ln|\Sigma_k| - 2\ln\pi_k), \tag{1.49}$$

$$r_0 = 1 - \int_{R^N} \max_{k \in S}(\pi_k n_N(x|\mu_k, \Sigma_k))dx. \tag{1.50}$$

The constructed Bayesian decision rule (1.49) can be presented in the form (1.34) with quadratic discriminant functions:

$$f_{ij}(x) = (x - \mu_i)^T \Sigma_i^{-1}(x - \mu_i) - (x - \mu_j)^T \Sigma_j^{-1}(x - \mu_j) + \tag{1.51}$$

$$+ \ln \frac{|\Sigma_i|\pi_j^2}{|\Sigma_j|\pi_i^2}, x \in R^N (1 \leq i < j \leq L),$$

and therefore it is called the *quadratic decision rule*.

If the assumption about the equality of covariance matrices for all L classes (so-called *Fisher model*)

$$\Sigma_1 = \Sigma_2 = \ldots = \Sigma_L = \Sigma \tag{1.52}$$

holds, then the Bayesian decision rule becomes linear:

$$d = d_0(x) = \arg\max_{k \in S}((\Sigma^{-1}\mu_k)^T x + \ln \pi_k - \mu_k^T \Sigma^{-1}\mu_k/2) \tag{1.53}$$

and it is characterized by linear discriminant functions

$$f_{ij}(x) = (\Sigma^{-1}(\mu_j - \mu_i))^T x + \frac{1}{2}(\mu_i^T \Sigma^{-1}\mu_i - \mu_j^T \Sigma^{-1}\mu_j) + \ln \frac{\pi_j}{\pi_i}. \tag{1.54}$$

The Bayesian decision rule (1.53) assumes a particularly simple form in the case of two classes ($L = 2$):

$$d = d_0(x) = \mathbf{1}(f(x)) + 1, x \in R^N, \tag{1.55}$$

where the discriminant function $f(x)$ is linear:

$$f(x) = b^T x + \beta, b = \Sigma^{-1}(\mu_2 - \mu_1), \tag{1.56}$$

$$\beta = \frac{1}{2}(\mu_1^T \Sigma^{-1}\mu_1 - \mu_2^T \Sigma^{-1}\mu_2) - \ln h.$$

The threshold h for the Bayesian decision rule is π_2/π_1, and it has other values for the decision rules considered in Section 1.3. The discriminant hypersurface Γ_0 (1.36) is the hyperplane in R^N:

$$\Gamma_0 = \{x : x \in R^N, f(x) = b^T x + \beta = 0\}. \tag{1.57}$$

To investigate the performance of two-class recognition by the decision rule (1.55), (1.56), let us find conditional error probabilities r_1, r_2 (1.23) and unconditional error probability r_0 (1.22). One can see from (1.48), (1.52) that the classes Ω_1, Ω_2 are distinguished by mathematical expectation vectors μ_1, $\mu_2 \in R^N$ only. A commonly used measure of their distinction is the value

$$\Delta = \sqrt{(\mu_2 - \mu_1)^T \Sigma^{-1}(\mu_2 - \mu_1)}, \tag{1.58}$$

which is called *interclass Mahalanobis distance.* With the increase of Δ the distinction between the classes Ω_1, Ω_2 increases.

Let us introduce the standard $N_1(0, 1)$ normal distribution function:

$$\Phi(z) = \frac{1}{\sqrt{2\pi}} \int_{-\infty}^{z} e^{-t^2/2} dt = 1 - \Phi(-z), z \in R^1. \tag{1.59}$$

Theorem 1.2 *Let the Gaussian model of observations (1.48), (1.52) hold for the classes Ω_1, Ω_2, and let Δ be the interclass Mahalanobis distance (1.58). Then for the linear decision rule (1.55), (1.56) the error probabilities are:*

$$r_i = \Phi(-\frac{\Delta}{2} + (-1)^i \frac{\ln h}{\Delta}), \ i = 1, 2; \tag{1.60}$$
$$r_o = \pi_1 \Phi(-\frac{\Delta}{2} - \frac{\ln h}{\Delta}) + \pi_2 \Phi(-\frac{\Delta}{2} + \frac{\ln h}{\Delta}).$$

Proof. According to (1.2), consider the discrete random variable $\nu \in S = \{1, 2\}$ with distribution $\{\pi_1, \pi_2\}$ indicating the true unobservable class number for a random observation $x \in R^N$. Then from (1.23), (1.55) we have:

$$r_1 = \mathbf{P}\{d_0(x) = 2 | \nu = 1\} = \mathbf{P}\{f(x) > 0 | \nu = 1\}.$$

According to (1.56), $f(x)$ is a linear combination of Gaussian random variables, and by the well-known theorem from multivariate statistical analysis (Anderson, 1958), (Tong, 1990), the conditional probability distribution of $f(x)$ under condition $\nu = i$ is also Gaussian:

$$\mathcal{L}\{f(x)|\nu = i\} = N_1(m_i, \zeta_i), m_i = b^T \mu_i + \beta, \zeta_i = b^T \Sigma b, i \in S. \tag{1.61}$$

Substituting the expression for b from (1.56) into (1.61) and using (1.58), we obtain the expression for the conditional variance

$$\zeta_i = (\mu_2 - \mu_1)^T \Sigma^{-1} (\mu_2 - \mu_1) = \Delta^2,$$

which does not depend on the class number $i \in S$. From (1.61) it follows that the random variable $(f(x) - m_i)/\Delta$ under condition $\nu = i$ has standard normal distribution $N_1(0, 1)$. This, together with (1.59), implies

$$r_1 \equiv \mathbf{P}\{\frac{f(x) - m_1}{\Delta} > -\frac{m_1}{\Delta} | \nu = 1\} = \Phi(\frac{m_1}{\Delta}). \tag{1.62}$$

Similarly,

$$r_2 = \Phi(-\frac{m_2}{\Delta}). \tag{1.63}$$

Now let us compute conditional mathematical expectations $\{m_i\}$ in (1.61) using (1.56), (1.58):

$$m_i = (\mu_2 - \mu_1)^T \Sigma^{-1} \mu_i + \frac{1}{2}(\mu_1^T \Sigma^{-1} \mu_1 - \mu_2^T \Sigma^{-1} \mu_2) - \ln h = \tag{1.64}$$

$$= (-1)^i \Delta^2 / 2 - \ln h, i \in S.$$

Substituting (1.64) into (1.62), (1.63) results in (1.60). Unconditional error probability r_0 in (1.60) can be found using the formula (1.22).

■

This theorem allows to express error probability for all six types of optimal decision rules mentioned in Section 1.3.

Corollary 1.3 *If the classes Ω_1, Ω_2 are equiprobable ($\pi_1 = \pi_2 = 1/2$), then the minimal error probability is*

$$r_0 = \Phi(-\frac{\Delta}{2}), \tag{1.65}$$

and it is attained by the linear decision rule (1.55), (1.56) at $h = 1$, which is equivalent to Bayesian decision rule (1.39), maximum likelihood decision rule (1.41), maximum posterior probability decision rule (1.42), minimax decision rule (1.44) and to Neyman - Pearson decision rule at $\epsilon = \Phi(-\Delta/2)$. Moreover, in this case the conditional error probabilities are equal: $r_1 = r_2 = r_0$.

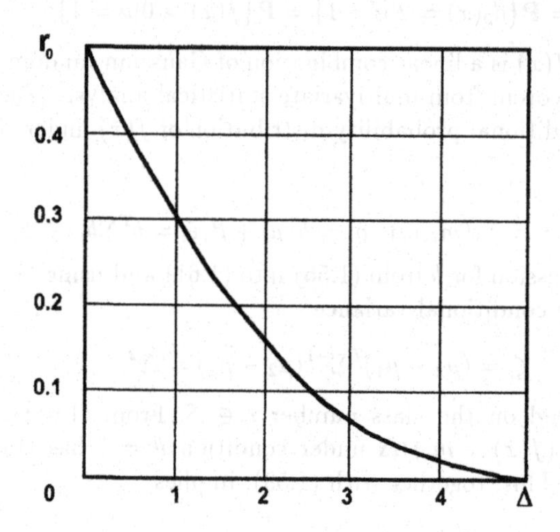

Figure 1.4: Error probability vs. interclass distance

The plot of monotone dependence (1.65) is shown at Figure 1.4. This figure and formula (1.65) are useful in practice for finding the critical value of interclass Mahalanobis distance that guarantees a preassigned error probability level γ, $0 < \gamma < 0.5$:

$$\Delta_\gamma = 2\Phi^{-1}(1-\gamma),$$

where $\Phi^{-1}(1-\gamma) = g_{1-\gamma}$ is the $(1-\gamma)$-quantile of the standard normal distribution $N_1(0, 1)$ (Bolshev and Smirnov, 1983). Let us list the critical values of Δ_γ for levels of γ often used in practice:

Level	0.2	0.1	0.05	0.01	0.005	0.001
MD	1.683	2.562	3.289	4.652	5.151	6.200

For the optimal quadratic decision rule (1.49), (1.50) in the case when the Fisher model (1.52) is distorted no simple analytical expression for error probability r_0 is known. In this case only the upper bound was found so far (using interclass Bhattacharya distance (Fukunaga, 1972)) :

$$r_0 \leq \sqrt{\pi_1\pi_2}\exp\{-1/4(\mu_2-\mu_1)^T(\Sigma_1+\Sigma_2)^{-1}(\mu_2-\mu_1)+$$
$$+1/2\ln\frac{|1/2(\Sigma_1+\Sigma_2)|}{\sqrt{|\Sigma_1||\Sigma_2|}}\}.$$

In addition, the book (Fomin *et al.*, 1986) gives some rather cumbersome expressions for characteristic functions of random variables $f_{ij}(x)$ (1.50), but the inverse Fourier transform for them is still unknown. Moreover, to specify a quadratic discriminant function, $N(N+1)/2$ parameters are required, whereas a linear discriminant function can be specified by N parameters only. For these reasons, the linear discriminant function is often used in pattern recognition problems (McLachlan, 1992), (Aivazyan *et al.*, 1989), (Fukunaga, 1972), (Anderson *et al.*, 1962), even in the cases of distorted Fisher condition (1.52). In (Anderson *et al.*, 1962) for the case of two equiprobable classes ($L=2, \pi_1=\pi_2$) an algorithm for the construction of linear discriminant function $F(x)$ (hyperplane) is presented; N coefficients of this hyperplane minimize the unconditional error probability r. The corresponding decision rule is:

$$d=d_7(x)=\mathbf{1}(F(x))+1, x\in R^N, \tag{1.66}$$

$$F(x)=b^Tx+\beta, b=(s\Sigma_1+(1-s)\Sigma_2)^{-1}(\mu_2-\mu_1),$$

$$\beta=-\frac{\sqrt{b^T\Sigma bb^T}\mu_1+\sqrt{b^T\Sigma bb^T}\mu_2}{\sqrt{b^T\Sigma b}+\sqrt{b^T\Sigma b}},$$

$$r=r(d_7)=\Phi(-\frac{u}{2}), u=u(s)=\frac{b^T(\mu_2-\mu_1)}{\frac{1}{2}(\sqrt{b^T\Sigma_1 b}+\sqrt{b_T\Sigma_2 b})}.$$

where the parameter $s\in(0,1)$ is chosen to maximize the value of u (the analogue of the Mahalanobis distance according to (1.65)):

$$u(s)\to\max_{0<s<1}. \tag{1.67}$$

It was found (Chernoff, 1972) that the solution of maximization problem (1.67) is reduced to finding a root of the following equation:

$$\left((s\Sigma_1+(1-s)\Sigma_2)^{-1}(\mu_2-\mu_1)\right)^T(s^2\Sigma_1-(1-s)^2\Sigma_2)\times$$

$$\times\left((s\Sigma_1+(1-s)\Sigma_2)^{-1}(\mu_2-\mu_1)\right)=0.$$

If $\Sigma_1=\Sigma_2$, then decision rule (1.66) turns into the investigated decision rule (1.55), (1.56). Note additionally that decision rule (1.66) is inapplicable in the pattern recognition problems where the classes are distinguished by the covariance matrices Σ_1, Σ_2 only and mathematical expectations μ_1, μ_2 have equal or close values.

1.5 Optimal Decision Rules in Discrete Space of Feature Variables

Until now in constructing and investigating optimal pattern recognition algorithms, we assumed the continuous observation model (when the observed feature vector $x = (x_j) \in R^N$ may be any point of the N-dimensional Euclidean space). However in practice the components of the vector x are often discrete variables, so the values of x are from discrete space:

$$X = \{u_1, u_2, \dots, u_K\}, 2 \le K \le \infty,$$

where $u_k = (u_{kj}) \in X$ is the k-th possible value for the observed vector of feature variables ($k = 1, \dots, K$). Let us state now the problem of construction of an optimal pattern recognition algorithm (1.17), (1.21) for discrete observation model.

Assume that the observations to be classified are recorded in the N-dimensional discrete space X with prior probabilities $\pi_1, \dots, \pi_L$. An observation from the class Ω_i (the true class number is $\nu = i$) is a random vector x with conditional discrete probability distribution

$$\mathbf{P}\{x = z | \nu = i\} = P_i(z), z \in X, i \in S, \tag{1.68}$$

$$\sum_{k=1}^{K} P_i(u_k) = 1.$$

The classification loss matrix W is given. The problem is to find a decision rule that minimizes risk (1.21). Similarly to (1.25), denote

$$f_k(z) = \sum_{l=1}^{L} \pi_l P_l(z) w_{lk}(z), z \in X, k \in S. \tag{1.69}$$

Theorem 1.3 *Suppose that the observation model is discrete and the loss matrix $W = (w_{ij})$, the prior probabilities $\{\pi_i\}$, and the conditional discrete probability distributions of feature variables $\{P_i(\cdot)\}$ for L classes are given. Then the Bayesian decision rule optimal under minimum risk criterion is randomized and is specified by the critical functions:*

$$\chi_j^*(z) = 0, \text{ if } j \notin J = \{j_1, \dots, j_Q\}, \tag{1.70}$$

$$\chi_{j_t}^*(z) \in [0, 1], \sum_{t=1}^{Q} \chi_{j_t}^*(z) = 1, z \in X,$$

where

$$J = J(z) = \text{Arg} \min_{k \in S} f_k(z), Q = Q(z) = |J(z)|. \tag{1.71}$$

Moreover, the minimal Bayesian risk is

$$r_o = \sum_{k=1}^{K} \min_{j \in S} f_j(z_k). \tag{1.72}$$

Proof. Let $\{\chi_j(z)\}$ be arbitrary critical functions that determine a randomized decision rule. By (1.68), (1.69), the risk functional (1.21) for decision rule $\{\chi_j\}$ assumes the form

$$r = r(\{\chi_j\}) = \sum_{k=1}^{K} \rho(u_k), \tag{1.73}$$

where $\rho(\cdot)$ is the risk density defined by (1.29). The rest of the proof is like the proof of Theorem 1.1. The only difference is that according to (1.68), the event (1.32) for $Q > 1$ can have nonzero probability; therefore, the randomization over the set J should be performed according to (1.69), (1.70).

■

Theorem 1.3 implies that if there are points $z \in X$, for which the cardinality $Q(z) = |J(z)|$ of the set $J(z)$ is greater than 1 then the Bayesian decision rule (1.70), (1.71) is non-unique: the values $\chi_j^*(z)$, $j \in J$, may be chosen arbitrarily. It is advisable to use this arbitrariness to minimize the maximal conditional error probability ($\max\{r_1, \ldots, r_L\} \to \min$):

$$\min_{i \in S} \sum_{k=1}^{K} P_i(u_k)\chi_i^*(u_k) \to \max$$

or to minimize the variance of conditional error probabilities $\{r_i\}$:

$$\sum_{i=1}^{L} \pi_i (r_i - r_0)^2 \to \min .$$

Sometimes to simplify the pattern recognition algorithm, one may abandon the mentioned randomization.

Corollary 1.4 *In the family of equivalent (by the minimum risk criterion) optimal decision rules (1.70), (1.71) there exists a nonrandomized Bayesian decision rule*

$$d = d_0(z) = \arg\min_{k \in S} f_k(z), z \in X, \tag{1.74}$$

with conditional error probability

$$r_i = \mathbf{P}\{d_0(x) \neq i | \nu = i\} = 1 - \sum_{k=1}^{K} P_i(u_k)\delta_{i,d_0(u_k)}, i \in S. \tag{1.75}$$

In practice, one may encounter pattern recognition problems in which the features $\{x_j\}$ of the observed object are the frequencies of some events in a series of M experiments. An adequate data model (1.68) for such problems is the polynomial probability distribution:

$$X = \{u_1, \ldots, u_k\} = \{z = (z_1, \ldots, z_N)^T :$$

$$z_j \in \{0, 1, \ldots, M\}, \sum_{j=1}^{N} z_j = M\},$$

$$P_i(z; \theta_i) = \frac{M!}{z_1! \ldots z_N!} \prod_{j=1}^{N} \theta_{ij}^{z_j}, z = (z_j) \in X, \tag{1.76}$$

where the parameter $\theta_i = (\theta_{i1}, \ldots, \theta_{iN})^T$ of the i-th class is the vector of elementary probabilities $(0 \leq \theta_{ij} \leq 1, \sum_{j=1}^{N} \theta_{ij} = 1)$ of the events recorded in the experiment. The classes $\{\Omega_i\}$ differ by the values of the parameters $\{\theta_i\}$.

Corollary 1.5 *The Bayesian decision rule (1.74) for classification of discrete data defined by polynomial distribution (1.76) in the case of two classes ($L = 2$) and (0-1)-loss matrix (1.18) is the linear one:*

$$d = d_0(z) = \mathbf{1}(f(z)) + 1, z \in X, \tag{1.77}$$

$$f(z) = b^T z + \beta, b = (b_j) \in R^N, \; b_j = \ln \frac{\theta_{2j}}{\theta_{1j}} (j = 1, \ldots, N), \; \beta = \ln \frac{\pi_2}{\pi_1}.$$

Proof. Substituting (1.18) into (1.69), (1.74) we obtain the Bayesian decision rule:

$$d = d_o(z) = \arg \max_{k \in \{1,2\}} (\pi_k P_k(z)) = \arg \max_k \ln(\pi_k P_k(z)).$$

Now, using (1.76), after some transformations we obtain (1.77). ■

Thus, similarly to the case of Gaussian model with equal covariance matrices (investigated in Section 1.4), in this case the Bayesian discriminant function is also linear, and it differs from (1.56) only by the values of coefficients b and β.

The exact values of the conditional error probabilities and the Bayesian error probability $r_0 = \pi_1 r_1 + \pi_2 r_2$ for the Bayesian decision rule (1.77) can be found by numerical methods using (1.75). Unfortunately, the sums in (1.75) cannot be expressed in analytical form. Approximate values of r_0, r_1, r_2 can be found using the Gaussian approximation of the linear statistics $f(x)$ in (1.77):

$$\mathcal{L}\{f(x) | \nu = i\} \approx N_1(M m_i, M \zeta_i),$$

$$m_i = \sum_{j=1}^{N} \theta_{ij} \ln \frac{\theta_{2j}}{\theta_{1j}},$$

$$\zeta_i = \sum_{j=1}^{N} \theta_{ij} \ln^2 \frac{\theta_{2j}}{\theta_{1j}} - m_i^2 > 0 (i = 1, 2).$$

Reasoning in the same way as in the proof of Theorem 1.2, we can find approximations for the case $\pi_1 = \pi_2$:

$$r_i \approx \Phi\left(-\sqrt{M}\frac{(-1)^i m_i}{\sqrt{\zeta}}\right), \quad r_o = \frac{r_1 + r_2}{2}.$$

In particular, in the case of symmetry in θ_1, θ_2 we have $m_2 = -m_1$, $\zeta_2 = \zeta_1 = \zeta$ (this is always the case, if $N = 2$), and we come to a simple formula:

$$r_o \approx \Phi\left(-\sqrt{M}\frac{K(1,2)}{2\sqrt{\zeta}}\right), \tag{1.78}$$

where

$$K(1,2) = \sum_{j=1}^{N} (\theta_{1j} - \theta_{2j}) \ln \frac{\theta_{1j}}{\theta_{2j}} > 0$$

is interclass *Kullback divergence* (Kullback, 1967) for two discrete probability distributions $\theta_1 = (\theta_{1j})$ and $\theta_2 = (\theta_{2j})$. It can be seen from the formula (1.78) that with larger interclass distance $K(1,2)$, larger number M of experiments, and less variance ζ the accuracy of pattern recognition is higher.

1.6 Regular Families of Probability Distributions

As it is seen from the previous Sections 1.1 – 1.5, optimal decision rules minimizing risk functional depend on the conditional N-variate distribution functions $F_1(\cdot), \ldots, F_L(\cdot)$ of the feature vector. In applied pattern recognition problems the distribution function $F_i(x)$, $x \in R^N$ $(i = 1, 2, \ldots, L)$ is usually unknown and assumed to belong to some family (set) of N-variate distribution functions and is estimated by some statistical procedure using a random sample.

Consider two main kinds of families of N-variate distribution functions used in statistical pattern recognition theory. The family of distribution functions for each class Ω_i $(i = 1, 2, \ldots, L)$ is assigned in the same way, therefore we shall omit the class index i from notations throughout this section.

1. Regular Parametric Family

Let us assume that the distribution function $F(x;\theta)$, $x \in R^N$, $\theta \in \Theta \subseteq R^m$, depends on the vector of m parameters θ. In other words, the distribution function is assumed to be an element of m-parametric family of N-variate distribution functions :

$$\mathbf{F} = \left\{F(x;\theta),\, x \in R^N \,:\, \theta \in \Theta\right\}.$$

This family induces the family of probability measures on the Borel σ-algebra $\mathbf{B} = \mathbf{B}(R^N)$:

$$\mathcal{P} = \{\mathbf{P}_\theta(B),\, B \in \mathbf{B} \,:\, \theta \in \Theta\},$$

where

$$\mathbf{P}_\theta(B) = \int_B dF(x;\theta).$$

Let the unknown parameter value θ be estimated by some statistical estimator $\hat{\theta} \in \Theta^*$ for a random sample $z_1, \ldots, z_n \in R^N$ of size n from the distribution $F(x;\theta)$ (Θ^* denotes the closure of Θ). Let the estimator $\hat{\theta}$ belong to the large family of minimum contrast estimators, which includes ML-estimator, LS-estimator, Huber estimator, etc. :

$$\hat{\theta} = \arg\min_{\theta \in \Theta^*} \frac{1}{n} \sum_{j=1}^{n} g(z_j;\theta),$$

where $g(z;\theta) : R^N \times \Theta^* \to R$ is the so-called *contrast function* (full description of these estimators is given in Section 3.1).

We introduce now some regularity conditions for the functions $F(\cdot)$, $g(\cdot)$ under which $\hat{\theta}$ is a consistent statistical estimator that admits a stochastic expansion in powers of $1/\sqrt{n}$. We shall use the *Chibisov regularity conditions of order k* (Chibisov, 1973). To formulate them we need the following notations. Let $\alpha = (\alpha_1, \ldots, \alpha_m)$ be an m-vector with integer nonnegative components, $\langle\alpha\rangle = \alpha_1 + \ldots + \alpha_m$. For any function $u(\theta)$ let

$$u^\alpha(\theta) = \frac{\partial^{\langle\alpha\rangle} u(\theta)}{\partial\theta_1^{\alpha_1} \ldots \partial\theta_m^{\alpha_m}}$$

be the partial derivative of the order $\langle\alpha\rangle$. Let us introduce an m-vector and $m \times m$-matrix respectively :

$$g^{(1)}(x;\theta) = \left(\frac{\partial g(x;\theta)}{\partial\theta_p}\right),\; g^{(2)}(x;\theta) = \left(\frac{\partial^2 g(x;\theta)}{\partial\theta_p \partial\theta_q}\right),\; p, q = 1, \ldots, m.$$

Assume the notations: $\mathbf{E}_\theta$ is expectation with respect to the probability measure $\mathbf{P}_\theta$;

$$a(\theta) = \mathbf{E}_\theta\left\{g(z_1;\theta)\right\},\; a^{(1)}(\theta) = \mathbf{E}_\theta\left\{g^{(1)}(z_1;\theta)\right\},\; A(\theta) = \mathbf{E}_\theta\left\{g^{(2)}(z_1;\theta)\right\}.$$

Then the *Chibisov regularity conditions* of order k have the following form.

(I) The mapping $\theta \to \mathbf{P}_\theta$ is continuous on Θ : if $|\theta - \theta_0| \to 0$, then

$$\sup_{B \in \mathbf{B}} |\mathbf{P}_\theta(B) - \mathbf{P}_{\theta_0}(B)| \to 0.$$

(II) $\forall x \in R^N$ the contrast function $g(x;\theta)$ is continuous w.r.t. $\theta \in \Theta^*$.

(III) $\forall \theta \in \Theta$ there exist $s > 2$ and a neighborhood U_θ of the point θ such that the function $\mathbf{E}_{\theta_0}\left\{|g(z_1;\theta_0)|^{s/2}\right\}$ converges uniformly w.r.t. $\theta_0 \in U_\theta$.

(IV) $\forall \theta \in \Theta$ there exists a neighborhood U_θ such that for any $U \subset U_\theta$ and any compact set $K \subset U_\theta$ the function $\mathbf{E}_{\theta_0}\left\{\left|\inf_{\theta \in U} g(z_1;\theta)\right|^{s/2}\right\}$ converges uniformly w.r.t. $\theta_0 \in K$.

(V) $\forall \theta_0 \in \Theta$ there exists its neighborhood U_{θ_0} such that the derivatives $g^\alpha(x;\theta)$ exist and are continuous w.r.t. $\theta \in U_{\theta_0}$ $\mathbf{P}_{\theta_0}$-almost everywhere for every α such that $1 \leq \langle\alpha\rangle \leq k+1$.

(VI) $a^{(1)}(\theta) = 0$, $\theta \in \Theta$.

(VII) The matrix $A(\theta)$ is positive definite; moreover, its minimal eigenvalue is separated from zero.

(VIII) There exists $t \geq 2$ such that $\mathbf{E}_\theta\left\{|g^{(\alpha)}(z_1;\theta)|^t\right\} < \infty$ for every $\alpha : 1 \leq \langle\alpha\rangle \leq k+1$, and all these moments converge uniformly w.r.t. $\theta \in K$.

(IX) $\forall \theta \in K$ there exist its neighborhood U_θ and a function $R(x;\theta)$ such that if $\theta_1, \theta_2 \in U_\theta$, $\langle\alpha\rangle = k+1$, then

$$|g^{(\alpha)}(x;\theta_1) - g^{(\alpha)}(x;\theta_2)| \leq R(x;\theta)|\theta_1 - \theta_2|,$$

and the random variable $R^{s/2}(z_1;\theta)$ is uniformly integrated w.r.t. $\mathbf{P}_{\theta_0}$, $\theta_0 \in K$.

2. Regular Nonparametric Family

A nonparametric family of distrbution functions is necessary in the situations where no parametric form of N-variate distribution function $F(x)$ is known, and its nonparametric estimators are constructed for a random sample $z_1, \ldots, z_n \in R^N$ (see Chapter 4).

We will assume that the distribution function $F(x)$ is absolutely continuous w.r.t. the Lebesgue measure in R^N, and there exists the N-variate probability density function :

$$p(x) = \frac{\partial^N F(x)}{\partial x_1 \ldots \partial x_N}, \; x \in R^N.$$

Let us define some regular nonparametric families of N-variate probability density functions $\mathcal{P} = \{p(\cdot)\}$ which are often used in statistical pattern recognition.

$\mathcal{P}_1$ is the family of N-variate probability density functions.

$\mathcal{P}_2$ is the family of N-variate probability density functions differentiable up to k-th order for some $k \geq 1$.

$\mathcal{P}_3$ is the family of bounded N-variate probability density functions with bounded derivatives up to order k.

$\mathcal{P}_4$ is the family of probability density functions $p(x)$ from $\mathcal{P}_3$ such that

$$\int_{R^N} p^l(x)dx < \infty$$

for some $l \geq 2$.

Let us note that the following chain of inclusions for these families holds :

$$\mathcal{P}_4 \subset \mathcal{P}_3 \subset \mathcal{P}_2 \subset \mathcal{P}_1.$$

Chapter 2

Violations of Model Assumptions and Basic Notions in Decision Rule Robustness

The chapter investigates the adequacy of classical (hypothetical) models of observed data described in Chapter 1. Main approaches for construction of decision rules based on training samples are compared for the situations with proir uncertainty. Classification and mathematical description of typical distortions of hypothetical models and data are given. Basic notions for decision rule robustness analysis, such as guaranteed risk, robustness factor, critical distortion level, robust decision rule, are defined.

2.1 Construction of Decision Rules Using Training Sample

In constructing optimal decision rules (Chapter 1) it was assumed that complete prior information about the alphabet $\{\Omega_1, \ldots, \Omega_L\}$ of classes is known:

1. The feature variable space $Y = X \subseteq R^N$ is given;

2. The loss matrix $W = (w_{ij})$ is given;

3. Exact probability characteristics of classes (prior probabilities $\{\pi_i\}$ and conditional probability density functions $\{p_i(\cdot)\}$ for continuous data model or conditional probability distributions $\{P_i(\cdot)\}$ for discrete data model) are known.

In this case, the optimality criterion was the minimum of risk functional r (1.21), and the optimal decision rule was the Bayesian decision rule defined by Theorems 1.1, 1.3. We shall say that we have *prior uncertainty* in the pattern recognition problem, if at least one of these three enumerated elements is unknown. In practical pattern recognition problems, probability characteristics of classes $\{\Omega_i\}$ are most often unknown.

Using an experimental data sample (statistics), we can reduce the prior uncertainty. Therefore we shall assume that in the case of prior uncertainty we observe a random sample consisting of n composed $(N+1)$-dimensional vectors:

$$A = \left\{ \begin{pmatrix} x_1 \\ \cdots \\ \tau_1 \end{pmatrix}, \begin{pmatrix} x_2 \\ \cdots \\ \tau_2 \end{pmatrix}, \ldots, \begin{pmatrix} x_n \\ \cdots \\ \tau_n \end{pmatrix} \right\}, x_j = (x_{jk}) \in R^N, \tau_j \in R^1,$$

$$j = 1, \ldots, n.$$

The sample A is obtained by a series of n independent experiments on observation of objects from all classes $\Omega_1, \ldots, \Omega_L$. The result of the j−th experiment is a composed $(N+1)$-vector $(x_j^T \vdots \tau_j)^T$, where x_j is a vector of random feature variables and τ_j is a value in some way associated with the class number $\nu_j \in S$ of the observed object. Sometimes the classes to which the observed data belong are fixed in advance. In these situations $\tau_1, \ldots, \tau_L$ are nonrandom. We shall assume here that the observations $x_1, \ldots, x_n$ are independent random vectors. The problem is to classify a random observation $x \in R^N$ independent of A.

The usage of a sample A for decision rule construction in prior uncertainty conditions is called *adaptation*, or *training;* the sample A is called *training sample,* and the corresponding decision rule $d = d(x; A)$, based on the training sample is called *adaptive decision rule* (ADR).

There are two different kinds of the training sample A, distinguished by the kind of functional dependence between the variables τ_j and ν_j $(j = 1, \ldots, n)$.

If $\tau_j = \nu_j$, $j = 1, \ldots, n$, then the sample A is called *classified training sample.* In other words, for each observation x_j the true class Ω_{ν_j} to which the observed object belongs is indicated. In this way, the sample A is partitioned into L nonintersecting subsamples $A_1, \ldots, A_L$:

$$A = \bigcup_{i=1}^{L} A_i, \; A_i \bigcap A_k = \emptyset \, (i \neq k),$$

where

$$A_i = \left\{ \begin{pmatrix} x_j \\ \cdots \\ \nu_j \end{pmatrix} : \begin{pmatrix} x_j \\ \cdots \\ \nu_j \end{pmatrix} \in A, \; \nu_j = i \right\}$$

is a training sample from class Ω_i. Let us agree to mark the elements of this sample by two indices:

$$A_i = \{x_{i1}, \ldots, x_{in_i}\},$$

where $x_{ik} = (x_{ikl}) \in R^N$ is the feature vector of the k-th observation for the object from class Ω_i and l is feature's number,

$$n_i = \sum_{j=1}^{n} \delta_{\nu_j, i}$$

is the size of the sample for class Ω_i ($n_1 + \ldots + n_L = n$). By the construction of the sample, the observations $x_{i1}, \ldots, x_{in_i}$ are independent. The subsamples $A_1, \ldots, A_L$ are also independent.

A sample A is called *unclassified training sample* if $\{\tau_j\}$ is absent in A or take values independent of $\{\nu_j\}$:

$$A = \{x_1, \ldots, x_n\}, x_j = (x_{jl}) \in R^N.$$

Samples of this kind are usually recorded in passive experiments, when there is no opportunity to plan the experiment or to learn the class numbers $\{\nu_j\}$ reliably by special investigations (see Chapter 7).

Thus, let the probability characteristics $\{\pi_i, p_i(\cdot)\}$ be unknown (without loss of generality, we use here the continuous model of data) and let a training sample A (classified or unclassified) be recorded. In this situation $\{\pi_i, F_i(\cdot)\}$ are unknown and risk functional (1.21) is underdetermined. Therefore the problem (1.17) of construction of an optimal classification algorithm turns into the problem of minimization of an underdetermined functional.

The latter problem is not considered in the classical theory of statistical inference. Let us briefly consider main principles of the construction of algorithms for classification of an observation x in presence of the training sample A.

2.1.1 Structural Minimization of Empirical Risk for Given Parametric Family of Decision Rules

Within this approach, proposed by Vapnik (Vapnik, 1974, 1979) the empirical classification risk, which is a function of decision rule parameters θ, is used as the optimality criterion. The restrictions for "complexity" ("capacity") of the family of decision rules under uniform (with respect to θ) convergence of the empirical risk to the true risk functional are found. Constructive methods of empirical risk minimization allowing to find the estimates of optimal decision rule parameters θ^* are suggested. This approach is implemented in the software system FOP (Vapnik, 1979). A deficiency of the approach stems from difficulties in selection of the parametric family of decision rules (the so-called *"rectifying space"*).

2.1.2 Statistical Hypothesis Testing with Parametric Model of Probability Distributions $\{p_i(\cdot)\}$ and Classified Training Sample Available

A complete system of L composite hypotheses $H_1, \ldots, H_L$ is formulated. The hypothesis H_i means that an observed random vector $x \in R^N$ to be classified and the elements of the sample A_i (from the class Ω_i) have the same probability distribution distinct from the distributions of all $L-1$ other training samples $A_1, \ldots, A_{i-1}, A_{i+1}, \ldots, A_L$.

The likelihood ratio test, traditional in statistical composite hypothesis testing theory, is used for testing these hypotheses $\{H_i\}$ and for classification of x.

The main deficiencies of this approach are the necessity of assignment of exact parametric probability model $\{p_i(\cdot)\}$ and high computational complexity of evaluation of the statistics for the likelihood ratio criterion, especially for multidimensional feature space R^N. Therefore this approach for $N > 1$ is realized for the Gaussian data model only (Anderson, 1958).

2.1.3 Decision Rules Minimizing Lower and Upper Bounds of Risk Density

In Section 1.3 it was found out that for a nonrandomized decision rule $d = d(x; A)$ the risk is of the form (for classified training sample A) :

$$r(d(\cdot); \{p_i(\cdot)\}) = \mathbf{E}\{w_{\nu d(x;A)}\} = \int_{R^{N(n+1)}} \rho(x, A; d(\cdot), \{p_i(\cdot)\}) dx dA, \tag{2.1}$$

where

$$\rho = \rho(x, A; d(\cdot), \{p_i(\cdot)\}) = \sum_{l=1}^{L} \pi_l p_l(x) w_{ld(x;A)} \prod_{i=1}^{L} \prod_{j=1}^{n_i} p_i(x_{ij}) \tag{2.2}$$

can be called *risk density;* $\{p_i(\cdot)\} \subset P$ are the distribution densities from a given set P. It can be seen from Theorem 1.1 that for given $\{p_i(\cdot)\}$ the Bayesian decision rule minimizes the risk density $\rho(\cdot)$ at each point $(x, A) \in R^{N(n+1)}$.

In the situation considered here, the densities $\{p_i(\cdot)\}$ are unknown. Therefore the risk r and its density $\rho(\cdot)$ cannot be used as measures of optimality for the adaptive decision rule. Let the functions $\rho_{\pm}(x, A; d(\cdot))$ be the upper and the lower bounds for risk density (2.2) respectively, so that the following inequalities hold for arbitrary densities $\{p_i(\cdot)\} \subset P$ (Kharin, 1978):

$$\rho_-(x, A; d(\cdot)) \leq \rho(x, A; d(\cdot), \{p_i(\cdot)\}) \leq \rho_+(x, A; d(\cdot)), \tag{2.3}$$

$$(x, A) \in R^{N(n+1)}.$$

By integrating (2.3) and using (2.1) we obtain a two-sided estimate of risk for arbitrary $\{p_i(\cdot)\} \subset P$:

$$r_-(d(\cdot)) \leq r(d(\cdot); \{p_i(\cdot)\}) \leq r_+(d(\cdot)),$$

where

$$r_{\pm}(d(\cdot)) = \int_{R^{N(n+1)}} \rho_{\pm}(x, A; d(\cdot)) dx dA \tag{2.4}$$

are the upper and the lower bounds for risk of adaptive decision rule $d(\cdot)$. We shall call the adaptive decision rule $d_+ = d_+(x; A)$ r_+*-optimal* if the upper bound for risk density is minimal:

$$d_+ = d_+(x; A) = \arg\min_{d\in S} \rho_+(x, A; d). \tag{2.5}$$

The r_-*-optimal adaptive decision rule* is defined similarly:

$$d_- = d_-(x; A) = \arg\min_{d\in S} \rho_-(x, A; d). \tag{2.6}$$

A variety of r_+- and r_--optimal adaptive decision rules arises from the variety of upper and lower bounds in (2.3) and the variety of loss matrices $W = (w_{ij})$ in (2.2). Let us construct two versions of the adaptive decision rule (2.5), (2.6) for a common practical situation (1.18) where the risk r means the error probability.

In the first version, ρ_+, ρ_- are the corresponding least upper and greatest lower bounds (with respect to $\{p_i(\cdot)\}$). Let

$$P\{\nu = i \mid x\} = \frac{\pi_i p_i(x)}{p(x)}; p(x) = \sum_{i=1}^{L} \pi_i p_i(x)$$

denote the unconditional probability density function of the random observation to be classified. Then from (2.2), (1.18) we obtain:

$$\rho_\pm = \pm \max_{\{p_i(\cdot)\}\subset P} \left(\pm \left(\prod_{i=1}^{L} \prod_{j=1}^{n_i} p_i(x_{ij}) \right) p(x)(1 - \mathbf{P}\{\nu = d \mid x\}) \right),$$

$$d_+ = d_+(x; A) =$$

$$= \arg\min_{d\in S} \max_{\{p_i(\cdot)\}\subset P} ((1 - \mathbf{P}\{\nu = d \mid x\}) p(x) \prod_{i=1}^{L} \prod_{j=1}^{n_i} p_i(x_{ij})). \tag{2.7}$$

Let us sketch a method for approximate solution of the complicated optimization problem (2.7).

First, we find the maximum likelihood estimates (ML-estimates) $\{\tilde{p}_i(\cdot)\}$, for which the maximum for the following expression is attained:

$$p(x) \prod_{i=1}^{L} \prod_{j=1}^{n_i} p_i(x_{ij}) \to \max_{\{p_i(\cdot)\}\subset P}.$$

Then using $\{\tilde{p}_i(\cdot)\}$ we formulate the problem:

$$1 - \mathbf{P}\{\nu = d \mid x\} \to \min_d,$$

which has an evident solution:

$$d_+ = d_+(x; A) = \arg\max_{l\in S}(\pi_l \tilde{p}_l(x)). \tag{2.8}$$

In the second version of this approach to adaptive decision rule construction under prior uncertainty conditions, we use the following bounds $\rho_\pm(\cdot)$ in (2.3) :

$$\rho_\pm = p(x)\prod_{i=1}^{L}\prod_{j=1}^{n_i} p_i(x_{ij}) \mp \min_{\{p_i(\cdot)\}\subset P}\left(\pm\pi_d p_d(x)\prod_{i=1}^{L}\prod_{j=1}^{n_i} p_i(x_{ij})\right).$$

Note that the first term of $\rho_\pm$ depends on $\{p_i(\cdot)\}$ only and does not depend on d, and this dependence is inessential because of the normalization condition:

$$r_\mp(d) = 1 \mp \int_{R^{N(n+1)}} \max_{\{p_i(\cdot)\}\subset P}\left(\pm\pi_d p_d(x)\prod_{i=1}^{L}\prod_{j=1}^{n_i} p_i(x_{ij})\right) dx dA.$$

As a result, we find the r_--optimal adaptive decision rule:

$$d_- = d_-(x; A) = \arg\max_{d\in D} \max_{\{p_i(\cdot)\}\subset P}\left(\pi_d p_d(x)\prod_{i=1}^{L}\prod_{j=1}^{n_i} p_i(x_{ij})\right). \tag{2.9}$$

Note that the particular case of the adaptive decision rule (2.9) for the Gaussian family of densities P was proposed by (Anderson, 1958).

2.1.4 Plug-in Decision Rules (PDR)

Let $\{\hat{\pi}_i, \hat{p}_i(\cdot)\}$ be any statistical estimators of probability characteristics $\{\pi_i, p_i(\cdot)\}$ based on the training sample A. The *plug-in decision rule* (Glick, 1972) is the adaptive decision rule $d = d_1(x; A)$ derived from the Bayesian decision rule (1.26), (1.25) by substitution of the estimators $\{\hat{\pi}_i, \hat{p}_i(\cdot)\}$ for unknown true probability characteristics $\{\pi_i, p_i(\cdot)\}$:

$$d = d_1(x; A) = \arg\min_{k\in S} \hat{f}_k(x),\ \hat{f}_k(x) = \sum_{l=1}^{L} \hat{\pi}_l \hat{p}_l(x) w_{lk}(x), x \in R^N. \tag{2.10}$$

Varying the loss matrix $W = (w_{lk})$ and kinds of statistical estimators $\{\hat{\pi}_i, \hat{p}_i(\cdot)\}$ we obtain a fairly rich family of plug-in decision rules. This family contains the adaptive decision rules derived by the above approaches 2.1.1–2.1.3. For example, the r_+-optimal adaptive decision rule (2.8) is a particular case of (2.10) for (0, 1)-loss matrix (1.18). For this reason, we shall place the main emphasis on the approach 2.1.4.

First of all, let us list some general consistency properties for the plug-in decision rule (Glick, 1972). Denote: $d = d(x; A)$ is an arbitrary adaptive decision rule, $d = d_1(x; A)$ is the plug-in decision rule (2.10), $d = d_0(x)$ is the Bayesian decision rule (1.26) for exactly known $\{\pi_i, p_i(\cdot)\}$; r_0 is the Bayesian risk (1.27). According to (1.28), (1.29),

$$\mathrm{rc}(d(\cdot)) = \int_{R^N} f_{d(x;A)}(x)dx \tag{2.11}$$

is the conditional risk in classification of a random observation $x \in R^N$ by adaptive decision rule $d = d(x; A)$ when training sample A is fixed;

$$\hat{\mathrm{rc}}(d(\cdot)) = \int_{R^N} \hat{f}_{d(x;A)}(x)dx \tag{2.12}$$

is the statistical estimator of conditional risk (2.11) based on the training sample A *(plug-in risk)*. The adaptive decision rule $d = d(x; A)$ is called *strongly consistent* (in the sense of Bayesian risk) if the conditional risk for this decision rule converges to the Bayesian risk almost surely as the sample size n increases:

$$\mathrm{rc}(d(\cdot)) \xrightarrow{a.s.} r_0. \tag{2.13}$$

If the convergence in (2.13) is in probability, then this adaptive decision rule is called *consistent.*

Property 1. The plug-in decision rule (2.10) minimizes the conditional risk estimator (2.12):

$$d = d_1(x; A) = \arg\min_{d(\cdot)} \hat{\mathrm{rc}}(d(\cdot)), \tag{2.14}$$

where minimization is with respect to all possible measurable functions

$$d(\cdot): \; R^{N(n+1)} \to S.$$

Property 2. If the estimators $\{\hat{\pi}_i, \hat{p}_i(\cdot)\}$ are such that at each point $x \in R^N$ the following nonbiasedness condition holds:

$$\mathbf{E}\{\hat{f}_k(x)\} = f_k(x), k \in S,$$

then

$$\mathbf{E}\{\, \hat{\mathrm{rc}}(d_1(\cdot))\} \leq r_0 \leq \mathrm{rc}(d_1(\cdot)). \tag{2.15}$$

Consequently, $\hat{\mathrm{rc}}(d_1(\cdot))$ is a biased estimator of the plug-in decision rule risk.

Property 3. If the estimators of the probability characteristics $\{\pi_i, p_i(\cdot)\}$ are strongly consistent $(n \to \infty)$:

$$\hat{\pi}_i \xrightarrow{a.s.} \pi_i, \;\; \hat{p}_i(x) \xrightarrow{a.s.} p_i(x), \;\; x \in R^N,$$

and for any sample $A \in R^{Nn}$ the following inequalities hold:

$$\hat{p}_i(x) \geq 0, \int_{R^N} \hat{p}_i(x)dx = 1, \;\; i \in S,$$

then the plug-in risk $\hat{\mathrm{rc}}(d(\cdot))$ uniformly converges to the conditional risk $\mathrm{rc}(d(\cdot))$ almost surely as $n \to \infty$:

$$\sup_{d(\cdot)} \mid \hat{\mathrm{rc}}(d(\cdot)) - \mathrm{rc}(d(\cdot)) \mid \xrightarrow{a.s.} 0.$$

The convergencies in probability and in the mean-square sense hold as well.

Property 4. Under the conditions of Property 3 as $n \to \infty$ the plug-in Bayesian decision rule (2.10) is strongly consistent in the sense of Bayesian risk.

Property 5. Under the conditions of Property 3 as $n \to \infty$ the plug-in risk for the plug-in decision rule (2.10) is a strongly consistent estimator:

$$\hat{rc}(d_1(\cdot)) \xrightarrow{a.s.} r_0.$$

Note in conclusion that for estimation of prior probabilities $\{\pi_i\}$ from a classified training sample $A = \bigcup_{i=1}^{L} A_i$ the relative frequencies $\hat{\pi}_i = n_i/n\,(i \in S)$ are used, which are strongly consistent, unbiased and efficient estimators. For the estimation of probability density functions $\{p_i(\cdot)\}$ (and probability distributions $\{P_i(\cdot)\}$) the following methods are used:

- when the parametric probability family is known *a priori*:
  - maximum likelihood method and its generalization (method of minimum contrast);
  - method of moments;
  - method of χ^2 minimization;
  - Bayesian method;
- when nonparametric prior uncertainty takes place:
  - Rosenblatt - Parzen method;
  - k-Nearest Neighbor (k-NN) method.

2.2 Types of Distortions

The principles of construction of optimal decision rules from Chapter 1 and adaptive decision rules from Section 2.1 are based on some model assumptions, which can be violated in practice. A classification of main distortion types (which produce violations of assumptions) is sketched at Figure 2.1. Let us give brief mathematical descriptions for each distortion type indicated at Figure 2.1.

D.1. Small-sample Effects

This kind of distortions is typical for all statistical plug-in procedures. It arises from noncoincidence of statistical estimates $\{\hat{\pi}_i, \hat{p}_i(\cdot), \hat{P}_i(\cdot)\}$ of probability characteristics based on a finite training sample A of size $n < \infty$ and their true values $\{\pi_i, p_i(\cdot), P_i(\cdot)\}$. The plug-in decision rules described in Section 2.1 are asymptotically optimal as $n \to \infty$. For finite sample size n, however, random deviations of statistical estimates

$$\{\hat{\pi}_i - \pi_i, \hat{p}_i(\cdot) - p_i(\cdot), \hat{P}_i(\cdot) - P_i(\cdot)\}$$

can produce significant increase of risk (Glick, 1972), (Raudis, 1976, 1980), (Fomin *et al.*, 1986), (Fukunaga *et al.*, 1989), (McLachlan, 1992)). In existing statistical

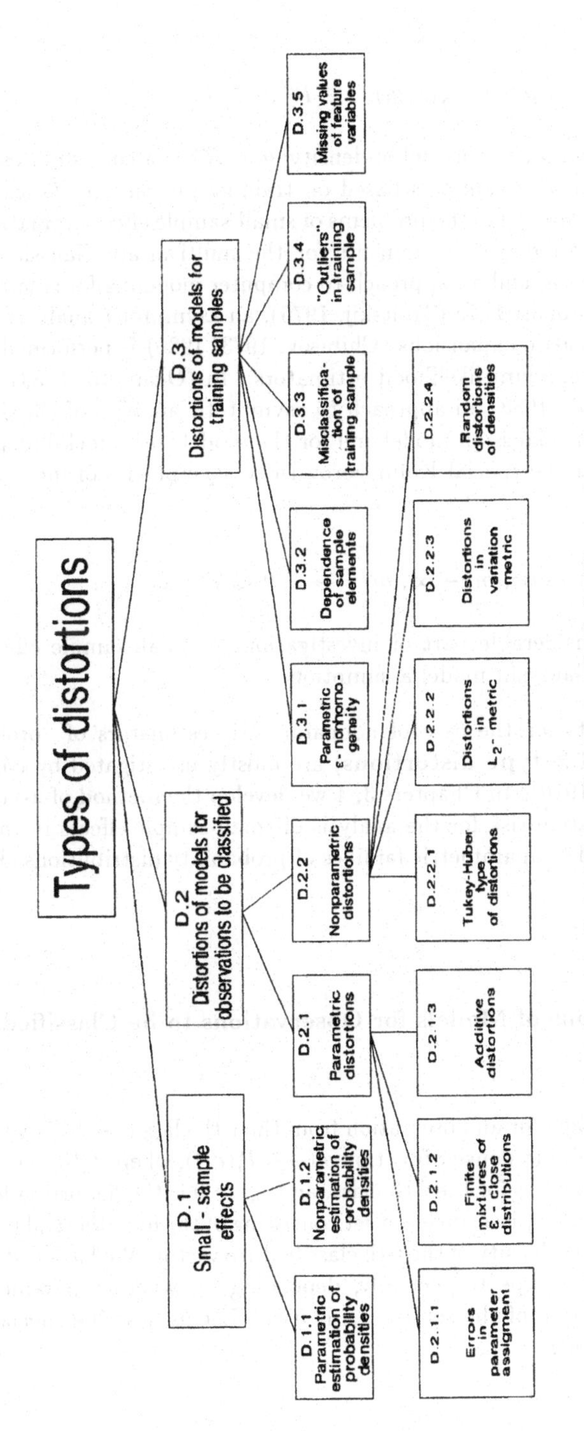

Figure 2.1. Classification of distortions types.

pattern recognition literature, the main attention is paid to small-sample effects when parametric estimators of probability density functions are used (**D.1.1–type distortions**):

$$\hat{p}_i(x) = q(x; \hat{\theta}_i),\ x \in R^N,\ i \in S,$$

where $q(\cdot)$ is a given parametric model of density; $\hat{\theta}_i \in R^m$ is a consistent estimator of the i-th class vector of parameters based on training sample. In (Glick, 1972), (Raudis, 1976), for the first time the problems of small-sample effect evaluation were mathematically formulated, risk increments for the multivariate Gaussian model were found in closed form, and an approach to computer modeling for small-sample effects detection was proposed. In (Troitskij, 1976), an asymptotic analysis of error probability using stochastic expansions (Chibisov, 1973, 1980) is performed for the case when $\{\hat{\theta}_i\}$ are maximum likelihood estimators. In (Okamoto, 1963), (Deev, 1970), (Meshalkin *et al.*, 1988) an approach to asymptotic analysis of classification error probability for the Gaussian model and for the model with block-independent feature variables under the special Kolmogorov-Deev asymptotics of increasing dimension:

$$n \to \infty,\ m \to \infty,\ m/n \to \lambda,\ 0 < \lambda < \infty$$

was developed. A considerable part of investigations of small-sample effects was performed under the Gaussian model assumption.

Small-sample effects for the case of nonparametric estimators of probability density functions (**D.1.2–type distortions**) are mostly investigated by computer experiments (Raudis, 1976). In Chapters 3, 4 we develop the method of asymptotic expansions of classification risk for the analysis of small-sample effects in the cases of both parametric and nonparametric families of probability distributions (Kharin, 1981–1983).

D.2. Distortions of Models for Observations to be Classified

A mathematical model for an observation from the i-th class ($i \in S$) is a random N-vector X_i with probability density function $p_i(\cdot) \in P_i(\epsilon_{+i})$, where $P_i(\epsilon_{+i})$ is a set of admissible distorted densities for the i-th class, and $\epsilon_{+i} \geq 0$ is the distortion level. If $\epsilon_{+i} = 0$, then $P_i(0) = \{p_i^0(\cdot)\}$ is a singleton set consisting of the hypothetical ("ideal", nondistorted) probability density of the i-th class feature vector. We have distortions of **D.2.1** type or **D.2.2** type respectively, depending on whether parametric or nonparametric description of the set $P_i(\epsilon_{+i})$ is used. Let us describe special cases of **D.2.1**-type distortions.

D.2.1. Parametric Distortions

D.2.1.1. Errors in Assignment of Parameter Values

Suppose that we have an m-parametric regular family of probability densities in R^N :

$$Q = \{q(x;\theta), x \in R^N : \theta \in R^m\},$$

$p_i^\circ(x) = q(x;\theta_i^\circ)$, and the set of admissible distorted densities for the i-th class has the following parametric description:

$$P_i(\epsilon_{+i}) = \{q(x;\theta_i^*), x \in R^N : \theta_i^* \in \Theta_i\}, i \in S,$$

where

$$\theta_i^\circ \in \Theta_i, \Theta_i \bigcap \Theta_j = \emptyset, \quad i,j \in S, \quad i \neq j.$$

With increasing distortion level ϵ_{+i} the diameter of the set $\Theta_i \subset R^m$ of admissible distortions of parameters increases. For example,

$$\Theta_i = \{\theta_i^* : (\theta_i^* - \theta_i^\circ)^T B_i(\theta_i^* - \theta_i^\circ) = \epsilon_i^2, 0 \leq \epsilon_i \leq \epsilon_{+i}\}$$

is a hyperellipsoid in R^m centered at θ_i^0;

$$\Theta_i = \{\theta_i^* = (\theta_{ij}^*) : \mid \theta_{ij}^* - \theta_{ij}^\circ \mid \leq \epsilon_i g_{ij}, \ j = 1, \ldots, m, \ 0 \leq \epsilon_i \leq \epsilon_{+i}\}$$

is a hyperparallelepiped in R^m. Here B_i is a given positive definite symmetric matrix and $\{g_{ij}\}$ are given positive numbers.

D.2.1.2. Finite Mixtures of ϵ-close Distributions

In this case,

$$P_i(\epsilon_{+i}) = \left\{ p_i(\cdot) : p_i(x) = \sum_{j=1}^{M_i} a_{ij} p_{ij}(x), 0 \leq a_{ij} \leq 1, \sum_{j=1}^{M_i} a_{ij} = 1 \right\}, \quad i \in S,$$

where $p_{i1}(x) \equiv p_i^\circ(x)$ is the hypothetical (assumed) probability density function of the i-th class feature vector, and $\{p_{i2}(x), \ldots\ldots, p_{iM_i}(x)\}$ is a set of $M_i - 1$ probability densities ϵ-close to $p_{i1}(x)$ in a given probabilistic metric $\rho(\cdot)$:

$$\rho(p_{ij}(\cdot), p_{i1}(\cdot)) = \epsilon_{ij}, 0 \leq \epsilon_{ij} \leq \epsilon_{+i}, \quad j = 2, \ldots, M_i.$$

This distorted model has an intuitive interpretation. The class Ω_i is in fact partitioned into $M_i \geq 2$ subclasses $\Omega_{i1}, \ldots, \Omega_{i,M_i}$:

$$\Omega_i = \bigcup_{j=1}^{M_i} \Omega_{ij}, \Omega_{ij} \bigcap \Omega_{ik} = \emptyset, \ \ j \neq k.$$

A subclass Ω_{ij} is described by the probability density function $p_{ij}(x)$ and is observed with probability a_{ij}; it is common in practice that

$$\max_{2 \leq j \leq M_j} a_{ij} \ll a_{i1}.$$

D.2.1.3. Additive Distortions

A random feature vector $X_i \in R^N$ from class Ω_i to be classified is represented as an additive mixture:

$$X_i = X_i^\circ + \epsilon_i Y_i, 0 \leq \epsilon_i \leq \epsilon_{+i},$$

where $X_i^\circ \in R^N$ is an unobservable (nondistorted) random vector of feature variables with density $q(x; \theta_i^\circ)$, and $Y_i \in R^N$ is a random vector of distortions, which is independent of X_i° and has unknown probability density $h(y)$, where $h(y)$ has fixed first and second order moments:

$$a = \mathbf{E}\{Y_i\}, B = \mathbf{Cov}\{Y_i, Y_i\}.$$

Note that round-off errors, which are common in practice (Broffit *et al.*, 1981), can be also investigated with the help of the **D.2.1.3** distortion type (see Section 5.5).

Let us describe particular cases of nonparametric **D.2.2**-type distortions.

D.2.2. Nonparametric Distortions

D.2.2.1. Tukey—Huber Type Distortions

The family of admissible distorted densities for the i-th class has a nonparametric description:

$$P_i(\epsilon_{+i}) = \{p_i(x), x \in R^N : p_i(x) = (1 - \epsilon_i)p_i^\circ(x) + \epsilon_i h_i(x),$$

$$0 \leq \epsilon_i \leq \epsilon_{+i}, h_i(x) \geq 0, \int_{R^N} h_i(x)dx = 1\}.$$

Thus, the distorted density $p_i(x)$ is a mixture of the hypothetical (expected) distribution $p_i^\circ(x)$ and an arbitrary distribution $h_i(x)$ describing data *contamination* by *outliers*. This distortion type, widespread in applications, was initially described by (Tukey, 1960) for the one-dimensional case only ($N = 1$) when the hypothetical distribution is Gaussian: $p_i^\circ(x) = n_1(x \mid \mu_i^\circ, \sigma_i^{\circ 2})$. (Huber, 1981) generalized this distortion type. The **D.2.2.1** type of distortions is the most popular one in robust statistics (Barnett *et al.*, 1978), (Hampel *et al.*, 1986), (Krasnenker, 1980), (Launer, 1979), (Ershov, 1978), (Zypkin, 1984).

D.2.2.2. Distortions in $\mathbf{L}_2$-metric

Let $\psi_i(x) : R^N \to R^1$ be a nonnegative normed weight function:

$$\psi_i(x) \geq 0, \int_{R^N} \psi_i(x) dx = 1.$$

Then

$$P_i(\epsilon_{+i}) = \left\{ p_i(x), x \in R^N : \int_{R^N} \frac{\left(p_i(x) - p_i^\circ(x)\right)^2}{\psi_i(x)} dx = \epsilon_i^2, 0 \leq \epsilon_i \leq \epsilon_{+i} \right\}$$

is the ϵ_{+i} - neighborhood (in $\mathbf{L}_2$-metric) of the hypothetical distribution $p_i^\circ(x)$. In mathematical statistics, the hypothetical distribution $p_i^\circ(x)$ is often used as the weight function, and in this special case we obtain the so-called χ^2-*metric* (see, e.g., (Borovkov, 1984)):

$$\rho = \rho(p_i, p_i^\circ) = \left(\int_{R^N} \frac{\left(p_i(x) - p_i^\circ(x)\right)^2}{p_i^\circ(x)} dx \right)^{1/2} = \left(\int_{R^N} \frac{p_i^2(x)}{p_i^\circ(x)} dx - 1 \right)^{1/2}. \quad (2.16)$$

Note that χ^2-metric is functionally related with widely used in pattern recognition interclass *Mahalanobis distance*

$$\Delta = ((\mu - \mu_0)^T \Sigma^{-1} (\mu - \mu_0))^{1/2}, \quad (2.17)$$

where $\mu, \mu_0 \in R^N$ are mean vectors for two Gaussian distributions of feature variables, and Σ is the common covariance matrix. It follows from (2.16), (2.17) that

$$\rho = (e^{\Delta^2} - 1)^{1/2}.$$

Figure 2.2 plots this monotonous dependence between $\ln \rho$ and Δ showing the correspondence of the χ^2-metric and the Mahalanobis metric.

D.2.2.3. Distortions in Variation Metric

This type of distortions looks like **D.2.2.2** type; the difference is only in the type of metric on the set of probability density functions. The family of admissible distorted densities for the i-th class has the following nonparametric form:

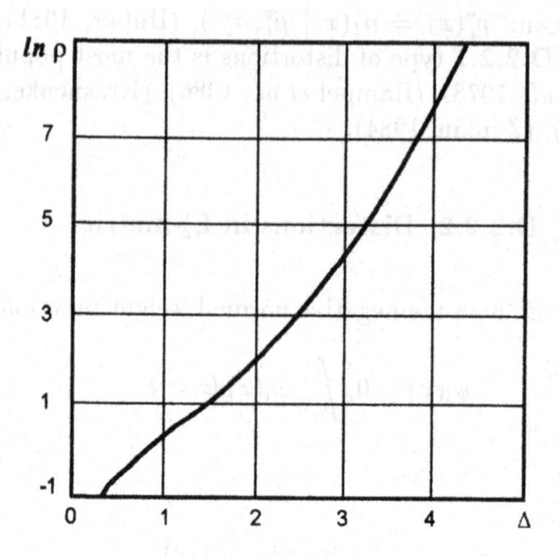

Figure 2.2: Plot of $\ln\rho$ dependence on Δ

$$\mathcal{P}_i(\epsilon_{+i}) = \{p_i(x), x \in R^N : p_i(x) \geq 0, \int_{R^N} p_i(x)dx = 1,$$

$$\rho(p_i, p_i^0) = \epsilon_i, 0 \leq \epsilon_i \leq \epsilon_{+i}\},$$

where

$$\rho = \rho(p_i, p_i^0) = \frac{1}{2}\int_{R^N} |p_i(x) - p_i^0(x)|dx$$

is probabilistic variation metric.

D.2.2.4. Random Distortions of Densities

In this case, the distorted probability distribution density of feature vector for the i-th class is represented in the form:

$$p_i(x) = p_i^\circ(x) + U_i(x), x \in R^N, i \in S, \tag{2.18}$$

where $U_i(x)$ is a random field with the following first and second order moments:

$$\mathbf{E}\{U_i(x)\} = A_i(x), \quad \mathbf{Cov}\{U_i(x), U_i(x')\} = \epsilon_i B_i(x, x'), 0 \leq \epsilon_i < \epsilon_{+i},$$

subject to the constraints:

$$U_i(x) \geq -p_i^\circ(x), \quad \int_{R^N} U_i(x)dx = 0.$$

The distortions of type (2.18) may appear, for example, when the probability density function is estimated from a random sample.

D.3. Distortions of Models for Training Samples

In constructing adaptive decision rules, the following classical model of the training sample A_i from the class Ω_i is traditionally used in pattern recognition theory: $A_i = \{x_{i1}, \ldots, x_{in_i}\} \subset R^N$ is a sequence of jointly independent random vectors with the same probability density $p_i^\circ(x)$, $i \in S$. In practice, this model is often distorted (Krzanowski, 1977), (Subrahmaniam *et al.*, 1978), (Kocherlakota *et al.*, 1987), (Tiku *et al.*, 1989).

D.3.1. Parametric ϵ-nonhomogeneity of Sample

The hypothetical probability density $p_i^\circ(x) = q(x; \theta_i^\circ)$ is an element of a regular m-parametric family Q , and the training sample A_i is nonhomogeneous. A sample value $x_{ij} \in R^N$ is a random vector with probability distribution density $q(x; \theta_{ij})$, where $\theta_{ij} \in \Theta_i$ is an unknown parameter value, and the set Θ_i of admissible parameter distortions for class Ω_i is defined in the same way as in the description of **D.2.1.1**-type distortions. The parametric ϵ-nonhomogeneity of the training sample is produced, for example, by instability of experiment conditions when this sample is being recorded.

D.3.2. Interdependence of Sample Elements

The training sample A_i from the class $\Omega_i (i \in S)$ is a sequence of dependent random vectors with the same marginal distribution density $p_i^\circ(x)$ (McLachlan, 1976), (Lawoko, *et al.*, 1988)). In Chapter 6 the following kinds of statistical dependence of training sample elements are defined and used:

- stationary vector time series,
- Markov sequence,
- vector autoregressive time series.

D.3.3. Misclassification of Training Sample

In the training sample A_i of size n_i only $(1 - \epsilon_i) \cdot 100\%$ observations are from $\Omega_i (i \in S)$, and the remaining $\epsilon_i \cdot 100\%$ observations are from $L - 1$ alien classes $\bigcup_{j \neq i} \Omega_j$; $0 \leq \epsilon_i \leq \epsilon_{+i}$. This situation is often appears in practice, for example, in biomedical data processing (Kox, 1974), (Launer, 1979), (Kharin, 1983). It is produced by "gross fool" and "teacher nonideality" in the process of statistical data archiving (McLachlan, 1972), (Aitchinson *et al.*, 1976), (Lachenbruch, 1979), (Chittineni, 1980), (Krishnan, 1988).

D.3.4. Outliers in Training Sample

The training sample $A_i (i \in S)$ from the i-th class is contaminated by *outliers* (Aivazyan *et al.*, 1981, 1989), (Tiku *et al.*, 1986); for the description of outliers it is convenient to use the Tukey-Huber model, as for **D.2.2.1**-type distortions of the observations to be classified.

This situation is often met in pattern recognition practice. For illustration, we present Figures 2.3, 2.4. They contain computer scatter diagrams of medical data collected at the Belarussian Research Institute of Onkology and Medical Radiology for *diagnostics of lung cancer*. The data were processed by the software package ROSTAN for robust statistical analysis at the Belarussian State University (see (Kharin *et al.*, 1994), (Abramovitch, Kharin, and Mashevskij, 1993)). There are $L = 2$ classes: $\Omega_1 = \{\text{patient has lung cancer}\}$, $\Omega_2 = \{\text{patient has chronical lung disease}\}$, which are observed in the 46-dimensional feature space.

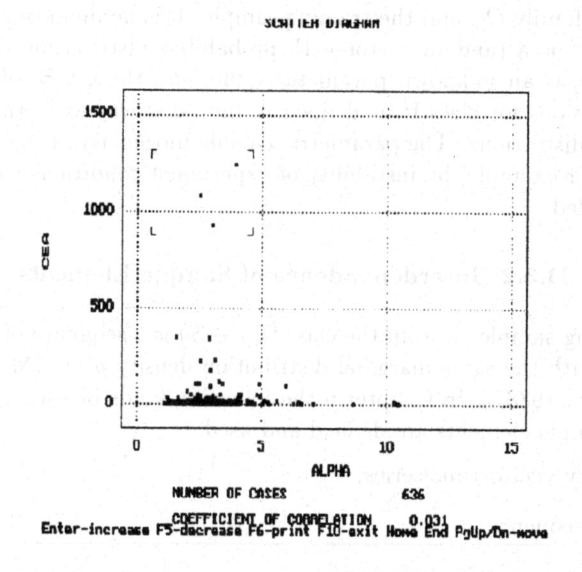

Figure 2.3: Outliers in a sample from Ω_1

Figure 2.3 presents a sample of 636 patients from Ω_1 in two-dimensional informative feature subspace: {ALPHA, CEA}. Figure 2.4 presents a composite sample of 934 patients from $\Omega_1 \cup \Omega_2$ in feature subspace: {CEA, NSE}. The description of the features ALPHA, CEA, NSE is given in Section 6.4, where the application problem of lung cancer recognition is considered.

Some outliers on the scatter diagrams are explained by presence of patients with terminal stages of cancer and by individuality of patients.

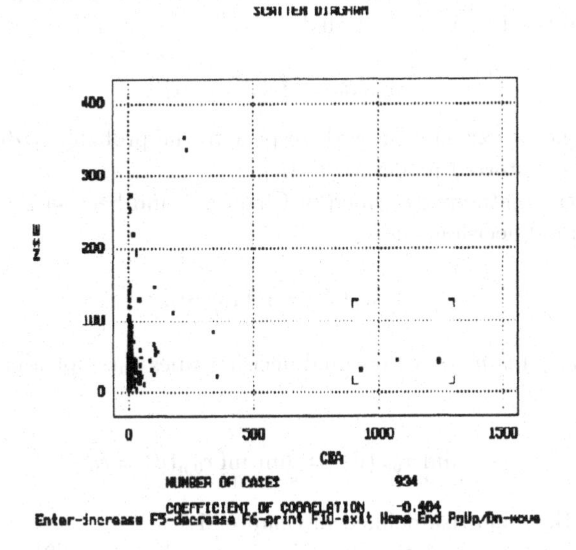

Figure 2.4: Outliers in composite sample

D.3.5. Missing Values of Feature Variables

There are $\epsilon_i \cdot 100\%$ of distorted observations in the training sample A_i for class $\Omega_i (i \in S)$, and each of them misses one or more values of feature variables ($0 \leq \epsilon_i \leq \epsilon_{+i}$). These "blanks" are produced in practice by many reasons (Little, 1987), (Hufnagel, 1988):

- measuring device fault during measurement;
- impossibility of measurement for some of feature variables;
- "nonideality" of statistical archive formation, etc.

2.3 Decision Rule Robustness Characteristics

Let $d = d_0(x; A)$ be an arbitrary decision rule constructed under some hypothetical model M_0, where $d \in S$ is the number of the class to which the feature vector $x \in R^N$ will be assigned and A is a training sample of size n used for the construction of the decision rule. Let M_ϵ denote an arbitrary admissible distorted data model for one of distortion types from Section 2.2 and let M_{ϵ_+} denote the set of admissible distorted data models, where $\epsilon_+ = \max_{i \in S} \epsilon_{+i}$ is the maximal admissible value for the distortion level. As in Chapter 1, the classification performance of the decision rule

$d_0(\cdot)$ in a situation where experimental data are fitted to distorted model $M_\epsilon \in M_{\epsilon+}$ will be characterized by risk functional:

$$r_{\epsilon,n}(d_0) = \mathbf{E}\{w_{\nu,d_0(x;A)}\}, \tag{2.19}$$

where $\mathbf{E}\{\cdot\}$ denotes expectation with respect to the probability distribution corresponding to the distorted model $M_\epsilon \in M_{\epsilon+}$.

Note that the methods described in Chapter 1 and Section 2.1 allow us to construct the optimal decision rule:

$$r_{0,n}(d_0) = \inf_{d(\cdot)} r_{0,n}(d), \tag{2.20}$$

or asymptotically (as $n \to \infty$) optimal decision rules $d_0(\cdot)$ for a given hypothetical model M_0 :

$$\lim_{n\to\infty} r_{0,n}(d_0) = \lim_{n\to\infty} \inf_{d(\cdot)} r_{0,n}(d) = r_0. \tag{2.21}$$

Here r_0 is the Bayesian risk defined in Chapter 1.

Distortions result in violation of optimality conditions (2.20), (2.21). Let us call the functional

$$r_{+n} = r_{+n}(d_0) = \sup_{M_\epsilon \in M_{\epsilon+}} r_{\epsilon,n}(d_0) \tag{2.22}$$

the *guaranteed (upper) risk* for the decision rule $d_0(\cdot)$ in presence of distortions $M_\epsilon \in M_{\epsilon+}$. Stability of decision rule $d_0(\cdot)$ will be evaluated by the *absolute risk increment*:

$$\beta_{+n} = \beta_{+n}(d_0) = r_{+n}(d_0) - r_0, \tag{2.23}$$

by the *robustness factor* (when $r_0 > 0$):

$$\kappa_{+n} = \kappa_{+n}(d_0) = (r_{+n}(d_0) - r_0)/r_0, \tag{2.24}$$

and by the *asymptotic robustness factor*:

$$\kappa_+ = \kappa_+(d_0) = \lim_{n\to\infty} \kappa_{+n}(d_0). \tag{2.25}$$

Since $M_0 \in M_{\epsilon+}$, (2.21) - (2.25) imply

$$\beta_{+n}(d_0) \geq 0, \kappa_{+n}(d_0) \geq 0, \kappa_+(d_0) \geq 0.$$

The smaller $\beta_{+n}, \kappa_{+n}, \kappa_+$, the more stable the decision rule $d_0(\cdot)$.

Let us say that a decision rule $d_0(\cdot)$ is *asymptotically robust decision rule of order* $\nu(d_0) = \nu_0$, if ν_0 is the largest positive integer for which

$$\kappa_+(d_0) = o\left(\epsilon_+^{\nu_0+1}\right). \tag{2.26}$$

It means that for the situations with fixed distortion level ϵ_+ and increasing training sample size n the guaranteed risk differs from the Bayesian risk only by the value of order $o\left(\epsilon_+^{\nu_0+1}\right)$. If $\nu_0 = +\infty$, that is, if

$$\lim_{n \to \infty} \kappa_{+n}(d_0) = 0, \tag{2.27}$$

then $d_0(\cdot)$ is referred to as ***asymptotically absolutely robust decision rule.***

Bearing in mind (1.21), let

$$r^* = \sum_{i,j=1}^{L} \pi_i w_{ij} / L \tag{2.28}$$

denote the risk value for the "worst" decision rule that makes decisions by equiprobable coin tossing. Similarly to (Hampel *et al.*,, 1986), the ***breakdown point*** for a decision rule $d = d(x; A)$ is the maximal value ϵ_+^* of distortion levels $\{\epsilon_{+i}\}$ for which the guaranteed risk r_{+n} is equal to r^* :

$$\epsilon_+^* = \min\{\alpha : \max_{\{0 \le \epsilon_{+i} \le \alpha\}} r_{+n}(d) = r^*\}. \tag{2.29}$$

The practical importance of this breakdown point ϵ_+^* is in fact the following: when the distortion levels $\{\epsilon_{+i}\}$ reach the critical value ϵ_+^*, then there are conditions under which the application of the decision rule $d(\cdot)$ leads to decisions as bad as for the "equiprobable coin tossing" decision rule.

Let δ, δ' be any given positive numbers characterizing the desirable level of decision rule robustness. We shall say that the decision rule $d(\cdot)$ satisfies the ***absolute δ-robustness condition,*** if

$$\beta_{+n}(d) = r_{+n}(d) - r_{0n}(d) \le \delta, \tag{2.30}$$

and it satisfies the ***relative δ'-robustness condition,*** if

$$\kappa_{+n}(d) = (r_{+n}(d) - r_{0n}(d))/r_{0n}(d) \le \delta'. \tag{2.31}$$

Passing to the limit as $n \to \infty$, similarly to (2.25), it is possible to define asymptotic versions of δ-robustness conditions (2.30), (2.31).

A decision rule $d = d^*(x; A)$ with the minimal value of guaranteed risk for all admissible distortions:

$$r_{+n}(d^*) = \inf_{d(\cdot)} r_{+n}(d) \tag{2.32}$$

is referred to as ***robust (with respect to distortions M_{ϵ_+}) decision rule.***

Note in conclusion that when searching for the robust decision rule one can extend the family of decision rules within which the rules with minimal guaranteed risk are sought by the randomized decision rules (see Chapter 5).

Chapter 3

Robustness of Parametric Decision Rules and Small-sample Effects

This chapter is devoted to analysis of risk robustness with respect to small-sample effects for commonly used plug-in decision rules derived from the Bayesian decision rule by substitution of unknown multivariate probability densities by their parametric estimates. These parametric estimates are constructed using the family of minimum contrast estimators of parameters (including ML-estimators, LS-estimators, etc.). We develop the method of asymptotic expansion for risk functional and construct general asymptotic expansions and approximations for the robustness factor. High accuracy of these approximations is demonstrated by computer experiments.

3.1 Asymptotic Risk Expansion for Decision Rules Using Minimum Contrast Estimators

In this section we develop the asymptotic risk expansion method (Chibisov, 1973, 1980), (Kharin, 1983) to evaluate the robustness of the parametric plug-in decision rule with respect to small-sample effects.

Assume that a separable m-parametric family of probability densities $\mathcal{P}$ is defined in the feature space R^N :

$$\mathcal{P} = \left\{p(x;\theta), x \in R^N : \theta \in \Theta \subseteq R^m\right\};$$

$$\theta_1^0 = (\theta_{1j}^0), \dots, \theta_L^0 = (\theta_{Lj}^0) \in \Theta,\ j = 1, \dots, m,$$

are L different fixed points in the parametric space Θ. Let the observations of L classes $\Omega_1, \dots, \Omega_L$ appear in R^N with prior probabilities $\pi_1, \dots, \pi_L$. An observation x from Ω_i is a random vector $X_i \in R^N$ with probability density $p(x;\theta_i^0)(i \in S)$. True parameter values $\{\theta_i^0\}$ are unknown.

There exists a classified training sample A:

$$A = (A_1^T \vdots A_2^T \vdots \ldots \vdots A_L^T) \in R^{nN};$$

$$A_i^T = (x_{i1}^T \vdots \ldots \vdots x_{in_i}^T) \in R^{n_i N}$$

is a random training sample consisting of n_i observations from class Ω_i; $n = n_1 + \ldots + n_L$ is the total sample size; $A_1, \ldots, A_L$ are jointly independent. The loss matrix $W = (w_{ik})$ is given. A random feature vector $X \in R^N$ from one of classes $\{\Omega_i\}$ observed independently from A is subject to classification.

According to Theorem 1.1, for a known true value of the composed parameter vector $\theta^{oT} = (\theta_1^{oT} \vdots \ldots \vdots \theta_L^{oT})$ the classification risk (expected losses) is minimized for the Bayesian decision rule (1.34):

$$d = d_0(x) = \sum_{i=1}^{L} i \mathbf{I}_{V_i^o}(x) = \sum_{i=1}^{L} i \prod_{k \neq i}^{L} \mathbf{1}(f_{ki}^0(x)), x \in R^N, \tag{3.1}$$

where $V_i^0 \subset R^N$ is the region of the Bayesian decision $d = i$ making,

$$f_{ki}^0(x) = \sum_{l=1}^{L} c_{lki} p(x; \theta_l^0), \qquad c_{lki} = \pi_l (w_{lk} - w_{li}) \tag{3.2}$$

is the Bayesian discriminant function for the pair of classes Ω_i, Ω_k $(i \neq k)$. The Bayesian risk is

$$r_0 = \sum_{i=1}^{L} \pi_i \mathbf{E} \left\{ w_{i, d_0(X_i)} \right\}.$$

According to (3.1),

$$r_0 = \sum_{k=1}^{L} \int_{R^N} \mathbf{I}_{V_k^o}(x) f_k^0(x) dx, \; f_k^0(x) = \sum_{i=1}^{L} \pi_i w_{ik} p(x; \theta_i^0). \tag{3.3}$$

Since $\{\theta_i^0\}$ are unknown, the plug-in decision rule (2.10) is used:

$$d = d(x; A) = \sum_{i=1}^{L} i \mathbf{I}_{V_i}(x) = \sum_{i=1}^{L} i \prod_{k \neq i} \mathbf{1}(f_{ki}(x; \hat{\theta})), x \in R^N, \tag{3.4}$$

where $\hat{\theta}^T = (\hat{\theta}_1^T \vdots \ldots \vdots \hat{\theta}_L^T)$ is the statistical estimator for θ^0 and

$$f_{ki}(x; \hat{\theta}) = \sum_{l=1}^{L} c_{lki} p(x; \hat{\theta}_l) \tag{3.5}$$

is the estimator for $f_{ki}^0(x)$ (with respect to sample A) derived by substitution of the minimum contrast estimator (MC-estimator) $\hat{\theta}_l$ (with respect to sample A_l) for θ_l^0 in (3.2) $(l \in S)$.

Let us define the family of MC-estimators $\hat{\theta}_l$, according to (Pfanzagl, 1969), (Chibisov, 1973). Denote: $\bar{\Theta}$ is the closure for Θ, $\bar{R}$ is the extended real line. A family of Borel functions $\{g(\cdot;\theta') : \theta' \in \bar{\Theta}\}$ defining Borel transformation $R^N \to \bar{R}$ is called *family of contrast functions* for the density family $\mathcal{P}$, if

$$\mathbf{E}_{\theta_i^0}\{g(X_i;\theta_i)\} = \int_{R^N} g(x;\theta_i)p(x;\theta_i^0)dx > \mathbf{E}_{\theta_i^0}\{g(X_i;\theta_i^0)\}, \tag{3.6}$$

for any $\theta_i^0 \in \Theta, \theta_i \in \bar{\Theta}, \theta_i \neq \theta_i^0$, where $\mathbf{E}_{\theta_i^0}\{\cdot\}$ denotes expectation with respect to distribution $p(\cdot;\theta_i^0)$.

Consider the following function:

$$L_i(\theta_i) = \sum_{j=1}^{n_i} g(x_{ij};\theta_i)/n_i, \qquad \theta_i \in \bar{\Theta}. \tag{3.7}$$

The *MC-estimator* of the parameter θ_i^0 (with respect to sample A_i) is the statistic ($i \in S$):

$$\hat{\theta}_i = \arg\min_{\theta_i \in \bar{\Theta}} L_i(\theta_i). \tag{3.8}$$

Under fairly weak regularity conditions (Chibisov, 1973), (see also Section 1.6), MC-estimators are consistent.

The family of MC-estimators is sufficiently rich and contains many well known estimators. Some examples:

1) If

$$g(x;\theta_i) = -\ln p(x;\theta_i), \tag{3.9}$$

then the MC-estimator turns into the maximum likelihood estimator (ML-estimator);

2) If θ_i is a location parameter and

$$\mathbf{E}_{\theta_i^0}\left\{\nabla_{\theta_i^0} g(X_i;\theta_i^0)\right\} = \mathbf{O}_m,$$

$\nabla^2_{\theta_i^0} g(x_i;\theta_i^0)$ is a nonnegative definite matrix, then we obtain the family of M-estimators well known in robust statistics (Huber, 1981), (Hampel *et al.*, 1986).

Let us construct an asymptotic expansion of the classification risk for the plug-in decision rule (3.4), (3.5), (3.8). Let us define the conditional risk for this plug-in decision rule at fixed training sample A:

$$\mathrm{rc}(A) = \sum_{i=1}^{L} \pi_i \mathbf{E}_{\theta_i^0}\left\{w_{i,d(X_i;A)} \mid A\right\} \tag{3.10}$$

and the unconditional risk, which according to (3.4) assumes the form:

$$r = \mathbf{E}\{\mathrm{rc}(A)\} = \mathbf{E}\left\{\sum_{j=1}^{L} \int_{R^N} \mathbf{I}_{V_j}(x) f_j^0(x)dx\right\}. \tag{3.11}$$

Denote:

$$n_* = \min_{i \in S} n_i, \tau_i = n_i^{-1/2}, \tau_* = n_*^{-1/2}, g_i = g(X; \theta_i);$$

$$g_i' = \nabla_{\theta_i} g(X; \theta_i) = (g_{ij}),$$

which is an m-vector-column;

$$g_i'' = \nabla^2_{\theta_i} g(X; \theta_i) = (g_{ijk}), \tag{3.12}$$

which is an $m \times m$-matrix;

$$a_i = a_i(\theta_i) = \mathbf{E}_{\theta_i}\{g_i\}, \quad a_i' = \mathbf{E}_{\theta_i}\{g_i'\}, \quad a_i'' = \mathbf{E}_{\theta_i}\{g_i''\},$$

$$\xi_i = \xi_i(\theta_i) = \tau_i^{-1}(L_i(\theta_i) - a_i(\theta_i)), \ \overset{\circ}{g_i'} = g_i'|_{\theta_i = \theta_i^0}.$$

Similarly, we shall use the following notations:

$$\overset{\circ}{g_i''}, \overset{\circ}{a_i''}, \xi_i', \overset{\circ}{\xi_i'}, p_i', p_i'', p_i''', \overset{\circ}{p_i}, \overset{\circ}{p_i'}, \overset{\circ}{p_i''}, \overset{\circ}{f_{kj}}, f_{kj}', f_{kj}'', f_{kj}'''.$$

We shall assume that the family $\mathcal{P}$ satisfies the Chibisov regularity conditions (Section 1.6) of order $k = 1$. These conditions are typical when MC-estimators are used. Let us formulate some auxiliary statements.

Lemma 3.1 *If Chibisov regularity conditions hold and the third order moments for* $\overset{\circ}{g_i'}$ *exist, then the random deviation*

$$\Delta\theta_i = (\Delta\theta_{ij}) = \theta_i - \theta_i^0 (j = 1, \ldots, m)$$

of the MC-estimator has third order moments, and the following asymptotic expansions hold:

- *for the bias:*

$$b_i = (b_{ij}) = \mathbf{E}\{\Delta\theta_i\} = \mathcal{O}(\tau_i^2)\mathbf{1}_m (i \in S), \tag{3.13}$$

- *for the covariance matrix:*

$$\mathbf{E}\{\Delta\theta_i \Delta\theta_k^T\} = \delta_{ik}\tau_i^2 V_i + \mathcal{O}(\tau_*^3)\mathbf{1}_{m \times m} \quad (i, k \in S), \tag{3.14}$$

$$V_i = (\overset{\circ}{a_i''})^{-1} \mathbf{E}_{\theta_i^0}\left\{\overset{\circ}{g_i'} \overset{\circ}{g_i'^T}\right\} (\overset{\circ}{a_i''})^{-1} = (v_{ijl}) \quad (j, l = 1, \ldots, m),$$

- *for the third order moments:*

$$\mathbf{E}\{|\Delta\theta_{ij_1}\Delta\theta_{kj_2}\Delta\theta_{lj_3}|\} = \mathcal{O}(\tau_*^3) \quad (j_1, j_2, j_3 = 1, \ldots, m; i, k, l \in S). \tag{3.15}$$

According to (3.11), the risk evaluation problem necessitates the analysis of the integrals of form (3.4) of random unit step functions. In this connection, we formulate some properties of these functionals and their derivatives. Let us introduce the generalized Dirac δ-function (Gelfand, 1959): $\delta(z) = d\mathbf{1}(z)/dz$, and its generalized k-th order derivative :

$$\delta^{(k)}(z) = d^{k+1}\mathbf{1}(z)/dz^{k+1}.$$

Lemma 3.2 *If $a = a(x), x \in R^N, a \in R^1$ is a differentiable function and*

$$\mathrm{mes}_N \{x :\mid \nabla a(x) \mid = 0\} = 0,$$

where mes *is the Lebesgue measure, then for any integrable function $\psi(x)$,*

$$\int_{R^N} \psi(x)\delta^{(k)}(a(x))dx = (-1)^k \frac{d^k}{dV^k} \int_{a(x)=V} \psi(x) \mid \nabla a(x) \mid^{-1} ds_{N-1} \mid_{V=0},$$

where the integral on the right side is the surface integral over the $(N-1)$-dimensional surface $a(x) = V$.

Lemma 3.3 *Let $a_1(x), \ldots, a_k(x)$ be differentiable functions $R^N \to R^1$ $(1 \leq k < N)$, and*

$$[a_1(x), \ldots, a_k(x)] = \sqrt{\det(\nabla^T a_i(x) \nabla a_j(x))} \quad (i, j = 1, \ldots, k).$$

If

$$\mathrm{mes}_N \{x : [a_1(x), \ldots, a_k(x)] = 0\} = 0,$$

then

$$\int_{R^N} \psi(x) \prod_{i=1}^{k} \delta(a_i(x))dx = \int_{\substack{a_1(x)=0, \\ \ldots \\ a_k(x)=0}} \psi(x)[a_1(x), \ldots, a_k(x)]^{-1} ds_{N-k}.$$

Lemma 3.4 *Under the conditions of Lemma 3.3 (for $k = 2$)*

$$\int_{R^N} \psi(x)\delta'(a_1(x))\delta(a_2(x))dx =$$

$$= -\frac{d}{dV} \int_{\substack{a_1(x)=V, \\ a_2(x)=0}} \psi(x)[a_1(x), a_2(x)]^{-1} ds_{N-2} \Big|_{V=0}.$$

Lemma 3.5 *If the conditions of Lemma 3.3 hold and $X \in R^N$ is a random vector with probability density $\psi(x)$, then the joint distribution density for the statistics $a_1 = a_1(X), \ldots, a_k = a_k(X)$ is given by the formula*

$$p_{a_1,\ldots,a_k}(V_1,\ldots,V_k) = \int_{\substack{a_1(x)=V_1,\\ \cdots \\ a_k(x)=V_k}} \psi(x)[a_1(x),\ldots,a_k(x)]^{-1} ds_{N-k}.$$

Lemmas 3.2–3.5 represent special properties of integrals with generalized δ-functions; they are proved in (Kharin, 1983) using the results from (Higgins, 1975) and (Shilov, 1972).

Let us assume the notations:

$$\overset{\circ}{\Gamma}_{tj} = \left\{ x : \overset{\circ}{f}_{tj}(x) = 0 \right\} \subset R^N$$

is a Bayesian discriminant hypersurface for the pair of classes $\{\Omega_t, \Omega_j\}$; $\overset{\circ}{\Gamma}{}'_{tj} \subseteq \overset{\circ}{\Gamma}_{tj}$ is the part of the hypersurface $\overset{\circ}{\Gamma}_{tj}$ that is the boundary of the regions V_t^0, V_j^0; $\Gamma_{tj} = \{x : f_{tj}(x;\theta) = 0\} \subset R^N$ is the hypersurface depending on parameters θ; $p_{f_{tjl}, f_{t'jl}, f_{t''jl}}(z_1, z_2, z_3)$ is joint probability distribution density of the statistics

$$f_{tjl} = f_{tj}(X_l), f_{t'jl} = f_{t'j}(X_l), f_{t''jl} = f_{t''j}(X_l),$$

where $j \neq t \neq t' \neq t''$, $j, l, t, t', t'' \in S$;

$$\overline{\Delta\theta_i} = \sqrt{n_i} V_i^{-1/2} \Delta\theta_i \tag{3.16}$$

is the normed deviation for the estimate $\hat{\theta}_i$ and $U_{\theta_i^0} \subset \Theta$ is a neighborhood of the point θ_i^0;

$$\alpha_{i\mu} = V_i^{1/2} \left(1/2 \sum_{j=1}^{L-1} \sum_{t=j+1}^{L} c_{itj} c_{\mu tj} \int_{\overset{\circ}{\Gamma}{}'_{tj}} \overset{\circ}{p}{}'_i \overset{\circ}{p}{}'^T_\mu \mid \nabla_x \overset{\circ}{f}_{tj}(x) \mid^{-1} ds_{N-1} \right) V_\mu^{1/2} \tag{3.17}$$

is an $(m \times m)$ - matrix $(i, \mu \in S)$.

Theorem 3.1 *Let Chibisov regularity conditions hold and let for any sufficiently small neighborhoods $U_{\theta_1^0}, \ldots, U_{\theta_L^0} \subset \Theta$ the partial derivatives*

$$p'(x;\theta_i),\ p''(x;\theta_i),\ p'''(x;\theta_i)$$

with respect to θ_i be uniformly bounded on the hypersurfaces

$$\Gamma_{tj} \subset R^N (\theta_i \in U_{\theta_i^0}, x \in \Gamma_{tj}, i, j, t \in S).$$

Moreover, suppose that $\{p(x;\theta_i)\}$ are triply differentiable with respect to x, $\{p'(x;\theta_i)\}$ are twice differentiable with respect to x, and $\{p''(x;\theta_i)\}$ are differentiable with respect to x. If for arbitrary $\{\theta_i \in U_{\theta_i^\circ}\}$, $j,l,t,t',t'' \in S$ $(j \neq t \neq t' \neq t'')$ the probability distribution densities

$$p_{f_{tjl}}(0) < \infty,\ p_{f_{tjl}f_{t'jl}}(0,0) < \infty,\ p_{f_{tjl}f_{t'jl}f_{t''jl}}(0,0,0) < \infty, \tag{3.18}$$

are bounded, then as $n_* \to \infty (n_1, \ldots, n_L \to \infty)$ the conditional classification risk (3.10) admits the following stochastic expansion:

$$\mathrm{rc}(A) = r_0 + \sum_{i,\mu=1}^{L} \frac{1}{\sqrt{n_i n_\mu}} (\overline{\Delta\theta_i})^T \alpha_{i\mu} (\overline{\Delta\theta_\mu}) + \zeta_{n*}, \tag{3.19}$$

$$\zeta_{n*} = n_*^{-3/2} \sum_{i,\mu,p=1}^{L} \sum_{s,\tau,q=1}^{m} \zeta_{is\mu\tau pq} \sqrt{(n_*/n_i)(n_*/n_\mu)(n_*/n_p)}\, \overline{\Delta\theta}_{is} \overline{\Delta\theta}_{\mu\tau} \overline{\Delta\theta}_{pq},$$

where $\zeta_{is\mu\tau pq}$ is a bounded random variable.

Proof. Denote

$$q(x;\theta) = \sum_{j=1}^{L} \overset{\circ}{f}_j(x) \prod_{k \neq j} \mathbf{1}(f_{kj}(x;\theta)). \tag{3.20}$$

Then according to (3.1), (3.3), (3.10) we have:

$$r_0 = \int_{R^N} q(x;\theta^0)dx,$$

$$\mathrm{rc}(A) = \int_{R^N} q(x;\hat{\theta})dx = \int_{R^N} q(x;\theta^0 + \Delta\theta)dx. \tag{3.21}$$

The Chibisov regularity conditions imply $\Delta\theta \overset{a.s.}{\to} \mathbf{O}_{mL}$, therefore there exists a value $\bar{n}_*$ such that $\hat{\theta}_i \in U_{\theta_i^\circ}$ $(i \in S)$ with probability one for $n_* > \bar{n}_*$. Apply now the Taylor formula to the integral in (3.21) that defines the function of $\{\theta_l\}$:

$$\mathrm{rc}(A) = r_0 + \sum_{i=1}^{L}\sum_{s=1}^{m} \Delta\theta_{is}(B_{is} + \sum_{\mu=1}^{L}\sum_{\tau=1}^{m} \Delta\theta_{\mu\tau}(B_{is\mu\tau} + \tag{3.22}$$

$$+ \sum_{p=1}^{L}\sum_{q=1}^{m} B_{is\mu\tau pq}\Delta\theta_{pq}));$$

$$B_{is} = \int_{R^N} \frac{\partial q(x;\theta^0)}{\partial \theta_{is}^\circ} dx,\ B_{is\mu\tau} = 1/2 \int_{R^N} \frac{\partial^2 q(x;\theta^0)}{\partial\theta_{is}^\circ \partial\theta_{\mu\tau}^\circ} dx,$$

$$B_{is\mu\tau pq} = 1/6 \int_{R^N} \frac{\partial^3 q(x;\bar{\theta})}{\partial\bar{\theta}_{is}\partial\bar{\theta}_{\mu\tau}\partial\bar{\theta}_{pq}} dx, \tag{3.23}$$

where $|\bar{\theta}_l - \theta_l^0| < |\Delta\theta_l|$, and therefore, $\bar{\theta}_l \in U_{\theta_l^0}$, $\bar{\theta}_l \overset{a.s.}{\to} \theta_l^0 \quad (l \in S)$.

Substituting (3.20) into (3.23) and using the following properties of the δ-function (Gelfand, 1959) : $\delta^{(k)}(-z) = (-1)^k \delta^{(k)}(z), z\delta(z) \equiv 0$, we obtain:

$$B_{is} = \int_{R^N} \sum_{j=1}^{L-1} \sum_{t=j+1}^{L} \overset{\circ}{f}_{tj}(x)\delta(\overset{\circ}{f}_{tj}(x))\, \overset{\circ}{f'}_{tjis}(x) \mathbf{I}_{V_j^\circ U V_t^\circ}(x)dx \equiv 0. \tag{3.24}$$

Here it was taken into account that according to (3.2), (3.3),

$$\overset{\circ}{f}_{tj}(x) = \overset{\circ}{f}_t(x) - \overset{\circ}{f}_j(x) = -\overset{\circ}{f}_{jt}(x).$$

Similarly, taking into account the property (Gelfand, 1959) $z\delta'(z) = -\delta(z)$, we have

$$B_{is\mu\tau} = 1/2 \sum_{j=1}^{L-1} \sum_{t=j+1}^{L} \int_{R^N} \overset{\circ}{f'}_{tjis} \overset{\circ}{f'}_{tj\mu\tau}\, \mathbf{I}_{V_j^\circ U V_t^\circ}(x)\delta(\overset{\circ}{f}_{tj}(x))dx.$$

According to (3.2), $\overset{\circ}{f'}_{tjis} = c_{itj}\, \overset{\circ}{p'}_{is}$, therefore by Lemma 3.2 at $k = 0$ we find:

$$B_{is\mu\tau} = 1/2 \sum_{j=1}^{L-1} \sum_{t=j+1}^{L} c_{itj} c_{\mu tj} \int_{\overset{\circ}{\Gamma}'_{tj}} \overset{\circ}{p'}_{is} \overset{\circ}{p'}_{\mu\tau} |\nabla \overset{\circ}{f}_{tj}(x)|^{-1} ds_{N-1}. \tag{3.25}$$

By similar transformation of $B_{is\mu\tau pq}$ in (3.23) with the use of Lemmas 3.2 - 3.4 we find the necessary and sufficient conditions for finiteness of $|B_{is\mu\tau pq}|$:

$$\sup_{\{\theta_i \in U_{\theta_i^\circ}\}} |B_{is\mu\tau pq}| < \infty, \tag{3.26}$$

which are satisfied by (3.19) and Lemma 3.5.

Now we substitute (3.24), (3.25) into (3.22) and express $\{\Delta\theta_i\}$ in terms of $\{\overline{\Delta\theta_i}\}$ according to (3.16). Using the notations (3.17) and

$$\zeta_{is\mu\tau pq} = \sum_{t,t',t''=1}^{m} B_{it\mu t' pt''} (V_i^{1/2})_{ts} (V_\mu^{1/2})_{t'\tau} (V_p^{1/2})_{t''q},$$

we obtain (3.19), and moreover, by (3.26),

$$\sup_{\{\theta_i \in U_{\theta_i^\circ}\}} |\zeta_{is\mu\tau pq}| < \infty. \tag{3.27}$$

■

The stochastic expansion (3.19) has an obvious interpretation. The conditional classification risk at a fixed training sample A is, according to (3.19), the sum of three terms:

$$\mathrm{rc}(A) = r_0 + \delta r(A) + \zeta_{n_*}. \tag{3.28}$$

The first term, i.e., the Bayesian risk r_0, represents the minimum admissible classification risk, which is attainable at exactly known true parameter values $\{\theta_i^0\}$ only. The second term

$$\delta r(A) = \sum_{i,\mu=1}^{L} (n_i n_\mu)^{-1/2} (\overline{\Delta\theta_i})^T \alpha_{i\mu} (\overline{\Delta\theta_i}) \tag{3.29}$$

with the help of (3.17) may be equivalently represented as

$$\delta r(A) = 1/2 \sum_{j=1}^{L-1} \sum_{t=j+1}^{L} \int_{\overset{\circ}{\Gamma}{}'_{tj}} \left(\sum_{i=1}^{L} c_{itj} n_i^{-1/2} \overset{\circ}{p}{}_i'^T V_i^{1/2} \overline{\Delta\theta_i} \right)^2 \times$$
$$\times \mid \nabla \overset{\circ}{f}_{tj}(x) \mid^{-1} ds_{N-1} \geq 0;$$

$\delta r(A) \geq 0$ is the nonnegative random increment of conditional classification risk induced by prior uncertainty: the true parameter values $\{\theta_i^0\}$ are unknown and estimated by the sample A. The third term ζ_{n_*} in (3.28) is the remainder of the stochastic expansion, and it is a random variable that converges to zero ($\zeta_{n_*} \overset{a.s.}{\to} 0$) sufficiently fast as $n_* \to \infty$.

Corollary 3.1 *The plug-in decision rule (3.4), (3.5), (3.8) is strongly consistent:*

$$\text{rc}(A) \overset{a.s.}{\to} r_0 \quad as \quad n_* \to \infty.$$

Now let us clarify the asymptotic probability distribution of the conditional risk rc(A). Let F denote the composed $(Lm \times Lm)$-matrix consisting of the following blocks ($(m \times m)$-matrices):

$$F_{i\mu} = \sqrt{\lambda_i \lambda_\mu} \alpha_{i\mu} \qquad (i, \mu = 1, \ldots, L), \tag{3.30}$$

where $\lambda_1, \ldots, \lambda_L \in [0, 1]$ are coefficients that are undefined at the moment.

Lemma 3.6 *The matrix F given by (3.30), (3.17) is a symmetric nonnegative definite matrix and*

$$\text{tr}(F) = 1/2 \sum_{j=1}^{L-1} \sum_{t=j+1}^{L} \int_{\overset{\circ}{\Gamma}{}'_{tj}} \mid \nabla \overset{\circ}{f}_{tj}(x) \mid^{-1} \sum_{i=1}^{L} \lambda_i c_{itj}^2 \overset{\circ}{p}{}_i'^T V_i \overset{\circ}{p}{}_i' ds_{N-1} \geq 0.$$

Proof is conducted by direct check of these properties. ■

Theorem 3.2 *If the conditions of Theorem 3.1 hold, $n_*/n_i \to \lambda_i (i \in S)$ as $n_* \to \infty$ and $\text{tr}(F) > 0$, then the asymptotic probability distribution of the normed random risk increment $n_*(\text{rc}(A) - r_0)$ coincides with the probability distribution of the random quadratic form:*

$$\mathcal{L}\{n_*(\mathrm{rc}(A) - r_0)\} \to \mathcal{L}\{\sum_{i=1}^{Lm} \chi_i \eta_i^2\}, \tag{3.31}$$

where $\eta_1, \ldots, \eta_{Lm}$ are independent Gaussian random variables with standard distribution $N_1(0,1)$ and $\chi_1 \geq \chi_2 \geq \ldots \geq \chi_{Lm} > 0$ are the eigenvalues of the matrix F.

Proof. By (3.28),

$$n_*(\mathrm{rc}(A) - r_0) = n_* \delta r(A) + n_* \zeta_{n_*}.$$

By Theorem 3.1, $n_* \zeta_{n_*} \xrightarrow{a.s.} 0$, therefore

$$\lim_{n_* \to \infty} \mathcal{L}\{n_*(\mathrm{rc}(A) - r_0)\} = \lim_{n_* \to \infty} \mathcal{L}\{n_* \delta r(A)\}. \tag{3.32}$$

By (3.29),

$$n_* \delta r(A) = \sum_{i,\mu=1}^{L} (\overline{\Delta\theta_i})^T (n_*/n_i)^{1/2} (n_*/n_\mu)^{1/2} \alpha_{i\mu} \overline{\Delta\theta_\mu}.$$

Under the conditions of Theorem 3.2 according to (3.30),

$$(n_*/n_i)^{1/2} (n_*/n_\mu)^{1/2} \alpha_{i\mu} \to F_{i\mu} (i, \mu \in S).$$

According to (Chibisov, 1980), the vector $\overline{\Delta\theta_i}$ is asymptotically normally distributed:

$$\mathcal{L}\{\overline{\Delta\theta_i}\} \to \mathcal{L}\{\Xi_i\} = N_m(\mathbf{O}_m, \mathbf{I}_m),$$

where $\Xi_1, \ldots, \Xi_L \in R^m$ are jointly independent Gaussian vectors with standard distribution. Therefore,

$$\mathcal{L}\{n_* \delta r(A)\} \to \mathcal{L}\{\Xi^T F \Xi\},$$

where $\Xi^T = (\Xi_1^T \vdots \ldots \vdots \Xi_L^T) \in R^{Lm}$ is a composed random vector with distribution $N_{Lm}(\mathbf{O}_{Lm}, \mathbf{I}_{Lm})$. By Lemma 3.6, there exists an orthogonal $(Lm \times Lm)$-matrix C such that

$$C^T F C = \chi = \mathrm{diag}\{\chi_1, \cdots, \chi_{Lm}\}, \quad F = C \chi C^T.$$

Then the random quadratic form is written as

$$\Xi^T F \Xi = \eta^T \chi \eta = \sum_{i=1}^{Lm} \chi_i \eta_i^2,$$

where $\eta = C^T \Xi$ and, by (Anderson, 1958), its distribution is $N_{Lm}(\mathbf{O}_{Lm}, \mathbf{I}_{Lm})$. Therefore (3.32) implies (3.31). ■

Corollary 3.2 *If* $\chi_1 = \ldots = \chi_{Lm}$, *then as* $n_* \to \infty$ *the normed random deviation of risk* $n_*(\operatorname{rc}(A) - r_0)/\chi_1$ *is asymptotically distributed according to the chi-square probability distribution law with* Lm *degrees of freedom.*

Note that in general case the distribution of the quadratic form in (3.31) can not be expressed by elementary functions in closed form, and it is convenient to use a computer program (Martynov, 1976) for its evaluation.

Now let us construct asymptotic expansions for unconditional risk defined by (3.11).

Theorem 3.3 *If the conditions of Theorem 3.1 hold and*

$$\mathbf{E}_{\theta_i^\circ}\{|\overset{\circ}{g}_{ij}|^3\} < \infty \quad (j = 1, \ldots, m, i \in S),$$

then as $n_* \to \infty$ *the following asymptotic expansion of unconditional risk for the plug-in decision rule (3.4), (3.5), (3.8) holds:*

$$r = r_0 + \sum_{i=1}^{L} \rho_i n_i^{-1} + \mathcal{O}(n_*^{-3/2}) \tag{3.33}$$

with the following coefficients $\{\rho_i\}$:

$$\rho_i = 1/2 \sum_{j=1}^{L-1} \sum_{t=j+1}^{L} c_{itj}^2 \int_{\overset{\circ}{\Gamma'}_{tj}} Q_i(x) \mid \nabla \overset{\circ}{f}_{tj}(x) \mid^{-1} ds_{N-1} \geq 0, \tag{3.34}$$

$$Q_i(x) = (\nabla_{\theta_i^\circ} p(x; \theta_i^0))^T V_i \nabla_{\theta_i^\circ} p(x; \theta_i^0) \geq 0, \tag{3.35}$$

where V_i *is the matrix given by formula (3.14).*

Proof. According to (3.11), $r = \mathbf{E}\{\operatorname{rc}(A)\}$. For $\operatorname{rc}(A)$ we use the stochastic expansion (3.19), (3.28), (3.29). The conditions of Lemma 3.1 are fulfilled, therefore by (3.14) - (3.16), the following mathematical expectations exist:

$$\mathbf{E}\{\overline{\Delta\theta_i}\,\overline{\Delta\theta_\mu^T}\} = \delta_{i\mu}\mathbf{I}_m + \mathcal{O}(\tau_*)\mathbf{1}_{m\times m},$$

$$\mathbf{E}\{\mid \overline{\Delta\theta_{is}}\,\overline{\Delta\theta_{\mu\tau}}\,\overline{\Delta\theta_{pq}} \mid\} < \infty.$$

By (3.19), (3.29), this fact implies the existence of the expectations

$$\mathbf{E}\{\delta r(A)\} = \sum_{j=1}^{L} n_i^{-1} \operatorname{tr}(\alpha_{ii}) + \mathcal{O}(n_*^{-3/2}), \quad \mathbf{E}\{\zeta_{n_*}\} = \mathcal{O}(n_*^{-3/2}). \tag{3.36}$$

Moreover, (3.17), (3.34), (3.35) together imply $\operatorname{tr}(\alpha_{ii}) = \rho_i$. Applying now the expectation operator $\mathbf{E}\{\cdot\}$ to (3.28) and using (3.36), we get (3.33).

■

Corollary 3.3 *The plug-in decision rule (3.4) is a consistent decision rule in the sense of unconditional risk convergence:* $r \to r_0$.

Corollary 3.4 *For the two-class case* ($L = 2$),

$$r = r_0 + \rho_1/n_1 + \rho_2/n_2 + \mathcal{O}(n_*^{-3/2}), \tag{3.37}$$

$$\rho_i = (\pi_i(w_{i1} - w_{i2}))^2 \int_{\overset{\circ}{\Gamma}_{12}} Q_i(x) \mid \nabla \overset{\circ}{f}_{12}(x) \mid^{-1} ds_{N-1}/2.$$

The stochastic expansion (3.19) allows to construct (as in Theorem 3.3) asymptotic expansions for higher order moments of conditional risk $\mathrm{rc}(A)$ (not only for its mean r) (see (Kharin and Duchinskas, 1981)). Let us illustrate this, for example, for the variance of conditional risk.

Theorem 3.4 *Under the conditions of Theorem 3.1, if*

$$\mathbf{E}_{\theta_i^\circ}\{|\overset{\circ}{g}_{ij}|^4\} < \infty \quad (j = 1, \ldots, m, i \in S),$$

then as $n_* \to \infty$ *the following asymptotic expansion of conditional risk variance holds:*

$$\mathbf{D}\{\mathrm{rc}(A)\} = \sum_{i=1}^{L}(2\,\mathrm{tr}(\alpha_{ii}^2))/n_i^2 + \sum_{i=1}^{L-1}\sum_{j=i+1}^{L}(4\,\mathrm{tr}(\alpha_{ij}\alpha_{ij}^T))/(n_i n_j) + \mathcal{O}(n_*^{-5/2}). \tag{3.38}$$

Proof is conducted in the same way as for Theorem 3.3 by using the stochastic expansion (3.19).

■

Note in conclusion that only $L+1$ main terms were evaluated in the risk expansion (3.33). By the method described here, if it is necessary, one can find next expansion terms of higher order (in comparison with $\mathcal{O}(n_*^{-1})$) and achieve the required accuracy of the asymptotic expansion.

3.2 Optimization and Robustness Analysis for Decision Rules Based on MC-estimators

In Section 3.1, in order to solve a pattern recognition problem under uncertainty conditions we used the plug-in decision rule (3.4), (3.5), (3.8). But there exists arbitrariness in choosing the contrast function $g(\cdot)$ that determines MC-estimators (3.7), (3.8). It means that in fact we have a family of plug-in decision rules:

$$D = \{d = d(x; A) : g(\cdot) \in G\},$$

where G is the set of admissible contrast functions (defined by the constraint (3.6)), and $d = d(x; A)$ is the plug-in decision rule (3.4), (3.5), (3.8) dependent on $g(\cdot)$.

For practice, it is interesting to find the plug-in decision rule $d^*(\cdot) \in D$ which is the most stable in D with respect to small-sample effects, and, according to (2.32), has minimal value of the unconditional risk r in D. First, let us evaluate a lower bound for risk r within the family D.

Theorem 3.5 *If in addition to the conditions of Theorem 3.3 there exist Fisher information matrices*

$$J_i = \mathbf{E}_{\theta_i^0}\{-\nabla^2_{\theta_i^0} \ln p(X_i; \theta_i^0)\} \quad (i \in S),$$

which are positive definite, then the following lower bounds for the coefficients of the asymptotic expansion (3.33) of unconditional risk hold:

$$\rho_i \geq \rho_i^* = \frac{1}{2}\sum_{j=1}^{L-1}\sum_{t=j+1}^{L} c^2_{itj} \int_{\overset{\circ}{\Gamma'}_{tj}} Q_i^*(x) \mid \nabla \overset{\circ}{f}_{tj}(x) \mid^{-1} ds_{N-1} \geq 0, \tag{3.39}$$

$$Q_i^*(x) = (\nabla_{\theta_i^0} p(x; \theta_i^0))^T J_i^{-1} \nabla_{\theta_i^0} p(x; \theta_i^0), \quad (i \in S). \tag{3.40}$$

Proof. Apply the information inequality (Ibragimov *et al.*, 1979) to the MC-estimator $\hat{\theta}_i$: for any $z \in R^m$,

$$z^T \mathbf{E}\{\Delta\theta_i(\Delta\theta_i)^T\} z \geq$$

$$z^T \left((\mathbf{I}_m + \nabla_{\theta_i^0} b_i(\theta_i^0))(n_i J_i)^{-1}(\mathbf{I}_m + \nabla_{\theta_i^0} b_i(\theta_i^0))^T \right) z,$$

where $\Delta\theta_i = \hat{\theta}_i - \theta_i^0$, and $b_i = b_i(\theta_i^0)$ is the bias of the estimator $\hat{\theta}_i (i \in S)$ determined by Lemma 3.1. Using (3.13), (3.14) we arrive at the following form of this inequality:

$$z^T V_i z - z^T J_i z \geq \mathcal{O}(\tau_i), z \in R^m.$$

Since the inequality holds as $\tau_i \to 0$ and V_i, J_i^{-1} do not depend on τ_i, we have

$$z^T V_i z \geq z^T J_i z. \tag{3.41}$$

For $z = \nabla_{\theta_i^0} p(x; \theta_i^0)$ it follows from (3.40), (3.41), (3.35) that

$$Q_i(x) \geq Q_i^*(x), x \in R^N (i \in S). \tag{3.42}$$

Multiplying both sides of (3.42) by $c^2_{itj} \mid \nabla \overset{\circ}{f}_{tj}(x) \mid^{-1} /2 \geq 0$, integrating with respect to $x \in \overset{\circ}{\Gamma'}_{tj}$ and summing over $t = j+1, \ldots, L$, $j = 1, \ldots, L-1$, we obtain (3.39), bearing in mind (3.34).

■

Corollary 3.5 *Up to $\mathcal{O}(n_*^{-3/2})$, the risk of the plug-in decision rule $d(\cdot) \in D$ is bounded from the below by r_-^*:*

$$r \geq r_-^*, \; r_-^* = \inf_{d(\cdot) \in D} r(d(\cdot)) = r_0 + \sum_{i=1}^{L} \rho_i^* n_i^{-1}. \tag{3.43}$$

Corollary 3.6 *Up to $O(n_*^{-3/2})$ the minimal risk value r_-^* is attained for the optimal plug-in decision rule $d^*(\cdot) \in D$ that uses maximum-likelihood estimators for $\{\hat{\theta}_i\}$.*

Proof. Choose the contrast function $g(\cdot)$ in the form (3.9). Then $\hat{\theta}_i$ is a ML-estimator. According to (3.12),

$$\overset{\circ}{a}_i'' = J_i, \ \mathbf{E}_{\theta_i^0}\{\overset{\circ}{g}_i' \overset{\circ}{g}_i'^T\} = J_i.$$

Substituting the above into (3.14), we find $V_i = J_i^{-1}$. Therefore we infer from (3.34), (3.35), (3.40) that the inequality (3.39) turns into equality: $\rho_i = \rho_i^* (i \in S)$. Moreover, (3.43) turns into equality too: $r = r_-^*$.

■

Let us recall the robustness factor

$$\kappa = \kappa(d(\cdot)) = (r(d(\cdot)) - r_0)/r_0 \geq 0, \tag{3.44}$$

where $r = r(d(\cdot))$ is the unconditional risk of the plug-in decision rule $d = d(x; A)$ under uncertainty conditions and r_0 is the Bayesian risk under the condition of full prior information. As it was noted in Section 2.3, κ is very important for applications, being a characteristic of relative robustness for the plug-in decision rule $d = d(x; A)$ with respect to small-sample effects. In the relative scale, the value of κ represents the increase of risk when prior uncertainty takes place and the plug-in decision rule $d = d(x; A)$ is used for classification. With lower values of κ, the plug-in decision rule is more accurate (for classification), more effectively it overcomes the existing prior uncertainty, and more stable it is with respect to small-sample effects.

It follows from Theorem 3.3 and (3.44) that for an arbitrary plug-in decision rule $d(\cdot) \in D$ the risk r and the robustness factor κ may be approximated as follows:

$$r \approx r_1 = r_0 + \sum_{i=1}^{L} \rho_i n_i^{-1}, \quad \kappa \approx \kappa_1 = \sum_{i=1}^{L} (\rho_i / r_0) n_i^{-1}. \tag{3.45}$$

The accuracy of these approximations is determined by the remainder term $O(n_*^{-3/2})$. We infer from Corollary 3.6 and (3.44) that the minimal values of r, κ attainable by the usage of the plug-in decision rule $d^*(\cdot) \in D$ with ML-estimators $\{\hat{\theta}_i\}$ may be approximated as follows:

$$\begin{aligned} r_-^* &\approx r_1^* = r_0 + \textstyle\sum_{i=1}^{L} \rho_i^* n_i^{-1}, \\ \kappa_-^* &\approx \kappa_1^* = \textstyle\sum_{i=1}^{L} (\rho_i^* / r_0) n_i^{-1}. \end{aligned} \tag{3.46}$$

These approximation formulas (3.45), (3.46) are convenient in practice for determining the sizes of training samples using the δ-robustness conditions from Section 2.3.

Let a number $\delta > 0$ be fixed. The training sample sizes $n_{\delta 1}, \ldots, n_{\delta L}$ are called *δ-admissible sample sizes,* if the robustness factor for these sample sizes does not exceed δ, i.e.,

$$\kappa^* \leq \delta. \tag{3.47}$$

If, in addition to the condition (3.47), the total size of the training sample is minimal:

$$n_{\bar{\delta 1}} + \ldots + n_{\bar{\delta L}} = \min\{n_{\delta 1} + \ldots + n_{\delta L}\}, \tag{3.48}$$

then such sample sizes $n_{\bar{\delta 1}}, \ldots, n_{\bar{\delta L}}$ are called *minimally δ-admissible sample sizes.*

The values $\{n_{\bar{\delta i}}\}$ are the minimal sizes of training samples $\{A_i\}$ which guarantee that the recognition risk exceeds the Bayesian risk by at most $\delta \cdot 100\%$.

Using the approximation (3.46) in (3.47), (3.48), one may state the optimization problem:

$$n_1 + \ldots + n_L \to \min,$$

$$\sum_{i=1}^{L} c_i/n_i \leq \delta, \; c_i = \rho_i^*/r_0.$$

Solving this problem we obtain approximation formulas for minimal δ-admissible sample sizes:

$$n_{\bar{\delta i}} \approx \left\lfloor \sum_{j=1}^{L} (\rho_i^* \rho_j^*)^{1/2}/(\delta r_0) \right\rfloor + 1 \qquad (i \in S). \tag{3.49}$$

In particular, if the expansion coefficients are equal: $\rho_1^* = \ldots = \rho_L^* = \rho^*$, then (3.49) assumes a simple form:

$$n_{\bar{\delta 1}} = \ldots = n_{\bar{\delta L}} = n_{\bar{\delta *}} = \lfloor L\rho^*/(\delta r_0) \rfloor + 1. \tag{3.50}$$

From (3.50) one can see that the minimal δ-admissible training sample size $n_{\bar{\delta *}}$ increases if the number of classes L and expansion coefficient ρ^* increase and if the Bayesian risk r_0 and admissibility level δ decrease.

Another problem of practical interest is the problem of evaluation of δ-admissible sample sizes from the probability distribution of random deviations of the conditional risk. This problem was treated within the empirical risk minimization theory (Vapnik *et al.*, 1974) with the usage of (too rough) conditions of the uniform convergence of sample frequencies. But here we shall use the asymptotic distribution of conditional risk determined in Theorem 3.2.

For the simplicity of exposition, assume that the training sample sizes $\{n_i\}$ are asymptotically equivalent, i.e., $n_*/n_i \to \lambda = 1 (i \in S)$. Denote:

$$G(z; F) = \mathbf{P}\{\sum_{i=1}^{Lm} \chi_i \eta_i^2 < z\}, \quad z \in R \tag{3.51}$$

is the probability distribution function of the random quadratic form $Q = \sum_{i=1}^{Lm} \chi_i \eta_i^2$, where $\{\chi_i \geq 0\}$ are the eigenvalues of the matrix F defined by (3.30), (3.17), and

$\eta_1, \ldots, \eta_{Lm}$ are independent random variables with standard Gaussian distribution $N_1(0,1)$. Let us find a *δ-admissible training sample size* $n_{\delta *}$ (the same for all classes) such that the conditional risk of the plug-in decision rule with sample sizes $n_i = n_{\delta *}$ $(i \in S)$ differs from the Bayesian risk at most by δ with probability larger than $1 - \gamma$ $(0 < \gamma < 1)$:

$$\mathbf{P}\{\operatorname{rc}(A) \le r_0 + \delta\} > 1 - \gamma.$$

From this condition we obtain by Theorem 3.2 the asymptotic expression:

$$n_{\delta *} = \lfloor G^{-1}(1-\gamma; F)/\delta \rfloor + 1, \tag{3.52}$$

where $G^{-1}(1-\gamma; F)$ is the $(1-\gamma)$-quantile for the distribution function $G(\cdot; F)$. For comparison, we cite the estimate from (Vapnik *et al.*, 1974) for the linear decision rule:

$$n_{\delta **} = \frac{32N}{\delta^2}\left(1 - \frac{\ln(\gamma/6) - \delta^2/16}{N} - \ln\frac{\delta^2}{32}\right). \tag{3.53}$$

Comparing (3.52) and (3.53), it is obvious that for a fixed significance level γ, $n_{\delta *} = \mathcal{O}(1/\delta)$ and $n_{\delta **} = \mathcal{O}(1/\delta^2)$, and therefore, as $\delta \to 0$, the value (3.53) becomes an overestimate in comparison with the value (3.52). Therefore we recommend to use more exact estimate (3.52). If the matrix F is not diagonal, then it is necessary to use numerical methods (Martynov, 1976) for evaluation of the quantile $G^{-1}(1-\gamma; F)$ in (3.52). For its simplification we give an easily computable upper bound for $n_{\delta *}$. Apply the Chebyshev inequality:

$$1 - G(\delta n_*; F) = \mathbf{P}\{Q > \delta n_*\} \le \mathbf{E}\{Q\}/(\delta n_*).$$

Since $\mathbf{E}\{Q\} = \sum_{i=1}^{Lm} \chi_i = \operatorname{tr}(F)$, we have

$$G(\delta n_*; F) \ge 1 - \operatorname{tr}(F)/(\delta n_*),$$

$$\mathbf{P}\{\operatorname{rc}(A) \le r_0 + \delta\} \ge 1 - \operatorname{tr}(F)/(\delta n_*).$$

Equating this probability to $1 - \gamma$, one can obtain

$$n_{\delta *} \le \operatorname{tr}(F)/(\delta\gamma), \tag{3.54}$$

where $\operatorname{tr}(F)$ is determined by Lemma 3.6.

Note in conclusion, if we use the formulas (3.50), (3.52), (3.54) in practice, we need to know the values of Bayesian risk r_0 and expansion coefficients $\{\rho_i\}$. Indeed, these values are contained in (3.50), (3.52) directly, and (3.54) contains the value $\operatorname{tr}(F)$, which can be expressed by $\{\rho_i\}$ according to Lemma 3.6 and formulae (3.34), (3.35) $(\lambda_i = 1)$:

$$\operatorname{tr}(F) = \sum_{i=1}^{L} \rho_i.$$

In this case, the inequality (3.54) assumes the form:

$$n_{\delta *} \leq \frac{1}{\delta \gamma} \sum_{i=1}^{L} \rho_i.$$

In practice, the values $\{\rho_i\}$, r_0 are unknown. Therefore, let us consider a problem of their statistical estimation.

Let us investigate, in general case, an arbitrary adaptive decision rule $d = d(x; A)$ with unconditional risk satisfying an asymptotic expansion:

$$r = \sum_{j=1}^{M} \beta_j \psi_j(n_1, \ldots, n_L) + o_1.$$

Here M is the number of expansion terms, o_1 is the remainder, $\{\psi_1(\cdot), \ldots, \psi_M(\cdot)\}$ is a known system of linearly independent monotone nonincreasing functions in L integer variables, $\{\beta_j\}$ are unknown parameters (coefficients) of the asymptotic expansion, which are known functionals w.r.t. unknown conditional probability density functions. For example, in the case of expansion (3.33) we have:

$$M = L + 1;\ \psi_1(n_1, \ldots, n_L) = 1,\ \psi_j(n_1, \ldots, n_L) = \frac{1}{n_{j-1}}\ (j = 2, \ldots, M);$$

$$\beta_1 = r_0,\ \beta_2 = \rho_1,\ \ldots,\ \beta_M = \rho_L;\ o_1 = \mathcal{O}(n_*^{-3/2}).$$

Let us assume that the investigated adaptive decision rule $d(x; \cdot)$ is subjected to $K \geq M$ experiments. The k-th experiment ($k = 1, \ldots, K$) is conducted in the following way. The set of L sample sizes is fixed:

$$n_1 = n_{k1}, \ldots, n_L = n_{kL},$$

and (either by an examination sample A' independent of the training sample A or by the jackknife method) an estimate for risk is calculated:

$$\hat{r}^{(k)} = \sum_{j=1}^{M} \beta_j \psi_j(n_{k1}, \ldots, n_{kL}) + \xi_k,$$

where ξ_k is the random error of estimation. Let us denote:

$$\psi_{kj} = \psi_j(n_{k1}, \ldots, n_{kL}),$$

$\Psi = (\psi_{kj})$ is the $(K \times M)$-matrix of experimental design,
$\hat{r} = (\hat{r}^{(k)})$ is the K-vector-column of risk estimates,
$\beta = (\beta_j)$ is the M-vector-column of expansion parameters,
$B = \text{diag}\{b_{kk}\}$ is the diagonal $(K \times K)$-matrix of weights $b_{kk} > 0$.
If experiments with an adaptive decision rule are designed in such a way that the matrix $\Psi^T \Psi$ is nonsingular, then the weighted least square estimate of β is defined by the expression:

$$\hat{\beta} = (\Psi^T B^{-1} \Psi)^{-1} \Psi^T B^{-1} \hat{r}.$$

This method of statistical estimation of $r_0, \{\rho_j\}$ was implemented in computer program and used to solve some practical pattern recognition problems. The results of the application to pattern recognition problems in oncology are presented in Section 3.3.

3.3 Adaptive Decision Rule Robustness for Gaussian Observations

In this section we analyze the robustness of various plug-in decision rules using the results from Sections 3.1, 3.2 and computer experiments. This analysis is performed for the situation defined in Section 1.4, when the observations $x \in R^N$ are Gaussian random vectors.

Gaussian Model and its Special Cases

Suppose that there are two $(L = 2)$ equiprobable classes $\Omega_1, \Omega_2; \mathcal{P} = \{p(x)\}$ is the family of N-variate nonsingular Gaussian densities: $p(x) = n_N(x \mid \mu, \Sigma)$; the loss matrix is (1.18) (here the risk is the classification error probability). Note that the Gaussian family $\mathcal{P}$ satisfies the Chibisov regularity conditions (Section 1.6), and we may use the decision rule from Sections 3.1, 3.2.

Denote

$$p_i(x) = n_N(x \mid \mu_i^0, \Sigma_i^0) \tag{3.55}$$

the Gaussian distribution density of the random vector of feature variables X_i from the class Ω_i, where $\mu_i^0 = (\mu_{ik}^0) \in R^N$ is the true mean vector and $\Sigma_i^0 = (\sigma_{ijk}^0)$ is the true covariance matrix for class Ω_i $(i \in S = \{1, 2\})$.

Let us consider three special cases C_1, C_2, C_3.

Case $C_1 : \Sigma_1^0 = \Sigma_2^0 = \Sigma^0$.

In this case the classes Ω_1, Ω_2 are distinguished by the values μ_1^0, μ_2^0 only. According to (1.55) , the Bayesian decision rule for this situation is the linear one:

$$d = d_0(x) = \mathbf{1}(b^{0T} x + \beta_0) + 1, x \in R^N, \tag{3.56}$$

$$b = (b_j) = (\Sigma^0)^{-1}(\mu_2^0 - \mu_1^0),\ \beta_0 = (\mu_1^0 + \mu_2^0)^T (\Sigma^0)^{-1} (\mu_1^0 - \mu_2^0)/2,$$

and the classification error probability is

$$r_0 = \Phi(-\Delta/2), \tag{3.57}$$

where

$$\Delta = \left((\mu_2^0 - \mu_1^0)^T (\Sigma^0)^{-1} (\mu_2^0 - \mu_1^0) \right)^{1/2}$$

is the Mahalanobis distance.

Case $C_2 : \mu_1^0 = \mu_2^0 = \mu^0$.

In this case the classes Ω_1, Ω_2 are distinguished only by covariances Σ_1^0, Σ_2^0, and the Bayesian decision rule is the quadratic one:

$$d = d_0(x) = \mathbf{1}((x-\mu^0)^T((\Sigma_1^0)^{-1} - (\Sigma_2^0)^{-1})(x-\mu^0) +$$

$$+ \ln(|\Sigma_1^0| / |\Sigma_2^0|)) + 1. \tag{3.58}$$

For an arbitrary N the expression for the Bayesian error probability is extremely bulky. We give it only for $N = 1$:

$$r_0 = 1/2 - (\Phi(\gamma((\ln \gamma^2)/(\gamma^2 - 1))^{1/2}) - \Phi(((\ln \gamma^2)/(\gamma^2-1))^{1/2})), \tag{3.59}$$

where

$$\gamma = (\sigma_{111}/\sigma_{211})^{1/2} \tag{3.60}$$

is the root-mean-square deviation ratio for the classes Ω_1, Ω_2.

Case $C_3 : \Sigma_i^0 = \mathrm{diag}\{\sigma_{ijj}^0\} \quad (j = 1, \ldots, N)$.

In this case the components of the random observation vector X_i are independent and can be distinguished for different classes both by the variances and the mean values. The Bayesian decision rule is the quadratic one:

$$d = d_0(x) = \mathbf{1}\left(\sum_{j=1}^N \left(\frac{(x_j-\mu_{1j})^2}{\sigma_{1jj}} - \frac{(x_j-\mu_{2j})^2}{\sigma_{2jj}} + \ln\frac{\sigma_{1jj}}{\sigma_{2jj}}\right)\right) + 1. \tag{3.61}$$

Using Plug-in Decision Rule with ML-estimators

At first, consider the case C_1 when the matrix Σ^0 is known, and $\{\theta_i^0 = \mu_i^0 : i \in S\}$ are the unknown parameters. According to Corollary 3.6, the minimal value of the error probability is attained for the adaptive decision rule $d = d(x; A)$ obtained from (3.56) by substitution of the ML-estimator, which is the sample mean:

$$\hat{\mu}_i = (\hat{\mu}_{ij}) = \sum_{k=1}^{n_i} x_{ik}/n_i \quad (i \in S) \tag{3.62}$$

for the true value μ_i^0. By Theorem 3.5 one can obtain an asymptotic expansion for the unconditional risk for this rule. By (3.55), (3.56) we may evaluate the characteristics present in the formulas (3.39), (3.40) for expansion coefficients $\{\rho_i^*\}$:

$$c_{i21} = 0.5, \quad \overset{\circ}{\Gamma}{}'_{21} = \overset{\circ}{\Gamma}_{21} = \{x : b^{0T}x + \beta_0 = 0\};$$

$$\forall x \in \overset{\circ}{\Gamma}_{21} \quad |\nabla_x \overset{\circ}{f}_{21}(x)| = p_1(x)(((\mu_2^0 - \mu_1^0)^T(\Sigma^0)^{-2}(\mu_2^0 - \mu_1^0))^{1/2}/2, \tag{3.63}$$

$$Q_i^*(x) = (x - \mu_i^0)^T(\Sigma^0)^{-1}(x - \mu_i^0)p_1^2(x), \quad x \in \mathring{\Gamma}_{21} .$$

As a result,

$$r^* = r_1^* + \mathcal{O}(n_*^{-3/2}), r_1^* = r_0 + \rho_1^*/n_1 + \rho_2^*/n_2, \tag{3.64}$$

$$\rho_1^* = \rho_2^* = (4(2\pi)^{1/2}\Delta)^{-1}(N - 1 + \Delta^2/4)\exp(-\Delta^2/8) = \rho^{(1)}.$$

According to (3.45), the robustness factor is

$$\kappa_1^* = \frac{\exp(-\Delta^2/8)}{4\sqrt{2\pi}\Delta\Phi(-\Delta/2)}(N - 1 + \Delta^2/4)(1/n_1 + 1/n_2). \tag{3.65}$$

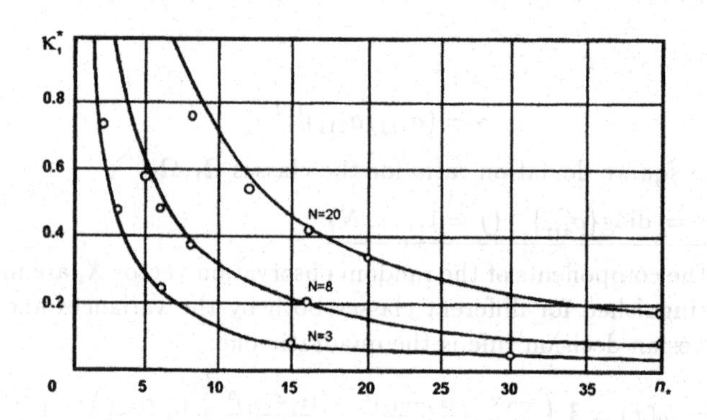

Figure 3.1: $\kappa_1^*(n_*, N)$ vs. sample size

The plots of dependence (3.65) of κ_1^* on sample size n_* for $n_1 = n_2 = n_*$, $\Delta = 2.56$ ($r_0 = 0.1$), and for different observation space dimensions $N = 3; 8; 20$ are given at Figure 3.1. Small circles near the curves mark the tabulated values κ^* obtained in (Raudis, 1976) by computer numerical integration of (3.11). The comparison of the plots κ_1^* and the tabulated values κ^* suggests that the remainder term in the expansion (3.33) of risk and in the approximation formulas (3.45) has small influence even for sample sizes $n_* \leq N$, and the approximation accuracy in (3.45) is sufficiently high.

Note that the risk expansion (3.64) agrees with the expansion, well-known in discriminant analysis, constructed by (Okamoto, 1963) for the Gaussian model (by a method based on special properties of Gaussian distribution) and with the results of (Krzysko, 1983). The passage to the limit in (3.65) under the Kolmogorov–Deev asymptotics:

$$n_i, N \to \infty, \ N/n_i \to \text{const} \ , \Delta \to \text{const}$$

results in the principal term of the Meshalkin–Serdobolskij expansion (Meshalkin *et al.*, 1978). Additionally, for the case C_1 we may give the expressions for minimal δ-admissible sample sizes, derived by (3.50), (3.57), (3.64):

$$n_{\delta}^{-} = n_{\delta_1}^{-} = n_{\delta_2}^{-} = \left\lfloor \frac{(N-1+\Delta^2/4)\exp(-\Delta^2/8)}{2(2\pi)^{1/2}\Delta\Phi(-\Delta/2)\delta} \right\rfloor + 1. \tag{3.66}$$

Figure 3.2 plots the dependence of minimal δ-admissible sample size n_{δ}^{-} on Mahalanobis distance Δ for fixed admissibility level $\delta = 0.1$ and for the following values of dimension of the observation space: $N = 1, 2, 5$.

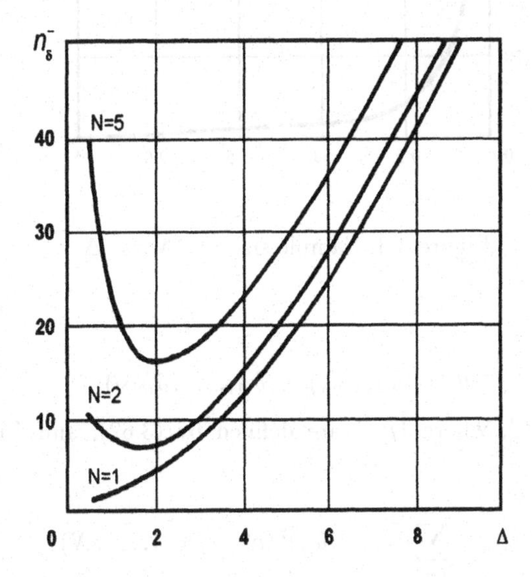

Figure 3.2: Minimal δ-admissible sample size $n_{\delta}^{-}(\Delta, N)$ vs. Mahalanobis distance

Figure 3.3 plots the dependence of relative δ-admissible sample size $n_{\delta *}/N$ on Δ for $N >> 1$ (derived by the formulas (3.52), (3.64) and Theorem 3.2) for $\delta = 0.19$, $\gamma = 0.05$. Here the dashed line indicates Vapnik's estimate $n_{\delta **}/N$ (Vapnik *et al.*, 1974) determined by (3.53). Vapnik's estimate does not use any information about the probability distribution of observations, and one can see at Figure 3.3 that it is excessively overestimated.

Consider now the case C_3, when the $(2N)$-vector

$$\theta_i^0 = (\mu_{i1}^0, \ldots, \mu_{iN}^0, \sigma_{i11}^0, \ldots, \sigma_{iNN}^0)\ (i \in S)$$

is the unknown vector of density parameters $p_i(x)$.

According to Corollary 3.6, the minimal value of the classification error probability is attained for the adaptive decision rule $d = d(x; A)$ produced from (3.61) by the substitution of ML-estimators

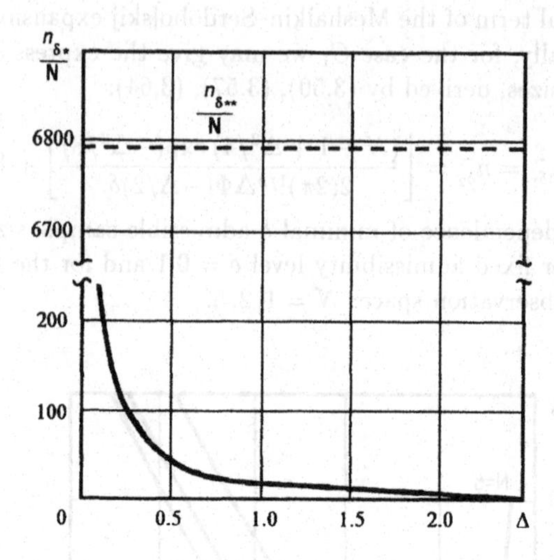

Figure 3.3: Sample size $n_{\delta *}/N$ vs. Δ

$$\hat{\theta}_i = (\hat{\mu}_{i1}, \ldots, \hat{\mu}_{iN}, \hat{\sigma}_{i11}, \ldots, \hat{\sigma}_{iNN})$$

for the true value θ_i^0, where $\{\hat{\mu}_{ij}\}$ are defined by (3.62), and $\{\hat{\sigma}_{ijj}\}$ are sample variances:

$$\hat{\sigma}_{ijj} = \sum_{k=1}^{n_i} (x_{ikj} - \hat{\mu}_{ij})^2 / n_i \quad (j = 1, \ldots, N). \tag{3.67}$$

Let us obtain the expansion of risk for this decision rule for the case $\Sigma_1^0 = \Sigma_2^0 = \Sigma^0 = \mathrm{diag}\{\sigma_{jj}^0\}$ similarly to the previous case:

$$Q_i^*(x) = 1/2(N + \sum_{j=1}^{N} (x_j - \mu_{ij}^0)^4 / (\sigma_{jj}^0)^2),$$

$$\rho_1^* = \rho_2^* = \frac{\exp(-\Delta^2/8)}{4(2\pi)^{1/2}\Delta}(2N - 3 + 3\Delta^2/4 + \nu/2(\Delta^4/16 - 3\Delta^2/2 + 3)) = \rho^{(2)}, \tag{3.68}$$

where

$$\nu = 1/\Delta^4 \sum_{j=1}^{N} \frac{(\mu_{1j} - \mu_{2j})^4}{\sigma_{jj}^{02}}, \quad 1/N \leq \nu \leq 1.$$

Let us compare the error probabilities $r^{(1)}, r^{(2)}$ for the two considered adaptive decision rules for the case

$$\Sigma_1^0 = \Sigma_2^0 = \Sigma^0 = \mathrm{diag}\{\sigma_{jj}^0\}$$

and for the same sample sizes n_1, n_2. These adaptive decision rules are distinguished by levels of prior uncertainty: in the first case the covariance matrices are known and only $2N$ parameters are estimated. In the second case these matrices are also unknown, and $4N$ parameters are estimated. The relative increase of error probability for the second adaptive decision rule with respect to the first one may be determined using (3.68), (3.64):

$$g = g(N, \Delta, \nu) = \frac{r^{*(2)} - r_0}{r_1^{*(1)} - r_0} =$$

$$\frac{(N - 2 + \Delta^2/2 + \nu(\Delta^4/16 - 3\Delta^2/2 + 3))}{N - 1 + \Delta^2/4}.$$

It may be shown that $g(N, \Delta, \nu) \geq 1$ for $\nu \in [1/N, 1]$.

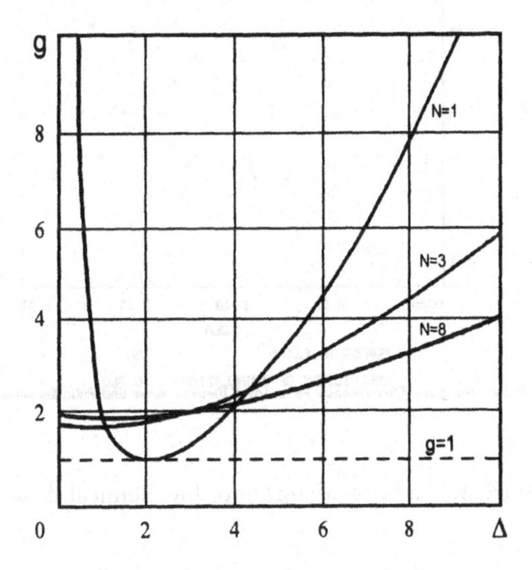

Figure 3.4: $g(n, \Delta, \nu)$ vs. Mahalanobis distance

Figure 3.4 plots the dependence of the coefficient g on Mahalanobis distance Δ for $N = 1; 3; 8$ at $\nu = 1/N$. (The equality $\nu = 1/N$ corresponds to the case when all N components of the observation vector possess the same "discrimination power":

$$(\mu_{2j}^0 - \mu_{1j}^0)^2 / \sigma_{jj}^0 = \text{const}\ , j = 1, \ldots, N).$$

It can be seen from these plots that:

1. The coefficient g increases monotonously as Δ increases for $N \geq 3$;

2. $g \to 2$ as $N \to \infty$ for fixed arbitrary Δ. This well matches with the fact that the second adaptive decision rule has two times more independent unknown parameters.

In conclusion, we illustrate the method of estimation of the δ-admissible sample size n_{δ}^{-} described in Section 3.2 for a real application problem of recognition of the earliest cancer stage by biochemical blood tests (see (Abramovich, Kharin, and Mashevskij, 1993)). In this investigation, the experimental data were collected from two groups of rats.

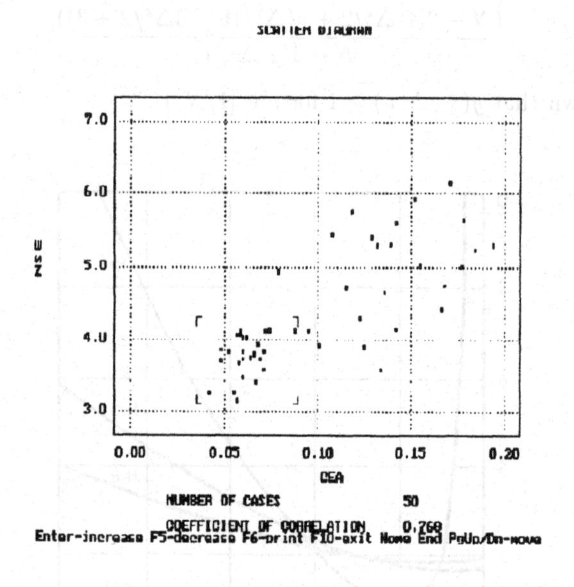

Figure 3.5: Scatter diagram of biochemical data

The first group consisted of 25 healthy rats and presented the class Ω_1. The second group (class Ω_2) consisted of 25 rats vaccinated by alveolar liver cancer. Biochemical blood tests were conducted on the third day after vaccination at the earliest stage of cancer. For each rat two biochemical tests were performed and two features were recorded: x_1 is the concentration of cancer-embryonic antigen (CEA); x_2 is the concentration of neuron-specific enalaza (NSE). Figure 3.5 presents a computer visualization of total sample data for 50 experimental rats in two-dimensional feature space (x_1, x_2). Note that all calculations in this investigation were performed by the software package ROSTAN (RObust STatistical ANalysis) published by the Belarussian State University (see (Kharin, 1994)).

The hypothetical Fisher model C_1 of statistical data was used in the investigation. According to the results of Section 3.2, the dependence of error probability r on training sample sizes $n_1 = n_2 = n_0$ was taken in the form:

$$r = \theta_1 + \frac{\theta_2}{n_0} + \xi,$$

where θ_1 and θ_2 are independent coefficients, ξ is a random error. The existing sample was divided into two subsamples: the examination subsample of size 20 (10 observations from each of two classes) and the training subsample of size 30 (15 observations from each class). A series of $K = 14$ computer experiments for the construction of the Fisher linear discriminant functions and the classification of the examination subsample was performed by the method from Section 3.2 and software package ROSTAN. The results of these experiments are presented in Table 3.1.

Table 3.1: Results of $K = 14$ Experiments

k	1	2	3	4	5	6	7
$n_0^{(k)}$	3	4	4	4	5	5	5
$r^{(k)}$	0.35	0.05	0.05	0.20	0.20	0.05	0.05
k	8	9	10	11	12	13	14
$n_0^{(k)}$	5	5	5	5	10	10	15
$r^{(k)}$	0.10	0.20	0.05	0.05	0.10	0.05	0.05

Table 3.2: Forecast of error probability

n_0	5	10	15	20	40	60	80	100
r	0.110	0.065	0.050	0.043	0.031	0.028	0.026	0.025

Table 3.3: δ-admissible sample sizes

δ	0.1	0.2	0.3	0.4	0.5	0.6	0.7	0.8	0.9	1.0
n_δ^-	226	113	76	57	46	38	33	29	26	23

Here $r^{(k)}$ is the error rate in the classification of the examination subsample at the k-th experiment. The experiments with equal sample size n_0 differ in sets of observations used for training.

To estimate the coefficients θ_1, θ_2, the weighted least squares method was used:

$$\sum_{k=1}^{K} (r^{(k)} - \theta_1 - \frac{\theta_2}{n_0^{(k)}})^2 b_k \to \min_{\theta_1, \theta_2}. \tag{3.69}$$

The weight factor $b_k > 0$ was chosen to be a function inversely proportional to the variance of conditional risk (3.38):

$$b_k = n_k^2, \quad k = 1, \ldots, K.$$

Solving the minimization problem (3.69), we obtained the following dependence of error probability r on training sample size n_0 :

$$r = 0.02 + \frac{0.45}{n_0}. \tag{3.70}$$

Formula (3.70) allows to forecast the error probability for a given sample size (see Table 3.2), and to calculate the δ–admissible sample size (see Table 3.3).

Chapter 4

Robustness of Nonparametric Decision Rules and Small-sample Effects

This chapter is devoted to the same problems as Chapter 3, but for the situations where no parametric model of probability distributions is known and nonparametric decision rules (Rosenblatt-Parzen, k-nearest neighbor) are used for classification. We find optimal values for smoothness parameters that optimize the robustness factor. We compare stability of parametric and nonparametric decision rules.

4.1 Robustness of Nonparametric Rosenblatt – Parzen Decision Rules

In constructing adaptive decision rules (Chapter 3) we assumed that probability distributions $\{p_i(\cdot)\}$ are any elements of a given parametric family of densities $\mathcal{P}$. Unfortunately in pattern recognition applications this prior information is usually absent. Therefore it is necessary to investigate the situations with nonparametric prior uncertainty: $\mathcal{P}$ is a nonparametric family of probability distributions. Let us formulate the mathematical model that corresponds to the distortion type **D**.1.2 (small-sample effects under nonparametric estimates of probability distributions), see Chapter 2.

Assume that a nonparametric family $\mathcal{P} = \{p(x)\}$ of triply differentiable probability densities is defined in feature space R^N and $p_1(\cdot), p_2(\cdot) \in \mathcal{P}$ are fixed probability distribution densities from this family. Random observations from two classes Ω_1, Ω_2 are recorded in R^N with prior probabilities π_1, π_2 $(\pi_1 + \pi_2 = 1)$. An observation from Ω_i is a random feature vector $X_i = (X_{il}) \in R^N$ with unknown density $p_i(x)$ $(i \in S = \{1, 2\})$. (The results may be generalized to the case of $L > 2$ classes, but this leads to bulky expressions.) A classified training sample A is observed: $A^T = (A_1^T \vdots A_2^T)$, where $A_i^T = (z_{i1}^T \vdots \ldots \vdots z_{in_i}^T)$ is a random sample of n_i observations from Ω_i; A_1 and A_2 are independent. An observation to be classified is a random vector $X \in R^N$ assigned to $\Omega_1 \cup \Omega_2$ and observed independently of A.

If $\{p_i(x)\}$ were given, then according to Corollary 1.2 of Theorem 1.1 the minimal risk r_0 would be attained for the Bayesian decision rule:

$$d = d_0(x) = \mathbf{1}(G_0(x)) + 1, x \in R^N, d \in S = \{1,2\}, \tag{4.1}$$

where $G_0(x)$ is the Bayesian discriminant function:

$$G_0(x) = c_2 p_2(x) - c_1 p_1(x),$$

$$c_1 = \pi_1(w_{12} - w_{11}) > 0, \; c_2 = \pi_2(w_{21} - w_{22}) > 0. \tag{4.2}$$

Since $\{p_i(x)\}$ are unknown, the family of nonparametric plug-in decision rules (Fix *et al.*, 1953), (Van Ryzin, 1965), (Zhivogljadov *et al.*, 1974), (Raudis, 1974) is used:

$$d = d(x; A) = \mathbf{1}\,(G(x; A)) + 1, \tag{4.3}$$

$$G(x; A) = c_2 \,\hat{p}_2\,(x) - c_1\,\hat{p}_1\,(x), x \in R^N,$$

where $\hat{p}_i\,(x)$ is the nonparametric Rosenblatt–Parzen (of kernel type) estimator (Parzen, 1962), (Rosenblatt, 1965), (Epanechnikov, 1969), (Loftsgarden *et al.*, 1974), (Vapnik *et al.*, 1974), (Devroye *et al.*, 1985) of the probability density $p_i(\cdot)$ by the sample A_i :

$$\hat{p}_i(x) = \frac{1}{n_i|H_i|}\sum_{j=1}^{n_i} K\left(H_i^{-1}(x - z_{ij})\right), x = (x_l) \in R^N. \tag{4.4}$$

Here $K(x) = K_1(x_1)\cdot\ldots\cdot K_N(x_N)$; $K_l(y), y \in R^1$ is a univariate kernel of the estimate, i.e., a nonnegative bounded differentiable even function, $K_l(|y|)$ is a non-increasing function of $|y|$, and

$$\int_{-\infty}^{+\infty} y^m K_l(y)dy < \infty \quad (m > 0), \qquad \int_{-\infty}^{+\infty} y^2 K_l(y)dy = 1\;;$$

$H_i = \operatorname{diag}\{h_{il}\}$ is a diagonal matrix of order N. The diagonal elements $h_{il} > 0$ of this matrix are called *smoothness parameters.* If

$$h_{il} = h_{il}(n_i) \to 0, \qquad n_i|H_i| \to \infty, \tag{4.5}$$

as $n_i \to \infty$, then the estimator (4.4) is consistent (Epanechnikov, 1969).

The following problems are to be solved:

1: Evaluation of risk of the plug-in decision rule (4.3), (4.4) in terms of training sample sizes $\{n_i\}$;

2: Finding optimal smoothness parameters $\{h_{il}^*\}$ for which the plug-in decision rule robustness is maximal, i. e., the risk value is minimal for fixed sample sizes;

3: Determining sample sizes that guarantee a given level of robustness.

These problems are particular cases of general plug-in decision rule robustness analysis problems from Section 2.3.

Note that Problem 1 was investigated in (Raudis, 1976) using the representation of the probability density of the discriminant function by means of Gram–Charlier series. But an analytical dependence of risk on $\{n_i\}$ was not found even for classification of Gaussian observations. In (Zhivogljadov *et al.*, 1974), (Raudis, 1976) Problems 1 and 2 were investigated by computer modeling of the plug-in decision rule and tabulating risk as a function of $\{n_i\}$, $\{h_{il}\}$. As the number of varying parameters is too large, this approach is not efficient.

To solve Problems 1 – 3, we shall apply, as in Chapter 3, the method of asymptotic risk expansions.

Let us transform the unconditional risk of the plug-in decision rule (4.3)

$$r = \sum_{i=1}^{2} \pi_i \mathbf{E}\{w_{i,d(X_i;A)}\}$$

to the form

$$r = \pi_1 w_{11} + \pi_2 w_{21} - \int_{R^N} \mathbf{E}\{\mathbf{1}(G(x;A))\} G_0(x) dx. \tag{4.6}$$

If we assume in (4.6) that $G(x;A) \equiv G_0(x)$, then we obtain the Bayesian risk for the Bayesian decision rule (4.1) :

$$r_0 = \pi_1 w_{11} + \pi_2 w_{21} - \int_{R^N} \mathbf{1}(G_0(x)) G_0(x) dx. \tag{4.7}$$

Assume the notations :

$$k = \prod_{l=1}^{N} \int_{-\infty}^{+\infty} K_l^2(y) dy, \qquad p_{il}^{(2)}(x) = \frac{\partial^2 p_i(x)}{\partial x_l^2};$$

$$o_1 = o\left(\max_{i,j,l,s} \{h_{ij}^2 h_{ls}^2, (n_i |H_i|^{-1})\} \right);$$

$\Gamma = \{x : G_0(x) = 0\}$ is the Bayesian discriminant surface; ds_{N-1} is the differential element of Γ;

$$\alpha_i = \int_\Gamma \frac{p_i(x)}{|\nabla G_0(x)|} ds_{N-1}, \qquad \beta_{ijl} = \int_\Gamma \frac{p_{ij}^{(2)}(x) p_{il}^{(2)}(x)}{|\nabla G_0(x)|} ds_{N-1} \quad (i \in S),$$

$$\beta_{3jl} = \int_\Gamma \frac{p_{1j}^{(2)}(x) p_{2l}^{(2)}(x)}{|\nabla G_0(x)|} ds_{N-1} \; (j, l = 1, \ldots, N); \tag{4.8}$$

$\xi_i = \hat{p}_i(x) - p_i(x)$ is a random variable (x is fixed) with probability density $p_{\xi_i}(z)$; $p_{\xi_i}^{(j)}(z)$ is the j-th order derivative of $p_{\xi_i}(z)$ $(z \in R^1)$.

First, let us investigate statistical properties of the estimator of discriminant function $G_0(x)$ defined by (4.3).

Lemma 4.1 *If $\{p_i(\cdot)\}$ are triply differentiable, then the random deviation $g(x;A) = G(x;A) - G_0(x)$ of discriminant function has moments up to the third order, and*

$$\mathbf{E}\{g^3(x)\} = o_1, \tag{4.9}$$

$$\mathbf{E}\{g^2(x;A)\} = k\sum_{i=1}^{2}\frac{c_i^2}{n_i|H_i|}p_i(x) + \frac{1}{4}\left(\sum_{j=1}^{N}\sum_{i=1}^{2}(-1)^{i+1}c_i h_{ij}^2 p_{ij}^{(2)}(x)\right)^2 + o_1.$$

Proof. By construction of ξ_1, ξ_2, they are independent and

$$g(x;A) = c_2\xi_2 - c_1\xi_1, \tag{4.10}$$

$$\mathbf{E}\{g^3(x;A)\} = -c_1^3\mathbf{E}\{\xi_1^3\} + c_2^3\mathbf{E}\{\xi_2^3\} + 3c_1^2c_2\mathbf{E}\{\xi_1^2\}\mathbf{E}\{\xi_2\} -$$

$$-3c_1c_2^2\mathbf{E}\{\xi_1\}\mathbf{E}\{\xi_2^2\}.$$

According to the known result (Epanechnikov, 1969),

$$\mathbf{E}\{\xi_i\} = \sum_{j=1}^{N} p_{ij}^{(2)}(x)h_{ij}^2/2 + o(\max_j h_{ij}^2),$$

$$\mathbf{E}\{\xi_i^2\} = k(n_i|H_i|)^{-1} + \frac{1}{4}\left(\sum_{j=1}^{N} p_{ij}^{(2)}(x)h_{ij}^2\right) +$$

$$+o\left(\max_{j,l}\{h_{ij}^2h_{il}^2, (n_i|H_i|)^{-1}\}\right).$$

Similarly to (Epanechnikov, 1969), one can find that $\mathbf{E}\{\xi_i^3\} = o_1$. This implies the first expression in (4.9). In the same way, one can derive the second expression in (4.9).

■

Theorem 4.1 *Let the following conditions be fulfilled :*

a) *the probability densities $p_1(\cdot)$, $p_2(\cdot)$ are triply differentiable and the surface integrals $\{\alpha_i\}$, $\{\beta_{ijl}\}$ in (4.8) are bounded;*

b) $\int_{R^N} p_1^{i_1}(x)p_2^{i_2}(x)dx < \infty \quad \forall i_1, i_2 \in \{0,1,2,3,4\} : i_1 + i_2 \leq 4;$

c) *there are $i \in S$ and $n_i = \bar{n}_i$ such that*

$$|p_{\xi_i}^{(j)}(z)| < \infty, \quad z \in R^1, \quad x \in R^N (j = 0,1,2).$$

Then the following expansion for the risk of decision rule (4.3) is possible:

$$r = r_0 + \frac{1}{2}\sum_{i=1}^{2} c_i^2 \left(\frac{k\alpha_i}{n_i|H_i|} + \frac{1}{4}\sum_{j,l=1}^{N} \beta_{ijl} h_{ij}^2 h_{il}^2 \right) - $$
$$- \frac{c_1 c_2}{4} \sum_{j,l=1}^{N} \beta_{3jl} h_{1j}^2 h_{2l}^2 + o_1. \qquad (4.11)$$

Proof. Let $p_g(v)$ be the probability density of $g(x)$ at fixed $x \in R^N$ and let

$$Q(t) = \int_{R^N} \mathbf{E}\{\mathbf{1}(G_0(x) + t \cdot g(x;A))\} G_0(x) dx, \quad t \in R^1. \qquad (4.12)$$

Then the expression (4.6) for risk assumes the form:

$$r = \pi_1 w_{11} + \pi_2 w_{21} - Q(1). \qquad (4.13)$$

We shall show that under the conditions of the theorem the function $Q(t)$ is triply differentiable in $(0,1)$. Indeed, (4.12) implies

$$Q^{(3)}(t) = \int_{R^N} Q_1(x,t) G_0(x) dx,$$

$$Q_1(x,t) = \int_{-\infty}^{+\infty} v^3 p_g(v) \delta^{(2)}(G_0(x) + tv) dv, \qquad (4.14)$$

where $\delta^{(2)}(z)$ is the second order derivative of the delta–function (Gelfand, 1959). Using properties of the delta-function we find from (4.14) :

$$Q_1(x,t) = t^{-3}(6u p_g(u) + 6u^2 p_g^{(1)}(u) + u^3 p_g^{(2)}(u)) \qquad (4.15)$$

at point $u = -G_0(x)/t$.

According to (4.14), ξ_i is the sum of n_i independent identically distributed random variables that have the same distribution as

$$\eta_i = (n_i|H_i|)^{-1} K(H_i^{-1}(x - X_i)) - n_i^{-1} p_i(x).$$

Therefore $p_{\xi_i}(z)$ is the convolution of n_i densities $p_{\eta_i}(z)$. By well known "smoothing" properties of the convolution (Koroljuk, 1984), if the condition **c)** holds for some $n_i = \overline{n}_i$, then it is holds for all $n_i > \overline{n}_i$. It follows from (4.10) that $p_g(v)$ is the convolution of densities $\{c_i^{-1} p_\xi(z/c_i) : i \in S\}$. Again, the properties of convolution and condition **c)** together imply that $\forall x \in R^N \; |p_g^{(j)}(u)| \leq e_j < \infty \, (j = 0,1,2)$. Then from (4.15) we get:

$$|Q_1(x,t)| \leq t^{-3}(6e_0|u| + 6e_1|u|^2 + e_2|u|^3)$$

at point $u = -G_0(x)/t$. Moreover, according to (4.14),

$$|Q^{(3)}(t)| \leq 6t^{-4} \sum_{j=0}^{2} e_j t^{-j} \int_{R^N} |G_0(x)|^{j+2} dx,$$

$$|G_0(x)| \leq c_1 p_1(x) + c_2 p_2(x),$$

hence bearing in mind condition **b)** we have

$$|Q^{(3)}(t)| < \infty, t \in (0,1). \tag{4.16}$$

Let us apply the Taylor formula to $Q(t)$ in the neighborhood of point $t = 0$:

$$Q(t) = Q(0) + Q^{(1)}(0)t + Q^{(2)}(0)t^2/2 + Q^{(3)}(t_1)t^3/6, 0 < t_1 < t. \tag{4.17}$$

From (4.12) and the known properties (Gelfand, 1959) of the Dirac function: $z\delta(z) \equiv 0$, $z\delta^{(1)}(z) \equiv -\delta(z)$, it follows that

$$Q(0) = \int_{R^N} \mathbf{1}(G_0(x)) G_0(x) dx, \qquad Q^{(1)}(0) = 0,$$

$$Q^{(2)}(0) = -\int_{R^N} \mathbf{E}\{g^2(x)\} \delta(G_0(x)) dx. \tag{4.18}$$

By Lemma 3.2 and (4.14), (4.16), we have: $Q^{(3)}(t_1) = o_1$ for $t_1 \in (0,1)$. Therefore substituting (4.18) into (4.17) at $t = 1$ and using Lemma 3.2, Lemma 4.1, and (4.7), (4.8), (4.13) we obtain (4.11).

■

Let us analyze the feasibility of conditions of Theorem 4.1. In particular, conditions **b)** are fulfilled if the densities $\{p_i(\cdot)\}$ are bounded. Condition **c)** is not particularly restrictive. It is of the same type as the condition for the densities in the Local Limit Theorem (see, e.g., (Koroljuk, 1984)).

It can be seen from (4.11) that under the conditions (4.5) the decision rule (4.3) is consistent: $r \to r_0$ as $n_1, n_2 \to \infty$.

Usually the smoothness parameter is a power function of the sample size :

$$h_{ij} = b_{ij} n_i^{-\gamma_i}, \qquad \gamma_i > 0,\ b_{ij} > 0. \tag{4.19}$$

Corollary 4.1 *Under the conditions of Theorem 4.1 and (4.19), the following expansion holds :*

$$r = r_1 + o\left(\max_i\{n_i^{N\gamma_i - 1}, n_i^{-4\gamma_i}, n_1^{-2\gamma_1} n_2^{-2\gamma_2}\}\right), \tag{4.20}$$

$$r_1 = r_0 + \sum_{i=1}^{2} \left(q_i n_i^{N\gamma_i - 1} + q_{2+i} n_i^{-4\gamma_i} \right) + q_5 n_1^{-2\gamma_1} n_2^{-2\gamma_2},$$

$$q_i = \frac{k c_i^2 \alpha_i}{2 \prod_{j=1}^{N} b_{ij}}, \quad q_{2+i} = \frac{c_i^2}{8} \sum_{j,l=1}^{N} \beta_{ijl} b_{ij}^2 b_{il}^2 \quad (i \in S),$$

$$q_5 = -c_1 c_2 \sum_{j,l=1}^{N} \beta_{3jl} b_{1j}^2 b_{2l}^2 / 4.$$

It follows from (4.20) that to assure the convergence $r \to r_0$ the assumption $0 < \gamma_i < N^{-1} (i \in S)$ is required. Let us choose $\{\gamma_i\}$ to maximize the rate of this convergence. Consider the situation where n_1, n_2 are increasing with the same rate:

$$n_i = \lambda_i n, \ 0 < \lambda_i < 1, \ \lambda_1 + \lambda_2 = 1; \ n_1 + n_2 = n \tag{4.21}$$

is the total sample size. Substituting (4.21) into (4.20) and minimizing r_1 with respect to γ_1, γ_2 we obtain the following.

Corollary 4.2 *The fastest convergence of r_1 to r_0 as $n \to \infty$ is observed for $\gamma_1^* = \gamma_2^* = 1/(N+4)$:*

$$r_1 = r_0 + q \cdot n^{-4/(N+4)}, \tag{4.22}$$

$$q = \sum_{i=1}^{2} (q_i + q_{2+i}) \lambda_i^{-4/(N+4)} + q_5 (\lambda_1 \lambda_2)^{-2/(N+4)}.$$

Note that γ_1^*, γ_2^* coincide with the values $\{\gamma_i\}$ found in (Epanechnikov, 1969) by minimization of the integral mean–square error for density estimation and with the results of (Marron, 1983).

As it was noted before, the robustness factor defined by (2.24) is an important performance characteristic of adaptive decision rules. According to (4.22), we obtain for the nonparametric decision rule (4.3) :

$$\kappa = \kappa_1 + o(n^{-4/(N+4)}), \quad \kappa_1 = (q/r_0) n^{-4/(N+4)}. \tag{4.23}$$

In addition, it is interesting to note that for the parametric decision rule from Sections 3.1, 3.2 that uses MC-estimators, (3.45) implies $\kappa_1 = \rho n^{-1}$. Thus, a non-parametric decision rule produces risk that converges to r_0 slower than the risk of a parametric decision rule. This loss in convergence rate grows as the dimension N of the observation space increases.

The expansion (4.11) of risk leads to approximate formulas useful for the analysis of robustness of decision rule (4.3) with respect to small sample effects: $r \approx r_1$,

$\kappa \approx \kappa_1$. Only six main terms are evaluated in (4.11). If necessary, one can evaluate (similarly to Theorem 4.1) further expansion terms and increase the accuracy of these approximate formulas.

The approximate formulas for r and κ are convenient for evaluation of minimal δ-admissible training sample size n_δ (see the definition in Section 3.2):

$$n_\delta = \min\{n : \kappa(n) \leq \delta\}. \tag{4.24}$$

Substituting (4.23) into (4.24) and suppressing the remainder term, we find an approximation for n_δ :

$$n_\delta \approx \left\lfloor (q/(\delta r_0))^{\frac{N}{4}+1} \right\rfloor + 1, \tag{4.25}$$

where q is defined by (4.20), (4.22).

Let us consider a problem arising in practical applications of risk expansion (4.11). Sometimes the discriminant surface is of infinite area. In this case the surface integrals $\{\alpha_i, \beta_{ijl}\}$ from (4.8) may be unbounded for some probability distributions, and the condition **a)** of Theorem 4.1 may fail. In such situations it is advisable to construct an expansion of type (4.11) for a special functional "close" to the risk functional. Let $\delta > 0$ be an arbitrarily small fixed value. Let us define the *δ-risk* (proceeding from (4.6)) :

$$r_\delta = \pi_1 w_{11} + \pi_2 w_{21} - \int_{T_\delta} \mathbf{E}\{\mathbf{1}(G(x;A))\} G_0(x)dx, \tag{4.26}$$

where a bounded region $T_\delta \subset R^N$ is chosen to assure the proximity of r_δ to r :

$$|r_\delta - r| \leq \delta. \tag{4.27}$$

Thus δ determines the accuracy of approximation of risk r. The following result is useful for the construction of region T_δ.

Lemma 4.2 *The condition (4.27) holds, if T_δ satisfies the inequality*

$$\mathbf{P}\{X_i \in T_\delta\} \geq 1 - \delta/(c_1 + c_2) \qquad (i \in S). \tag{4.28}$$

Proof. From (4.6), (4.26) and (4.2) we have:

$$|r_\delta - r| \leq |\mathbf{E} \int_{R^N \setminus T_\delta} \mathbf{1}(G(x;A)) G_0(x)dx| \leq \sum_i^2 c_i \mathbf{P}\{X_i \notin T_\delta\}.$$

By (4.28), $\mathbf{P}(X_i \bar{\in} T_\delta) \leq \delta/(c_1 + c_2)$, therefore (4.27) follows.

■

The asymptotic expansion of δ-risk (4.26) retains the form (4.11), but the surface integrals in (4.8) are evaluated over the piece $\Gamma \bigcap T_\delta$ of the Bayesian discriminant surface.

Consider now the Problem 2 of smoothness parameter optimization in the situation (4.21). We shall determine optimal smoothness parameters $\{h^*_{ij}\}$ from the criterion of robustness factor minimization:

$$\kappa = \kappa(\{h_{ij}\}, n) \to \min, \tag{4.29}$$

where the minimum is sought with respect to $\{h_{ij}\}$ under the restrictions (4.5). Instead of $\kappa(\cdot)$ in (4.29), we shall use the main expansion term $\kappa_1(\{h_{ij}\}, n)$. This will give us asymptotically optimal values $\{h^*_{ij}\}$. They come closer to the values that satisfy (4.29) as the sample size n increases. According to Corollary 4.2, the maximal order of the convergence rate in $\kappa_1(\{h_{ij}\}, n) \to 0$ as $n \to \infty$ is achieved for

$$h_{ij} = b_{ij} n_i^{-1/(N+4)}.$$

Then by (4.20), (4.22) and (4.23) the problem (4.29) is transformed to the problem of finding optimal values $\{b^*_{ij}\}$:

$$q = q(\{b_{ij}\}) \to \min_{\{b_{ij}>0\}} . \tag{4.30}$$

The latter problem is equivalent to the problem of polynomial (of degree $2N + 4$) minimization with respect to $2N$ variables $\{b_{ij}\}$. Therefore it cannot be solved analytically in general case; it is necessary to use numerical methods, for example, the methods of approximating programming (Himmelblau, 1972). Let us consider some particular cases of problem (4.30).

Let $\sigma^2_{ij} = \mathbf{D}\{X_{ij}\} < \infty$ be the variance of the j-th component of feature vector X_i from class Ω_i. Computer experiments suggest that it is advisable to select $\{b_{ij}\}$ proportional to $\{\sigma_{ij}\}$:

$$b_{ij} = b \cdot \sigma_{ij} \qquad (j = 1, \dots, N, i = 1, 2), \tag{4.31}$$

where $b > 0$ is an unknown proportionality factor.

Denote

$$\tau_1 = \frac{k}{2} \int_\Gamma \sum_{i=1}^{2} c_i^2 \lambda_i^{-4/(N+4)} \left(2 \prod_{j=1}^{N} \sigma_{ij} \right)^{-1} p_i(x) |\nabla G_0(x)|^{-1} ds_{N-1},$$

$$\tau_2 = \frac{1}{8} \int_\Gamma \left(\sum_{j=1}^{N} \sum_{i=1}^{2} (-1)^i c_i \lambda_i^{-2/(N+4)} \sigma^2_{ij} p^{(2)}_{ij}(x) \right)^2 |\nabla G_0(x)|^{-1} ds_{N-1}. \tag{4.32}$$

Theorem 4.2 *Under the conditions of Theorem 4.1 and (4.21), the optimal values $b^*, q^* = q(\{b^* \cdot \sigma_{ij}\})$ exist, they are unique and defined by the expressions*

$$b^* = \left(\frac{N\tau_1}{4\tau_2} \right)^{1/(N+4)}, \qquad q^* = \left(1 + \frac{4}{N} \right) \cdot \left(\frac{N \tau_1 \tau_2^N}{4} \right)^{1/(N+4)}. \tag{4.33}$$

Proof. From (4.20), (4.22), (4.30)—(4.32) we obtain the following problem :

$$q(b) = \tau_1 b^{-N} + \tau_2 b^4 \to \min_{b>0} .$$

Since $\tau_1, \tau_2 > 0$, the unique minimum point can be found as the root of the nonlinear equation

$$\frac{dq(b)}{db} = 0.$$

■

Corollary 4.3 *If*

$$\sigma_{ij}^2 = \sigma_i^2, j = 1, \ldots, N, i \in S, w_{11} = w_{22} = 0, w_{12} = w_{21} = 1,$$

the classes are equiprobable ($\pi_1 = \pi_2 = 0.5$) *and the sample sizes are equal* ($\lambda_1 = \lambda_2 = 0.5$), *then*

$$b^* = \left(\frac{\frac{Nk}{2} \int_\Gamma (\sigma_1^{-N} p_1(x) + \sigma_2^{-N} p_2(x)) |\nabla G_0(x)|^{-1} ds_{N-1}}{\int_\Gamma \frac{(\mathrm{tr}(\sigma_2^2 \nabla^2 p_2(x) - \sigma_1^2 \nabla^2 p_2(x)))^2}{|\nabla G_0(x)|} ds_{N-1}} \right)^{1/(N+4)} . \tag{4.34}$$

Consider one more situation, where the variances of components X_1, X_2 can be *a priori* assumed to be equal for the classes Ω_1, Ω_2, i.e., $\sigma_{ij}^2 = \sigma^2$. In this situation, (4.34) assumes a simpler form:

$$b^* = \left(\frac{Nk}{8} \int_\Gamma \frac{p_1(x) + p_2(x)}{|\nabla G_0(x)|} ds_{N-1} \Big/ \int_\Gamma \frac{(\mathrm{tr} \nabla^2 G_0(x))^2}{|\nabla G_0(x)|} ds_{N-1} \right)^{1/(N+4)} . \tag{4.35}$$

The optimal values $\{b_{ij}^*\}$ found by (4.30) and (4.32)—(4.35) depend on $\{\sigma_{ij}^2, \alpha_i, \beta_{ijl}\}$, τ_1 and τ_2, which in turn depend on unknown densities $\{p_i(\cdot)\}$. For practical applications of $\{b_{ij}^*\}$ in the nonparametric decision rules (4.3), (4.4) it is necessary to estimate $\{\sigma_{ij}^2, \alpha_i, \beta_{ijl}\}, \tau_1$ and τ_2 on the basis of the observed sample A. The sample variances are consistent estimators for $\{\sigma_{ij}^2\}$. The problem of estimation of $\{\alpha_i, \beta_{ijl}\}, \tau_1$ and τ_2 may be reduced, for example, to the problem of Monte-Carlo estimation of the surface integrals in (4.8), (4.32). Another way to solve this estimation problem is as follows.

First, (4.8) and (4.32) imply that the values τ_1, τ_2 can be expressed in terms of $\{\alpha_i, \beta_{ijl}\}$. In order to estimate the coefficients $\{\alpha_i, \beta_{ijl}\}$ of risk expansion in (4.11) one can apply the method from (Kharin *et al.*, 1981).

Concluding, note that in (Epanechnikov, 1969), (Vapnik, 1974) and (Hand, 1982) the optimal values of $\{h_{ij}\}$ are determined by the criterion of mean-square density estimation error minimization. An evident drawback of this approach is that $\{h_{1j}\}$ and $\{h_{2j}\}$ are estimated independently, so that the information about mutual positions of the classes Ω_1, Ω_2 in the observation space R^N is ignored.

4.2 Robustness of Nonparametric k-Nearest Neighbor Decision Rules

In this section we investigate another type of nonparametric decision rule, namely, the *k-Nearest Neighbor (k–NN) decision rule* (Fix *et al.*, 1953), (Loftsgarden *et al.*, 1965), (Meshalkin, 1969), (Patrick, 1972), (Snapp *et al.*, 1994), under the same conditions as in Section 4.1. This decision rule is derived from the Bayesian decision rule by the replacement of unknown probability densities by their k–NN–estimators (Fix *et al.*, 1953), (Fukunaga *et al.*, 1973), (Loftsgarden *et al.*, 1965), (Mack *et al.*, 1975).

Within the framework of probability data model of observations from Section 4.1 let us consider the nonparametric decision rule (4.3) in which the Rosenblatt–Parzen estimators (4.4) are replaced by the generalized k-NN–estimators (Mack *et al.*, 1975):

$$\hat{p}_i(x) = \frac{1}{n_i \rho_i^N} \sum_{j=1}^{n_i} L_i\left(\frac{x - z_{ij}}{\rho_i}\right) \qquad (i \in S), \tag{4.36}$$

where $\rho_i = \rho_i(x; A_i)$ is the Euclidean distance between x and the k_i-th nearest neighbor of the point x within the set of training sample points A_i; the positive integer k_i, $2 \le k_i \le n_i$, i.e., "the number of neighbors", is a parameter of the estimator; $L_i(u)$, $u = (u_k) \in R^N$ is a bounded integrable weight function such that

$$\int_{R^N} L_i(u)du = 1, \quad \int_{R^N} |u|^3 |L_i(u)| du < \infty,$$

$$\int_{R^N} u_k L_i(u)du = 0, \quad k = 1, \ldots, N.$$

If

$$k_i = k_i(n_i) \to \infty, \qquad k_i(n_i)/n_i \to 0 \tag{4.37}$$

as $n_i \to \infty$, then the estimator (4.36) is consistent (Mack *et al.*, 1975). Note that in statistical pattern recognition more simple k-NN–estimators with uniform weight function (Fix *et al.*, 1953), (Loftsgarden *et al.*, 1965), (Meshalkin, 1969), (Patrick, 1972) are used:

$$L_i(u) = (2\pi^{N/2})^{-1} N\Gamma(N/2)\mathbf{I}_{[0,1)}(|u|). \tag{4.38}$$

It was discovered (Fix *et al.*, 1953), (Meshalkin, 1969) that as $n_1, n_2 \to \infty$ the limit value of risk for the decision rule (4.36), (4.38) is the Bayesian risk r_0. But for the practical use of the k-NN decision rule the knowledge of this fact is not enough (Patrick, 1972), (Raudis, 1976): it is necessary to evaluate the risk at finite sample sizes $\{n_i\}$. Furthermore, this decision rule contains undetermined parameters $\{k_i\}$. Therefore we consider here the following problems of analysis of adaptive decision rule stability with respect to finite sample effects:

1) evaluation of risk (for the decision rule with estimators (4.36)) as function of $\{n_i, k_i\}$;

2) finding optimal values $\{k_i^*\}$, for which the robustness factor is minimal at fixed values $\{n_i\}$;

3) statistical estimation of the optimal values $\{k_i^*\}$ and dependence of risk on $\{n_i\}$ on the basis of the observed sample A.

Note that among the papers devoted to k-NN–procedures most of them are devoted to investigation of statistical properties of k-NN–density estimators (Fix *et al.*, 1953), (Loftsgarden *et al.*, 1965), (Patrick, 1972), (Fukunaga *et al.*, 1973), (Mack *et al.*, 1975)). At the same time the problem 1) is not solved in general case and the problem 2) is replaced by a simpler but not equivalent problem: find the optimal value k^* under the criterion of mean–square density estimation error minimization. In (Patrick, 1972), an upper bound for the moment of the conditional risk was found; but the analytical form of its dependence on $\{n_i, k_i\}$ was not found. (This dependence was only illustrated for the simplest example of one-dimensional observations with triangular probability distribution.) We would also like to mention the paper (Snapp and Venkatesh, 1994), where Problem 1) is investigated for the case (4.38) by the multidimensional Laplace method of integration.

Similarly to Section 4.1, let us construct the asymptotic expansion of risk for the plug-in decision rule (4.3), (4.36).

Let us introduce the notations (for $i \in S$):

$$F_i(y; x) = \mathbf{P}\{|X_i - x| < y\},\ y > 0,\ x \in R^N \tag{4.39}$$

is the distribution function of the random variable $|X_i - x|$ for fixed x;

$$a_{i2} = \int_{R^N} L_i(u)du > 0,\ \ B_i(x) = (p_i(x))\nabla^2 p_i(x); \tag{4.40}$$

$$L_i = (L_{ikl}),\ L_{ikl} = \int_{R^N} u_k u_l L_i(u)du \quad (k, l = 1, \ldots, N); \tag{4.41}$$

$$o_1 = o\left(\max_i\{k_i^{-1}, (k_i n_i^{-1})^{4/N}\}\right).$$

Using the notations from Section 4.1, let us investigate the asymptotics of the statistics $g(x; A) = G(x; A) - G_0(x)$ first.

Theorem 4.3 *If the densities $\{p_i(\cdot)\}$ are bounded, triply differentiable and $1 - F_i(y; x) = o(y^{-\alpha})$ for some $\alpha > 0$ $(i \in S)$, then the random deviation $g(x; A)$ of the Bayesian discriminant function estimator has moments up to the third order, and*

$$\mathbf{E}\{|g(x; A)|^3\} = o_1, \tag{4.42}$$

$$\mathbf{E}\{g^2(x;A)\} = \pi^{N/2}/\Gamma((N+2)/2)\sum_{i=1}^{2} a_{i2}c_i^2 k_i^{-1} + 1/(4\pi^2\Gamma((N+2)/2))\times$$

$$\times\left(\sum_{i=1}^{2}(-1)^{i+1}c_i p_i^{1-2/N}(x)\,\mathrm{tr}(L_iB_i(x))(k_i/n_i)^{2/N}\right)^2 + o_1. \tag{4.43}$$

Proof. From (4.2), (4.36) under the notations (4.39)—(4.41) we get

$$g(x;A) = c_2\xi_2 - c_1\xi_1. \tag{4.44}$$

According to (4.36) and (4.39), ξ_2, ξ_1 are independent, therefore (4.44) implies:

$$\mathbf{E}\{g^2(x;A)\} = c_2^2\mathbf{E}\{\xi_2^2\} + c_1^2\mathbf{E}\{\xi_1^2\} - 2c_1c_2\mathbf{E}\{\xi_1\}\mathbf{E}\{\xi_2\} =$$

$$= \sum_{i=1}^{2} c_i^2\mathbf{D}\{\xi_i\} + \left(\sum_{i=1}^{2}(-1)^{i+1}c_i\mathbf{E}\{\xi_i\}\right)^2. \tag{4.45}$$

From (Mack *et al.*, 1975) and (4.39),

$$\mathbf{E}\{\xi_i\} = (\Gamma((N+2)/2))^{2/N}(\mathrm{tr}(L_iB_i(x))/(2\pi))(p_i(x))^{1-2/N}(k_i/n_i)^{2/N} +$$

$$+p_i(x)/(Nk_i)\int_{|u|=1} L_i(u)ds_{N-1} + o((k_i/n_i)^{2/N} + 1/k_i),$$

$$\mathbf{D}\{\xi_i\} = (p_i^2(x)/k_i)(\pi^{N/2}a_{i2}/(\Gamma((N+2)/2)) + o(1/k_i). \tag{4.46}$$

Substituting (4.46) into (4.45), by (4.39)—(4.41) we get (4.43). As it was made in (Mack *et al.*, 1975), one can show that $\mathbf{E}\{|\xi_i|^3\} = o_1$. Then similarly to (4.43), we obtain (4.42).

■

Denote (for $i, j \in S$):

$$b_i(x) = \nabla_x \ln p_i(x) \in R^N; \quad b(x) = |b_2(x) - b_1(x)|;$$

$$\Gamma = \{x : G_0(x) = 0\} \subset R^N \tag{4.47}$$

is the Bayesian discriminant surface in R^N; for surface integrals over Γ :

$$\alpha_i = \pi^{N/2}a_{i2}c_i/(2\Gamma((N+2)/2)\int_\Gamma p_i(x)/b(x)ds_{N-1} \geq 0, \tag{4.48}$$

$$\beta_{ij} = (-1)^{i+j}c_i^{1-2/N}c_j^{2/N}\Gamma^{4/N}((N+2)/2)\int_\Gamma \mathrm{tr}(L_iB_i(x))\,\mathrm{tr}(L_jB_j(x))\times$$

$$\times p_i^{1-4/N}(x)(b(x))^{-1}ds_{N-1}/(8\pi^2).$$

Theorem 4.4 *Assume that the conditions of Theorem 4.3 hold,*

$$\mathrm{mes}_N\{x : |\nabla_x G_0(x)| = 0\} = 0,$$

where mes *is the Lebesgue measure, and for some* $i \in S$ *the derivatives are bounded:*

$$|p_{\xi_i}^{(j)}(z)| < \infty, \quad z \in R^1, \; j \in \{0,1,2\}. \tag{4.49}$$

Then the risk of k-NN–decision rule (4.3), (4.36) admits the following asymptotic expansion:

$$r = r_1 + o_1,$$

$$r_1 = r_0 + \sum_{i=1}^{2} \alpha_i k_i^{-1} + \sum_{i,j=1}^{2} \beta_{ij} (k_i/n_i)^{2/N} (k_j/n_j)^{2/N}. \tag{4.50}$$

Proof is conducted using Theorem 4.3 in the same way as for Theorem 4.1. ■

Note that (4.49) imposes weak restrictions on the probability distribution densities $\{p_i(\cdot)\}$ and the weight functions $\{L_i(\cdot)\}$. It is found that even for the case of discontinuous weight functions $\{L_i(\cdot)\}$ the condition (4.49) is easily fulfilled.

Corollary 4.4 *If the weight functions* $\{L_i(\cdot)\}$ *are discontinuous and defined by (4.38), then to ensure (4.49) it is sufficient that*

$$|F_i^{(k)}(y;x)| = |\partial^k F_i(y;x)/\partial y^k| < \infty, \; y \in R^1, \; x \in R^N; k = 1,2,3. \tag{4.51}$$

The coefficients of the asymptotic expansion of risk are:

$$\alpha_i = c_i/2 \int_\Gamma p_i(x)/b(x) ds_{N-1} \geq 0, \tag{4.52}$$

$$\beta_{ij} = \frac{(-1)^{i+j} c_i^{1-2/N} c_j^{2/N} \Gamma^{4/N}((N+2)/2)}{8\pi^2 (N+2)^2} \int_\Gamma \frac{\mathrm{tr}(B_i(x))\,\mathrm{tr}(B_j(x))}{b(x)} p_i^{1-4/N}(x) ds_{N-1}.$$

Proof. Substituting (4.38) into (4.36), we represent k-NN–estimator in the form

$$\hat{p}_i(x) = ((k_i - 1)/(n_i \pi^{N/2}))\,(\Gamma(N/2+1))/Y_{(k_i)}^N, \tag{4.53}$$

where $Y_{(k_i)} = \rho_i$ is the k_i-th order statistics for a random sample $\{|x - z_{ij}| : j = 1, \ldots, n_i\}$ of size n_i from the distribution $F_i(y;x)$ defined in (4.39)—(4.41). From the theory of order statistics (David, 1975), we get the probability distribution density of $Y_{(k_i)}$:

$$p_{Y_{(k_i)}}(y) = n_i!/((k_i-1)!(n_i-k_i)!) F_i^{k_i-1}(y;x)(1 - F_i(y;x))^{n_i-k_i} F_i^{(1)}(y;x). \tag{4.54}$$

According to (4.39)—(4.41), (4.53), the random variable ξ_i is a functional transformation of $Y_{(k_i)}$. Then by rules of functional transformation of random variables we derive from (4.54) the probability density of ξ_i:

$$p_{\xi_i}(z) = (n_i\pi^{N/2})/((k_i-1)N\Gamma(N/2+1))\tau^{N+1}p_{Y_{(k_i)}}(\tau), z \in R^1, \tag{4.55}$$

where

$$\tau = \left((k_i-1)\Gamma(N/2+1)/(n_i\pi^{N/2}(z+p_i(x)))\right)^{1/N}.$$

By conditions of Theorem 4.3, we have from (4.54), (4.55):

$$\tau^{N+1}(1-F_i(\tau;x))^{n_i-k_i} = o(\tau^{N+1-\alpha(n_i-k_i)}) \to 0$$

under the asymptotics (4.37). Then (4.49) follows from (4.51), (4.54), (4.55) for $j=0$. Differentiating (4.55) j times with respect to z and analyzing this result (as in (4.55)) we obtain (4.49) for $j=1,2$.

■

From (4.50) it follows that the conditions (4.37) are sufficient for the consistency of the k-NN–decision rule (4.3), (4.36):

$$r \to r_0, \text{ as } n_1, n_2 \to \infty.$$

Consider now a situation where the parameters $\{k_i\}$ of decision rule (4.36) are power functions of sample sizes ($i=1,2$):

$$k_i = \lfloor b_i n_i^{\gamma_i} \rfloor, \quad \gamma_i > 0, \quad b_i > 0. \tag{4.56}$$

In this case the risk expansion given by Theorem 4.4 becomes the expansion in powers of values inverse to the sample sizes.

Corollary 4.5 *Under the conditions of Theorem 4.4 and (4.56) the following expansion holds:*

$$r = r_1 + o(\max_i\{n_i^{-\gamma_i}, n_i^{-4(1-\gamma_i)/N}, n_1^{-2(1-\gamma_1)/N} n_2^{-2(1-\gamma_2)/N}\}), \tag{4.57}$$

$$r_1 = r_0 + \sum_{i=1}^{2}\left(q_i n_i^{-\gamma_i} + q_{2+i} n_i^{-4(1-\gamma_i)/N}\right) + q_5 n_1^{-2(1-\gamma_1)/N} n_2^{-2(1-\gamma_2)/N},$$

$$q_i = \alpha_i/b_i, \quad q_{2+i} = \beta_{ii} b_i^{4/N} \quad (i \in S), \tag{4.58}$$

$$q_5 = (\beta_{12}+\beta_{21})(b_1 b_2)^{2/N}.$$

It can be seen from (4.57) that to ensure the convergence $r \to r_0$ it is necessary that $0 < \gamma_i < 1$ ($i \in S$). Using the arbitrariness of $\{\gamma_i\}$ we choose them to maximize the convergence rate for risk.

Corollary 4.6 *Under the condition (4.21) the fastest convergence of r_1 to r_0 is achieved for $\gamma_1^* = \gamma_2^* = 4/(N+4)$:*

$$r_1 = r_0 + qn^{-4/(N+4)},$$

$$q = \sum_{i=1}^{2}(q_i + q_{2+i})\lambda_i^{-4/(N+4)} + q_5(\lambda_1\lambda_2)^{-2/(N+4)}, \tag{4.59}$$

where $q_1, \ldots, q_5$ are defined in (4.58).

Let us analyze (4.59). First, notice that γ_1^*, γ_2^* coincide with the values $\{\gamma_i\}$ found in (Meshalkin, 1969) and (Fukunaga *et al.*, 1973) under the criterion of integral mean–square density estimation error minimization. Second, the expressions (4.59) are of the same kind as the expressions (4.22) for the nonparametric Rosenblatt–Parzen decision rule. The only difference is in coefficients $\{q_i\}, q$. Thus, the order of convergence rate of r to the Bayesian risk r_0 is the same both for the Rosenblatt–Parzen decision rule and for the k-NN decision rule.

Consider now the problems 2) and 3) stated at the beginning of this section. The problem 2) is solved partially by Corollary 4.6: the optimal growth rates $\{\gamma_i^*\}$ for parameters $\{k_i\}$ are found, but the optimal values for the coefficients $\{b_i\}$ in (4.56) are unknown.

We shall find optimal parameters $\{k_i^*\}$ of the k-NN decision rule under the criterion of unconditional risk minimization:

$$r = r(k_1, k_2) \to \min, \tag{4.60}$$

where the minimization is with respect to $k_i \in \{2, 3, \ldots, n_i\} (i \in S)$. Instead of r in (4.60), we shall use (as in Section 4.1 for the optimization of the Rosenblatt-Parzen decision rule) the main term of the asymptotic expansion for risk $r_1 = r_1(k_1, k_2)$ defined by (4.50), neglecting the remainder term o_1. In this situation, (4.60), (4.50) is a problem of nonlinear discrete optimization. For "small" sample sizes n_1, n_2 the optimal values k_1^*, k_2^* can be found by computer comparison of $(n_1 - 1)(n_2 - 1)$ values of the function $r_1 = r_1(k_1, k_2)$. For "large" sample sizes $\{n_i\}$ it is advisable to use special procedures of discrete optimization, for example the branch and bound method.

In addition, let us construct an approximate solution of the problem (4.50), (4.60). According to (4.50), the main term of the expansion for $r_1(k_1, k_2)$ is a differentiable function of the variables k_1, k_2. Therefore we temporarily remove the requirement of discreteness of variables, and using (4.50) we express the necessary condition for minimum of $r_1 (i \in S)$ as follows:

$$\frac{\partial r_1}{\partial k_i} = \frac{-\alpha_i}{k_i^2} + \frac{4\beta_{ii}}{Nn_i}\left(\frac{k_i}{n_i}\right)^{4/N-1} +$$

$$+ \frac{2(\beta_{12} + \beta_{21})}{Nn_i}\left(\frac{k_i}{n_i}\right)^{2/N-1}\left(\frac{k_{3-i}}{n_{3-i}}\right)^{2/N} = 0. \tag{4.61}$$

The system (4.61) is a system of nonlinear equations with respect to $\{k_i\}$. It cannot be solved in analytical form. Let us linearize it by Taylor formulas in the neighborhood of some starting point (k_1^0, k_2^0):

$$\left(\frac{k_i}{n_i}\right)^\gamma = \left(\frac{k_i^0}{n_i}\right)^\gamma \left(1 - \gamma + \gamma\frac{k_i}{k_i^0}\right), \quad \frac{1}{k_i^2} = \frac{1}{(k_i^0)^2}\left(3 - 2\frac{k_i}{k_i^0}\right) \quad (i \in S).$$

We take the values of $\{k_i^0\}$ as recommended in (Meshalkin, 1969):

$$k_i^0 = n_i^{4/(N+4)}, \tag{4.62}$$

which coincide with (4.56) for $b_i = 1, \gamma_i = \gamma_i^*(i = 1, 2)$. Denote (for $i \in S$):

$$D_i = \frac{-3\alpha_i}{(k_i^0)^2} + \frac{8\beta_{ii}}{Nn_i}(1 - 2/N)\left(\frac{k_i^0}{n_i}\right)^{4/N-1} +$$

$$+\frac{4(\beta_{12} + \beta_{21})}{Nn_i}(1 - 1/N)(1 - 2/N)\left(\frac{k_i^0}{n_i}\right)^{2/N-1}\left(\frac{k_{3-i}^0}{n_{3-i}}\right)^{2/N},$$

$$E_i = \frac{2\alpha_i}{(k_i^0)^2} - \frac{4\beta_{ii}}{Nn_i}(1 - 4/N)\left(\frac{k_i^0}{n_i}\right)^{4/N-1} -$$

$$-\frac{2(\beta_{12} + \beta_{21})}{Nn_i}(1 - 2/N)^2\left(\frac{k_i^0}{n_i}\right)^{2/N-1}\left(\frac{k_{3-i}^0}{n_{3-i}}\right)^{2/N},$$

$$F_i = \frac{8(\beta_{12} + \beta_{21})}{N^2 n_i}(1 - 1/N)\left(\frac{k_i^0}{n_i}\right)^{2/N}\left(\frac{k_{3-i}^0}{n_{3-i}}\right)^{2/N}.$$

Then solving the linearized system (4.61) we find:

$$k_i^* = \left\lfloor n_i^{4/(N+4)} \frac{E_{3-i}D_i - F_i D_{3-i}}{F_1 F_2 - E_1 E_2} \right\rfloor \quad (i \in S). \tag{4.63}$$

In practice it is impossible to immediately apply (4.63) to optimization of the k-NN decision rule because $\{k_i^*\}$ depend on the expansion coefficients $\{\alpha_i, \beta_{ij}\}$ from (4.50), which are defined by (4.48) as functionals of unknown densities $\{p_i(\cdot)\}$. Similarly to Section 4.1, the method from (Kharin *et al.*, 1983) may be used to estimate $\{\alpha_i, \beta_{ij}\}$.

4.3 Comparative Analysis of Robustness for Adaptive Decision Rules

Multivariate Gaussian data model is often used in pattern recognition theory when the effectiveness of decision rules is compared. Therefore we shall compare robustness of parametric and nonparametric ADR with respect to small-sample effects under Gaussian data model described in Section 3.3.

4.3.1 Application of Rosenblatt–Parzen ADR

As in Section 3.3, consider first the situation C_1 for $\Sigma^0 = \mathbf{I}_N$. (The last condition is equivalent to the assumption that Σ^0 is known, because one can ensure this condition by linear transformation of observations: $x' = (\Sigma^0)^{-1/2}x$.) We choose the Gaussian kernel and smoothness parameters in (4.4) according to (4.19) and Corollary 4.2:

$$K(y) = 1/(2\pi)^{1/2}e^{-y^2/2}, \quad h_{ij} = bn_i^{-1/(N+4)}.$$

According to (4.8), (3.56) the Bayesian discriminant hypersurface

$$\Gamma = \{x : (\mu_1^0 - \mu_2^0)^T x - (\mu_1^0 - \mu_2^0)^T(\mu_1^0 + \mu_2^0)/2 = 0\}$$

is a hyperplane, with infinite surface area. Therefore we shall construct an asymptotic expansion of δ-risk r_δ defined by (4.26) ($0 < \delta << 1$). Let us apply Lemma 4.2 taking the region T_δ to be the union of two spheres:

$$T_\delta = T_\delta^{(1)} \bigcup T_\delta^{(2)}, \quad T_\delta^{(i)} = \{x : (x - \mu_i^0)^T(x - \mu_i^0) \leq R_\delta^2\},$$

where R_δ^2 is the $(1 - \delta)$-quantile of the chi-square distribution with N degrees of freedom. Here $c_1 = c_2 = 0.5$, $\mathbf{P}\{X_i \in T_\delta^{(i)}\} = 1 - \delta$, and the condition (4.28) (and hence, (4.27)) is fulfilled.

Let us apply Theorem 4.1 for δ-risk r_δ. The surface integrals in (4.8) must be evaluated over the region $\Gamma \bigcap T_\delta$, which is an $(N - 1)$-dimensional sphere of radius $(R_\delta^2 - \Delta^2/4)^{1/2}$. It is centered at the center of the segment connecting the points $\mu_1^0, \mu_2^0 \in R^N$. According to (4.8),

$$\alpha_i = (2/\Delta)\,\mathrm{mes}_{N-1}(\Gamma \bigcap T_\delta),$$

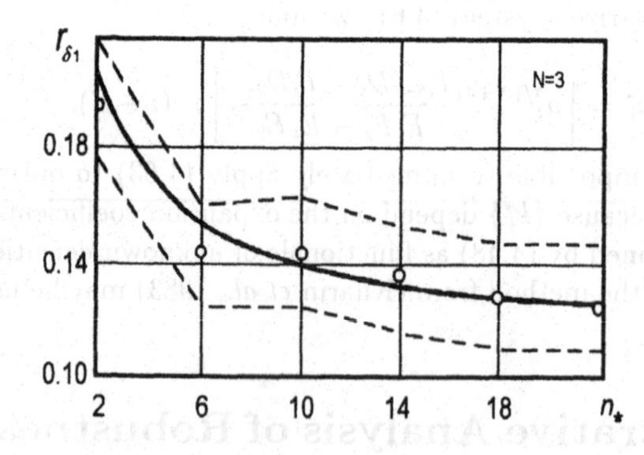

Figure 4.1: Risk vs. sample size for $N = 3$

where $\mathrm{mes}_{N-1}(\Gamma \bigcap T_\delta)$ is the $(N - 1)$-dimensional volume of $\Gamma \bigcap T_\delta$. The remaining terms in (4.11) vanish, since $\mathrm{tr}(\nabla^2 p_2(x)) = \mathrm{tr}(\nabla^2 p_1(x))$ for $x \in \Gamma$. Therefore the asymptotic expansion for δ-risk assumes the form

$$r_\delta = r_{\delta 1} + o(n_*^{-4/(N+4)}), \quad n_* = n_1 = n_2,$$

$$r_{\delta 1} = \Phi(-\Delta/2) + \left((R_\delta^2 - \Delta^2/4)^{(N-1)/2}\right) \times$$

$$\times \left(2^{N+1} b^N \pi^{1/2} \Delta \Gamma((N+1)/2)\right)^{-1} n_*^{-4/(N+4)}. \tag{4.64}$$

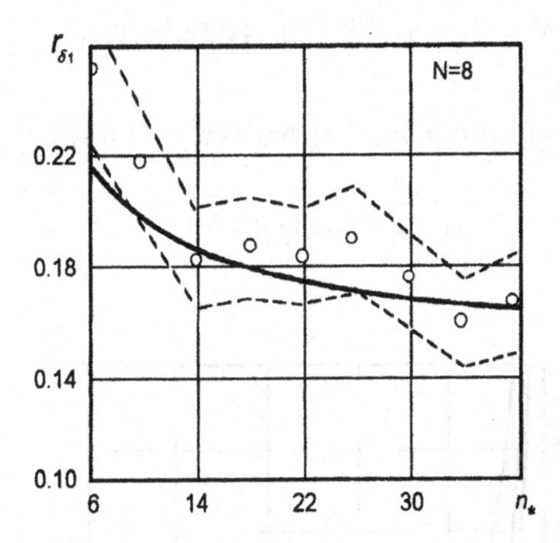

Figure 4.2: Risk vs. sample size for $N = 8$

Figure 4.1 plots the dependence of $r_{\delta 1}$ on n_* by solid line for $\Delta = 2.56$ (this means that $r_0 = 0.1$):

$$N = 3, \delta = 0.005, b = 1, \mu_1^0 = \mathbf{O}_N, \mu_2^0 = (\mu_{2i}), \mu_{2i} = \Delta/N^{1/2}, (i = 1, \ldots, N).$$

The points near this solid curve indicate risk values found by computer simulation of the adaptive decision rule. The dashed lines are the lower and upper 80%-confidence limits for risk. Similar results are presented at Figure 4.2 for $N = 8$, $\delta = 0.1$, $\Delta = 2.56$ (with 90%-confidence limits). It can be seen that the approximation accuracy for error probability (risk) is sufficiently high, and it increases with the increase of n_*. Note in addition, that in the considered situation $r_{\delta 1}$ depends monotonously on b. As $b \to \infty$, $r_{\delta 1}$ decreases, but the remainder term of (4.64) increases.

Consider now the nonparametric adaptive decision rule in the situation C_2 and demonstrate influence of smoothness parameters on error probability for $N = 1$. Denote:

$$h_i = h_{i1} = b\sigma_i n_*^{-1/5} \quad (i \in S); \gamma = (\sigma_{111}/\sigma_{211})^{1/2}.$$

Without loss of generality, $\gamma > 1$. The Bayesian risk in this situation is defined by (3.59). From Theorem 4.1 and (4.23) we may find the approximate formula for the robustness factor:

$$\kappa = \kappa_1 = ((1+\gamma)(\sqrt{2}b)^{-1} + \gamma^{-1/(\gamma^2-1)}(\ln^2\gamma)b^4)\times$$

$$\times\, (4r^0(\pi(\gamma^2-1)\ln\gamma)^{1/2})^{-1}n_*^{-4/5}. \tag{4.65}$$

Optimization with respect to b by Theorem 4.2 gives

$$b^* = ((1+\gamma)\gamma^{1/(\gamma^2-1)})/(4\sqrt{2}\ln^2\gamma))^{1/5},$$

$$\kappa_1 = \min_b \kappa_1(b) = 5n_*^{-4/5}/(16r_0(\pi(\gamma^2-1)\ln\gamma)^{1/2})\times$$

$$\times\, ((1+\gamma)^4\ln^2\gamma)/\gamma^{1/(\gamma^2-1)})^{1/5}. \tag{4.66}$$

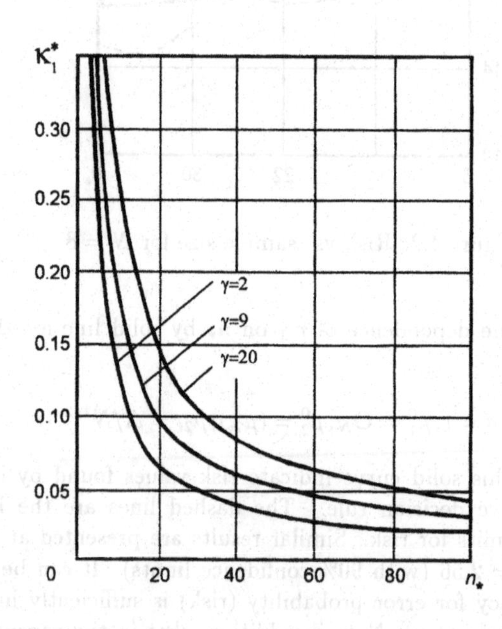

Figure 4.3: Robustness factor vs. sample size

Figure 4.3 plots the dependence of κ_1^* on sample size n_* for $\gamma = 2; 9; 20$.

The minimal δ-admissible sample sizes $n_{*\delta}(b^*)$ for both classes Ω_1, Ω_2 for $\delta = 0.2$ and different values of γ evaluated by (4.25), (4.66) at optimal values of smoothness parameter $h_i^* = b^*\sigma_i n_*^{-1/5}$, are:

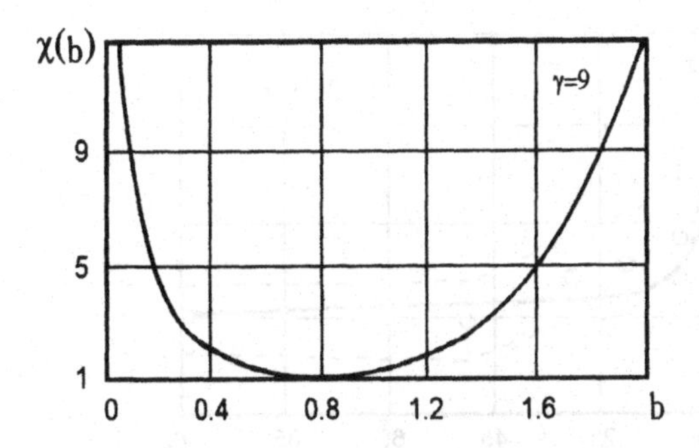

Figure 4.4: Plot of $\chi(b)$ dependence

γ	2	5	7	9	20
$n_{*\delta}(b^*)$	6	6	7	9	14

The dependence of the ratio of minimal δ-admissible sample sizes

$$\chi(b) = n_{*\delta}(b)/n_{*\delta}(b^*)$$

on b for $\gamma = 9$ ($r_0 = 0.101$) is presented at Figure 4.4. It can be seen that efficiency loss of the adaptive decision rule (i.e., error probability increment) induced by non-optimal choice of the parameter b can exceed the loss induced by the nonrepresentativity of the training sample (n_* has small value). For example, for $\gamma = 9$, $\delta = 0.2$ and optimal value $b^* = 0.83$ the error probability for the nonparametric ADR differs from the Bayesian error probability by at most 20%. If b is non-optimal, for example $b = 1.5$, then you need three times larger sample size: $n_{*\delta}(b^*) = 27$.

4.3.2 Application of k-NN Decision Rules

Let us assume that the weight function $L_i(\cdot)$ in (4.36) is a uniform function defined by (4.38). For the situation C_1 the computations by the formulas (4.40), (4.41), (4.47), (4.48) give ($i, j \in S$):

$$L_i = \mathbf{I}_N/(N+2),\ a_{i2} = (2\pi^{N/2})^{-1}N\Gamma(N/2),\ b(x) = |(\Sigma^0)^{-1}(\mu_2^0 - \mu_1^0)|,$$

$$\alpha_i = \exp^{-\Delta^2/8}/(4(2\pi)^{1/2}\Delta),\ \beta_{ij} = (-1)^{i+j}\beta_N,\ \beta_N > 0.$$

The expression of β_N is too bulky and is not presented here. The main term of the asymptotic expansion (4.50) is:

$$r_1 = r_0 + \exp(-\Delta^2/8)/(4(2\pi)^{1/2}\Delta)\left(\frac{1}{k_1} + \frac{1}{k_2}\right) +$$

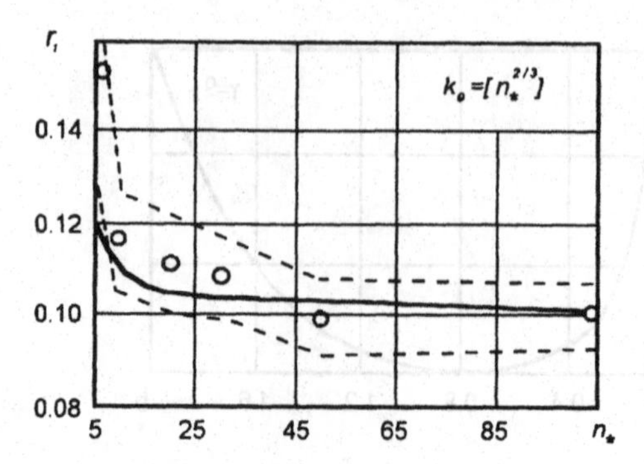

Figure 4.5: Risk vs. sample size

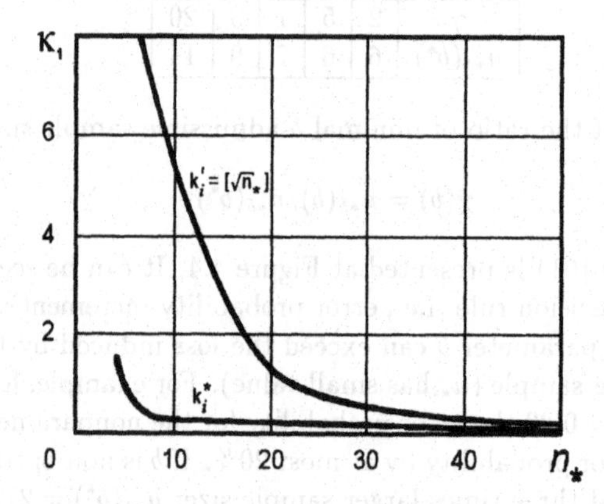

Figure 4.6: Robustness factor vs. sample size

$$+ \beta_N \left(\left(\frac{k_2}{n_2} \right)^{2/N} - \left(\frac{k_1}{n_1} \right)^{2/N} \right)^2 . \tag{4.67}$$

Figure 4.5 plots the dependence of error probability r_1 on n_* by the solid line for $n_1 = n_2 = n_*$, $k_1 = k_2 = \lfloor n_*^{4/(N+4)} \rfloor$, $N = 2$, $\Delta = 2.56 (r_0 = 0.1)$. The points shown near the curve are found by computer modeling of the adaptive decision rule. The dashed lines are the lower and upper 80%-confidence limits for the ADR error probability.

Consider the situation C_2, defined in Section 3.3. Let us analyze the influence of the parameters $\{k_i\}$ of the k-NN decision rule on the error probability. From (4.50), (4.59) we obtain the following approximate formula for the robustness factor:

$$\kappa = \kappa_1(k_1, k_2) = \left(\frac{2(2\pi(\gamma^2 - 1)\ln\gamma^2)^{1/2}\gamma^{1/(\gamma^2-1)})r_0}{k_1^{-1} + k_2^{-1} + ((a_1 k_1)^2 + (a_2 k_2)^2)^2} \right)^{-1}, \tag{4.68}$$

$$a_1 = ((\pi)^{1/2}\gamma^{1/(\gamma^2-1)})/(2(3)^{1/2}n_1)(1 - (\ln\gamma^2)/(\gamma^2 - 1))^{1/2},$$

$$a_2 = ((\pi)^{1/2}\gamma^{1/(\gamma^2-1)})/(2(3)^{1/2}n_2)(\gamma^2(\ln\gamma^2)/(\gamma^2 - 1) - 1)^{1/2}.$$

The problem (4.60) of optimal selection of $\{k_i\}$ is solved in closed form:

$$k_1^* = \left[(4a_1^4(1 + (a_2/a_1)^{2/3}))^{-1/5}\right]',$$

$$k_2^* = \left[(a_2/a_1)^{-2/3}(4a_1^4(1 + (a_2/a_1)^{2/3}))^{-1/5}\right]', \tag{4.69}$$

where $[z]'$ denotes rounding to the nearest integer.

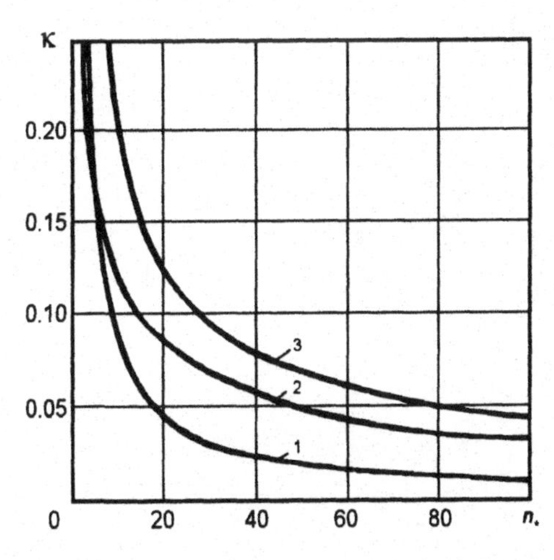

Figure 4.7: Plots of robustness factor $\kappa(n_*)$ for three decision rules

Figure 4.6 plots the dependence of κ_1 on sample size n_* for $n_1 = n_2 = n_*$, $\gamma = 9$ ($r_0 = 0.101$) for optimal parameter values $\{k_i^*\}$ defined in (4.69) and for values $\{k_i' = [(n_*)^{1/2}]'\}$ recommended in (Patrick, 1972). For the values $\{k_i^0\}$ determined by (4.62) (recommended in (Meshalkin, 1969)) the curve $\kappa_1(k_1^0, k_2^0)$ passes higher than the curve $\kappa_1(k_1', k_2')$. Considerable gain is observed in stability of pattern recognition when optimal values $\{k_i^*\}$ are used.

4.3.3 Performance Comparison for Parametric and Nonparametric ADR

Approximate formulas for risk functional and for robustness factor constructed in Sections 3.3, 4.3 allow to compare parametric and nonparametric adaptive decision rules and to give some recommendations on their practical use.

For example, consider the situation C_1, defined in Section 3.3, with $\pi_1 = \pi_2$, $n_1 = n_2 = n_*$, $N = 2$, $b = 1$, $\delta = 0.005$. For this case, Figure 4.7 plots the dependence of the robustness factor κ (relative increment of error probability) on n_* for three adaptive decision rules: optimal parametric adaptive decision rule (3.56), (3.62) (curve 1); nonparametric k-NN adaptive decision rule (curve 2) and nonparametric Rosenblatt–Parzen adaptive decision rule (curve 3). It can be seen from this figure that for "large" sample size $n_* > 20$ nonparametric k-NN adaptive decision rules and especially the Rosenblatt–Parzen adaptive decision rule lose in robustness in comparison with the parametric decision rule. For "small" sample sizes $n_* < 10$ (for the considered situation C_1) the k-NN decision rule is close in robustness to the parametric decision rule. Note in addition that in the situation C_2 the k-NN decision rule (with optimal values $\{k_i^*\}$) and the Rosenblatt–Parzen decision rule (with optimal values $\{h_{ij}^*\}$) demonstrated the same degree of classification robustness.

Chapter 5

Decision Rule Robustness under Distortions of Observations to be Classified

This chapter is devoted to problems of robust pattern recognition for common in practice types of distortions of multivariate observations subject to classification: Tukey-Huber type contaminations, additive distortions of observations (including round-off errors), distortions produced by mixtures of probability distributions, distortions defined by means of L_2-metric, or χ^2-metric, or variation metric, and random distortions of probability distributions. Using the asymptotic expansion method, we find estimates for the robustness factor and critical distortion levels ("breakdown points"). We also construct robust decision rules that minimize the guaranteed risk value.

5.1 Robustness to Tukey–Huber Type Contaminations

Suppose that random observations $X \in R^N$ from $L \geq 2$ classes $\Omega_1, \ldots, \Omega_L$ in the feature space R^N are recorded with prior probabilities $\pi_1, \ldots, \pi_L$ $(\pi_1 + \ldots + \pi_L = 1)$. An observation from the class Ω_i is a random vector $X_i \in R^N$ with N–variate probability density function $p_i(x)$. The loss matrix $W = (w_{ij})$ is given. The density $p_i(\cdot)$ admits Tukey–Huber type nonparametric (functional) distortions (type D.2.2.1) defined in Section 2.2 :

$$p_i(\cdot) \in \mathcal{P}_i(\epsilon_{+i}),$$

$$\mathcal{P}_i(\epsilon_{+i}) = \left\{ p_i(\cdot) : p_i(x) = (1 - \epsilon_i) p_i^0(x) + \epsilon_i h_i(x), x \in R^N, 0 \leq \epsilon_i \leq \epsilon_{+i} < 1 \right\}, \quad (5.1)$$

where $h_i(\cdot)$ is an arbitrary density of the contaminating distribution.

This type of distortions often happens in practice and has an obvious sense. The class Ω_i of observed objects is actually partitioned into two subclasses:

$$\Omega_i = \Omega_i^0 \cup \Omega_i^h, \; \Omega_i^0 \cap \Omega_i^h = \emptyset;$$

Ω_i^0 is the well-studied (most frequently observed) subclass; Ω_i^h is the seldomly observed and poorly investigated subclass. Observations from Ω_i^0 are characterized by a known (hypothetical, estimated) density $p_i^0(x)$, and observations from Ω_i^h are characterized by an unknown density $h_i(x)$. An object from Ω_i belongs to Ω_i^0 with probability $1-\epsilon_i$ and to Ω_i^h with probability ϵ_i.

In contrast to Section 2.3, here we shall consider a generalized family of randomized decision rules determined by a set of L critical functions:

$$\chi = (\chi_1(x), \ldots, \chi_L(x)),\ \chi_i(x) \geq 0, \sum_{i=1}^{L} \chi_i(x) = 1,\ x \in R^N. \tag{5.2}$$

The Borel function $\chi_i = \chi_i(x) : R^N \to [0,1]$ determines the probability of the event that when the feature vector $x \in R^N$ is registered the decision $d = i \in S$ is made. This means that with probability χ_i the observation $x \in R^N$ should be ascribed to class Ω_i. In particular, if $\chi_1(x), \ldots, \chi_L(x) \in \{0,1\}$, which means that there exists a Borel function $d = d(x) : R^N \to S$ such that

$$\chi_i(x) = \delta_{i,d(x)}, i \in S, x \in R^N,$$

where $\delta_{i,j}$ is the Kronecker symbol, then the decision rule (5.2) becomes nonrandomized. The risk functional for decision rule (5.2) is of the form

$$r = r(\chi; \{p_i\}) = \sum_{i=1}^{L} \pi_i \mathbf{E}\left\{\sum_{j=1}^{L} w_{i,j}\chi_j(X_i)\right\} =$$

$$= \sum_{i,j=1}^{L} \pi_i w_{i,j} \int_{R^N} p_i(x)\chi_j(x)dx. \tag{5.3}$$

If distortions are absent ($\epsilon_+ = \max_i \epsilon_{+i} = 0$), then, as it was shown in Chapter 1, the minimal risk r_0 is attained for the nonrandomized Bayesian decision rule:

$$\chi_i^0(x) = \delta_{i,d_0(x)},\ d_0(x) = \arg\min_{k \in S} f_k^0(x),$$

$$f_k^0(x) = \sum_{j=1}^{L} \pi_j p_j^0(x) w_{jk} \quad , x \in R^N, r_0 = r(\chi^0; \{p_i^0\}) \geq 0, \tag{5.4}$$

which is unique up to a zero Lebesgue measure set. The guaranteed (upper) risk for the decision rule χ is defined similarly to (2.22) :

$$r_+ = r_+(\chi) = \sup r(\chi; \{p_i\}), \tag{5.5}$$

where sup is taken with respect to all admissible distributions (5.1). Similarly to (2.32), we define *robust decision rule* $\chi^* = (\chi_1^*(x), \ldots, \chi_L^*(x))$ to be the decision rule that minimizes the guaranteed risk :

$$r_+^* = r_+(\chi^*) = \inf_{\chi} r_+(\chi). \tag{5.6}$$

The robustness level for the Bayesian decision rule χ^0 is characterized by the *absolute risk increment* $\beta_+(\chi^0) = r_+(\chi^0) - r_0 \geq 0$ and by the *robustness factor* (at $r_0 > 0$):

$$\kappa_+ = \kappa_+(\chi^0) = (r_+(\chi^0) - r_0)/r_0 \geq 0. \tag{5.7}$$

The less β_+, κ_+, the more stable the decision rule χ^0.

Let us denote: $w_{i+} = \max_j w_{ij}$, $\overline{w}_+ = \sum_{i=1}^{L} \pi_i w_{i+}$,

$$q(p_i^0; \chi_j) = \int_{R^N} p_i^0(x)\chi_j(x)dx, \quad r_{0i} = \sum_{j=1}^{L} w_{i,j} q(p_i^0, \chi_j^0) \tag{5.8}$$

is the conditional risk of the Bayesian decision rule for nondistorted observations from Ω_i $(i \in S)$.

Theorem 5.1 *If Tukey–Huber distortions (5.1) take place, then the guaranteed risk for the Bayesian decision rule (5.4) is*

$$r_+(\chi^0) = r_0 + \sum_{i=1}^{L} \epsilon_{+i}\pi_i(w_{i+} - r_{0i}). \tag{5.9}$$

Proof. From (5.1), (5.3), (5.4) for fixed ϵ_i bearing in mind (5.8) we obtain:

$$r(\chi^0; \{p_i\}) = r_0 - \sum_{i=1}^{L} \epsilon_i \pi_i r_{0i} + \sum_{i=1}^{L} \epsilon_i \pi_i \sum_{j=1}^{L} w_{ij} q(h_i; \chi_j^0).$$

As $h_i(\cdot)$ is an arbitrary density in R^N, then from (5.4), (5.8) we find:

$$\sup_{\{h_i(\cdot)\}} \sum_{j=1}^{L} w_{ij} q(h_i; \chi_j^0) = w_{i+},$$

therefore

$$\sup_{\{h_i(\cdot)\}} r(\chi^0, \{p_i\}) = r_0 + \sum_{i=1}^{L} \epsilon_i \pi_i (w_{i+} - r_{0i}).$$

Maximizing now this expression with respect to $\epsilon_i \in [0, \epsilon_{+i}]$ and using (5.5) we come to (5.9).

■

From (5.9) and definitions given in Section 2.3, one can obtain the following conditions of absolute and relative $\delta(\delta')$–robustness for the Bayesian decision rule (5.4):

$$\sum_{i=1}^{L} \epsilon_{+i}\pi_i(w_{i+} - r_{0i}) \leq \delta,$$

$$\left(\sum_{i=1}^{L} \epsilon_{+i}\pi_i(w_{i+} - r_{0i})\right) \cdot \left(\sum_{i=1}^{L} \pi_i r_{0i}\right)^{-1} \leq \delta'.$$

Corollary 5.1 *The breakdown point (2.29) for the Bayesian decision rule (5.4) is determined by the expression*

$$\epsilon_+^* = (r^* - r_0)/(\overline{w}_+ - r_0).$$

Corollary 5.2 *For equidistorted classes* ($\epsilon_{+i} = \epsilon_+, i \in S$),

$$\beta_+(\chi^0) = \epsilon_+(\overline{w}_+ - r_0), \ \kappa_+(\chi^0) = \epsilon_+(\overline{w}_+/r_0 - 1).$$

Corollary 5.3 *Under the conditions of Corollary 5.2 for the* $(0-1)$*–loss matrix (1.18) (when the risk* r *is in fact the error probability) the following expressions are valid :*

$$\beta_+(\chi^0) = \epsilon_+(1 - r_0), \quad \kappa_+(\chi^0) = \epsilon_+(r_0^{-1} - 1),$$

$$\epsilon_+^* = 1 - (L(1 - r_0))^{-1}. \tag{5.10}$$

Figure 5.1 plots the dependence (5.10) of the breakdown point ϵ_+^* on the Bayesian error probability r_0 and the number of classes L, which is important in practice. It can be seen from Figure 5.1 that, for example, if we have two classes ($L = 2$) and $r_0 = 0.2$, then the distortion level must not reach the critical value $\epsilon_+^* = 0.375$.

Note that by the same method as it was used in Theorem 5.1, one can find the greatest lower bound of risk for the Bayesian decision rule under Tukey–Huber distortions. In particular, under the conditions of Corollary 5.2 the value of error probability of the Bayesian decision rule for arbitrary distortions (5.1) is within the segment:

$$r_0 - \epsilon_+ r_0 \leq r(\chi^0; \{p_i\}) \leq r_0 + \epsilon_+(1 - r_0). \tag{5.11}$$

Risk variability zones, important for practical applications, are displayed at Figure 5.2 for the Bayesian decision rule under Tukey–Huber distortions (the vertically hatched area is for $\epsilon_+ = 0.1$ and the horizontally hatched one is for $\epsilon_+ = 0.2$).

Consider now the problem of robust decision rule synthesis.

Theorem 5.2 *Under distortions (5.1) the robust decision rule has the form:*

$$\chi_i^*(x) = \delta_{i,d_*(x)}, \quad d_*(x) = \arg\min_{k \in S} f_k(x),$$

$$f_k(x) = \sum_{j=1}^{L} \pi_j(1 - \epsilon_{+j})p_j^0(x)w_{j,k}, x \in R^N. \tag{5.12}$$

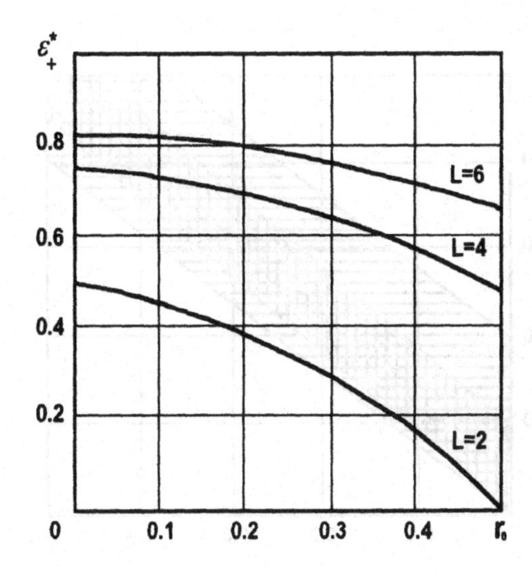

Figure 5.1: Breakdown point vs. Bayesian error probability

Proof. For any decision rule χ, similarly to (5.9), we have:

$$r_+(\chi) = \sum_{i=1}^{L} \pi_i \epsilon_{+i} w_{i+} + \int_{R^N} \sum_{j=1}^{L} \chi_j(x) f_j(x) dx.$$

By (5.6) and the constraints (5.2), we obtain from this expression a unique (up to a zero Lebesgue measure set) robust decision rule (5.12).

■

Corollary 5.4 *The Bayesian decision rule (5.4) is robust in the case of equidistorted classes.*

Proof. Since $\epsilon_{+j} = \epsilon_+$ does not depend on $j \in S$, the formulas (5.2), (5.4) result in:

$$f_k(x) = (1 - \epsilon_+) f_k^0(x), \quad d_+(x) = d_0(x), \quad x \in R^N.$$

■

Let us evaluate the guaranteed risk $r_+(\chi^*)$ for the robust decision rule constructed in Theorem 5.2 and compare it with guaranteed risk $r_+(\chi^0)$ for the Bayesian decision rule given by Theorem 5.1. Since both these decision rules are nonrandomized, we rewrite them to the form (1.34), convenient for investigation:

$$d = d_0(x) = \sum_{i=1}^{L} i \mathbf{I}_{V_i^0}(x), \quad \mathbf{I}_{V_i^0}(x) = \prod_{\substack{k=1, \\ k \neq i}}^{L} \mathbf{1}(f_{ki}^0(x)),$$

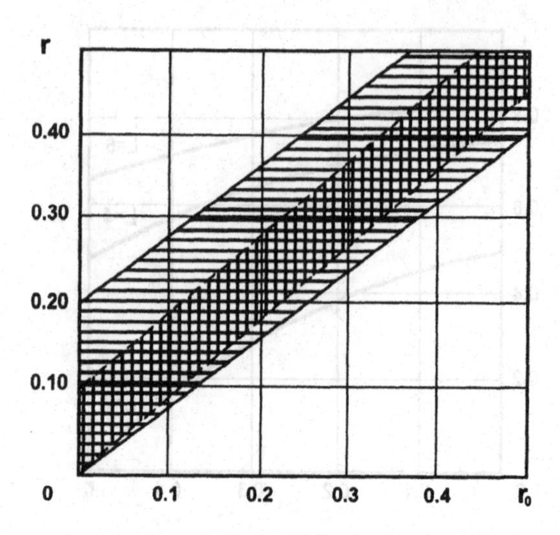

Figure 5.2: Risk variability zones for Bayesian decision rule under Tukey–Huber distortions

where V_i^0 is the region of Bayesian decision "$d = i$" making;

$$f_{ki}^0(x) = f_k^0(x) - f_i^0(x);$$

$$d = d^*(x) = \sum_{i=1}^{L} i\mathbf{I}_{V_i^*}(x), \quad \mathbf{I}_{V_i^*}(x) = \prod_{\substack{k=1,\\ k\neq i}}^{L} \mathbf{1}(f_{ki}(x)), \tag{5.13}$$

where V_i^* is the region of robust decision "$d = i$" making; $f_{ki}(x) = f_k(x) - f_i(x)$.

Assume the notations:

$$[f_{ij}^0(x), f_{kj}^0(x)] = |\nabla_x f_{ij}^0(x)|^2 |\nabla_x f_{kj}^0(x)|^2 - (\nabla_x^T f_{ij}^0(x) \nabla_x f_{kj}^0(x))^2;$$

$$g_i(x) = f_i(x) - f_i^0(x); \quad \Gamma_{ij} = \{x : f_{ij}^0(x) = 0\} \tag{5.14}$$

is an $(N-1)$–dimensional hypersurface in R^N which is the Bayesian discriminant hypersurface for the pair of classes Ω_i^0, Ω_j^0 $(i \neq j)$; Γ'_{ij} is the part of the surface Γ_{ij} which is the border of the region V_j^0;

$$\epsilon_+ = \max\{\epsilon_{+1}, \ldots, \epsilon_{+L}\}$$

is the maximal distortion level for all classes.

Lemma 5.1 *For the robust decision rule (5.13) and an arbitrary integrable function $\Psi(x), x \in R^N$ the asymptotic expansion in powers of ϵ_+ is valid for the following functional ($j \in S$):*

$$B_j = \int_{R^N} \Psi(x)\mathbf{I}_{V_j^*}(x)dx = \int_{R^N} \Psi(x)\mathbf{I}_{V_j^0}(x)dx+$$

$$+\int_{R^N} \Psi(x)\sum_{i\neq j} g_{ij}(x)\delta(f_{ij}^0(x)) \prod_{l\notin\{i,j\}} \mathbf{1}(f_{lj}^0(x))dx+$$

$$+\frac{1}{2}\int_{R^N} \Psi(x)\sum_{i\neq j}\Big(g_{ij}^2(x)\delta'(f_{ij}^0(x)) \prod_{l\notin\{i,j\}} \mathbf{1}(f_{lj}^0(x))+$$

$$+\sum_{k\neq j} g_{ij}(x)g_{kj}(x)\delta(f_{ij}^0(x))\delta(f_{kj}^0(x) \prod_{l\notin\{i,j,k\}} \mathbf{1}(f_{lj}^0(x))\Big)dx + \mathcal{O}(\epsilon_+^3),$$

where $g_{ij}(x) = g_i(x) - g_j(x) = \mathcal{O}(\epsilon_+)$ and $\delta(z), \delta'(z)$ are the generalized Dirac function and its derivative (Gelfand et al., *1959).*

Proof. Consider the following function in $L-1$ variables $t = (t_k),\ k \in S,\ k \neq j$:

$$B(t) = \int_{R^N} \Psi(x)\prod_{k\neq j}\mathbf{1}(f_{kj}^0(x) + t_k g_{kj}(x))dx.$$

According to (5.13), (5.14), the value of this function at $\{t_k = 1 : k \neq j\}$ is equal to B_j. By the Taylor formula in the neighborhood of the point $t = \mathbf{O}_{L-1}$ we have

$$B_j = B(t)|_{\{t_k=1\}} = B(\mathbf{O}_{L-1}) + \sum_{i\neq j}\left(\frac{\partial B(t)}{\partial t_i}+\right.$$

$$\left.\left.+\frac{1}{2}\sum_{k\neq j}\left(\frac{\partial^2 B(t)}{\partial t_i \partial t_k} + \frac{1}{3}\sum_{l\neq j}\frac{\partial^3 B(\bar{t})}{\partial \bar{t}_i \partial \bar{t}_k \partial \bar{t}_l}\right)\right)\right|_{t=\mathbf{O}_{L-1}}, \qquad (5.15)$$

where $\bar{t} = (\bar{t}_k), 0 < \bar{t}_k < 1 \quad (k \in S, k \neq j)$. By properties of the Dirac function (Gelfand *et al.*, 1959), $d\mathbf{1}(z)/dz = \delta(z), d\delta(z)/d(z) = \delta'(z)$. Therefore

$$\left.\frac{\partial B(t)}{\partial t_i}\right|_{t=\mathbf{O}_{L-1}} = \int_{R^N} \Psi(x)g_{ij}(x)\delta(f_{ij}^0(x)) \prod_{l\notin\{i,j\}} \mathbf{1}(f_{lj}^0(x))dx = \mathcal{O}(\epsilon_+).$$

Similarly we find other terms in (5.15). The last term in (5.15) is of order $\mathcal{O}(\epsilon_+^3)$; we take it to be the remainder. Substituting the obtained terms into (5.15) we conclude the proof.

■

Theorem 5.3 *If the hypothetical densities $\{p_i^0(x)\}$ are differentiable functions and the surface integrals* $(i,j,k,l,m \in S,\ i<j<k)$

$$a_{lmij} = \int_{\Gamma'_{ij}} p_l^0(x)p_m^0(x) \mid \nabla_x f_{ij}^0(x) \mid^{-1} ds_{N-1},$$

$$b_{lmijk} = \int_{\Gamma'_{ij}\cap\Gamma'_{kj}} f_i^0(x)p_l^0(x)p_m^0(x)[f_{ij}^0(x), f_{kj}^0(x)]^{-1} ds_{N-2}, \tag{5.16}$$

are finite, then the guaranteed risk for the robust decision rule (5.13) admits the following asymptotic expansion in powers of $\{\epsilon_{+i}\}$*:*

$$r_+(d^*) = r_+(d_0) - \sum_{l,m=1}^{L} \rho_{lm}\pi_l\pi_m\epsilon_{+l}\epsilon_{+m} + \mathcal{O}(\epsilon_+^3). \tag{5.17}$$

The expansion coefficients are:

$$\rho_{lm} = \sum_{i=1}^{L-1}\sum_{j=i+1}^{L}\Bigg((w_{lj}-w_{li})(w_{mj}-w_{mi})\left(\frac{a_{lmij}}{2} - \sum_{k=j+1}^{L} b_{lmijk}\right) -$$

$$- \sum_{k=j+1}^{L}(w_{lk}-w_{li})(w_{mk}-w_{mj})b_{lmijk}\Bigg). \tag{5.18}$$

Proof. According to (5.12)–(5.14) and (5.4),

$$r_+(d^*) = \sum_{i=1}^{L} \epsilon_{+i}\pi_i w_{i+} +$$

$$+\sum_{j=1}^{L}\left(\int_{R^N} f_j^0(x)\mathbf{I}_{V_j^*}(x)dx + \int_{R^N} g_j(x)\mathbf{I}_{V_j^*}(x)dx\right).$$

Apply Lemma 5.1 to the right side of this expression and collect the terms of order ϵ_+, ϵ_+^2 and $\mathcal{O}(\epsilon_+^3)$ using (5.3), (5.9):

$$r_+(d^*) = r_+(d_0) + \sum_{i=1}^{L-1}\sum_{j=i+1}^{L}\Bigg(\frac{1}{2}\int_{R^N} g_{ij}^2(x)\delta(f_{ij}^0(x)) \prod_{l\notin\{i,j\}} \mathbf{1}(f_{lj}^0(x))dx -$$

$$- \sum_{k=j+1}^{L}\int_{R^N} f_i^0\left(g_{ij}^2 + g_{ik}g_{jk}\right)\delta(f_{ij}^0)\delta(f_{jk}^0) \prod_{l\notin\{i,j,k\}} \mathbf{1}(f_{lj}^0(x))dx\Bigg) + \mathcal{O}(\epsilon_+^3). \tag{5.19}$$

Here we used the properties of the Dirac function (Gelfand *et al.*, 1959):

$$\delta'(-z) = -\delta'(z),\ z\delta'(z) = -\delta(z)$$

and the equalities:

$$f_{ij}^0(x) = -f_{ji}^0(x); \quad f_i^0(x) = f_j^0(x) \text{ for } x \in \Gamma_{ij}.$$

According to (5.14),

$$g_{ij}(x) = g_i(x) - g_j(x) = \sum_{l=1}^{L} \pi_l \epsilon_{+l}(w_{lj} - w_{li}) p_l^0(x). \tag{5.20}$$

Substitute (5.20) into (5.19) and use Lemma 3.3 about integrals with the Dirac function. Taking into account the notations (5.16), (5.18) we eventually obtain the expansion (5.17).

■

Corollary 5.5 *In the case of two classes ($L = 2$) the coefficients of the asymptotic expansion (5.17) are of the form:*

$$\rho_{lm} = (w_{l1} - w_{l2})(w_{m1} - w_{m2}) a_{lm12}/2. \tag{5.21}$$

Proof. For $L = 2$ the values $\{b_{lmijk}\}$ in (5.18) vanish.

■

Corollary 5.6 *If under the conditions of Corollary 5.5 we use the $(0-1)$–loss matrix (1.18), then the guaranteed (upper) value of the robust decision rule error probability satisfies the asymptotic expansion:*

$$r_+(d^*) = r_+(d_0) -$$

$$-\frac{1}{2}\int_{\Gamma_{12}} \left(\pi_1 \epsilon_{+1} p_1^0(x) - \pi_2 \epsilon_{+2} p_2^0(x)\right)^2 \mid \nabla_x f_{12}^0(x) \mid^{-1} ds_{N-1} + \mathcal{O}(\epsilon_+^3). \tag{5.22}$$

Practical importance of Theorems 5.1, 5.3 is in the fact that (5.9) and (5.17) produce a formula for approximate (up to $\mathcal{O}(\epsilon_+^3)$) computation of the guaranteed risk for the robust decision rule:

$$r_+(d^*) \approx r_0 + \sum_{i=1}^{L} \epsilon_{+i} \pi_i (w_{i+} - r_{0i}) - \sum_{l,m=1}^{L} \rho_{lm} \pi_l \pi_m \epsilon_{+l} \epsilon_{+m}. \tag{5.23}$$

For comparison of robustness for the Bayesian and robust decision rules we introduce, as in (5.7), the characteristics:

$$\kappa_0 = \kappa_+(\chi^0) = \frac{r_+(d_0) - r_0}{r_0}, \quad \kappa_* = \kappa_+(\chi^*) = \frac{r_+(d^*) - r_0}{r_0},$$

$$\kappa = \frac{r_+(d_0) - r_+(d^*)}{r_0}. \tag{5.24}$$

The value κ_0 indicates the increase of the Bayesian decision rule guaranteed risk with respect to the Bayesian risk r_0; κ_* is a similar characteristics for the robust decision rule; κ indicates the gain in the value of guaranteed risk when the robust decision rule is used instead of the Bayesian decision rule. The formulas (5.9), (5.17), (5.23), (5.24) allow us to evaluate these characteristics:

$$\kappa_0 \approx \sum_{i=1}^{L} \frac{\epsilon_{+i}\pi_i(w_{i+} - r_{0i})}{r_0}, \quad \kappa \approx \sum_{l,m=1}^{L} \epsilon_{+l}\epsilon_{+m}\frac{\rho_{lm}\pi_l\pi_m}{r_0},$$

$$\kappa_* \approx \sum_{i=1}^{L} \frac{\epsilon_{+i}\pi_i(w_{i+} - r_{0i})}{r_0} - \sum_{l,m=1}^{L} \epsilon_{+l}\epsilon_{+m}\frac{\rho_{lm}\pi_l\pi_m}{r_0}. \tag{5.25}$$

Suppose that $\delta > 0$ is fixed. It is very useful for practical applications to evaluate the critical values of distortion levels $\{\epsilon_{+i}\}$ at which the robust DR guaranteed risk gain value (with respect to the Bayesian DR) crosses the given level δ:

$$\kappa > \delta. \tag{5.26}$$

At an appropriate choice of δ, this inequality specifies the conditions when the robust decision rule is recommended to apply. Using (5.25) we obtain the explicit form of the inequality (5.26):

$$\sum_{l,m=1}^{L} \epsilon_{+l}\epsilon_{+m}\frac{\rho_{lm}\pi_l\pi_m}{r_0} > \delta. \tag{5.27}$$

In conclusion, let us consider the situation, common in applications, where $L = 2$, the $(0-1)$–loss matrix is used and the assumed hypothetical model of classes Ω_1^0, Ω_2^0 is the Gaussian model (of Fisher type) defined in Section 1.4:

$$p_i^0(x) = n_N(x \mid a_i, B)$$

is the assumed N–variate normal distribution density of observations from Ω_i^0; $a_i \in R^N$ is the mean vector; B is the nonsingular covariance matrix. According to (1.55) and Theorem 1.2, the Bayesian decision rule for the undistorted model $\{\Omega_i^0\}$ is linear:

$$d = d_0(x) = \mathbf{1}(b^T x - \gamma_0), \quad b = B^{-1}(a_2 - a_1),$$

$$\gamma_0 = \frac{(a_2 + a_1)^T B^{-1}(a_2 - a_1)}{2} + \ln\frac{\pi_1}{\pi_2}; \tag{5.28}$$

$$r_0 = 1 - \sum_{i=1}^{2} \pi_i \Phi\left(\frac{\Delta}{2} + \frac{(-1)^i}{\Delta}\ln\frac{\pi_1}{\pi_2}\right), \tag{5.29}$$

$$\Delta = \sqrt{(a_2 - a_1)^T B^{-1}(a_2 - a_1)}.$$

By Theorem 5.1,

$$r_+(d_0) = r_0 + \sum_{i=1}^{2} \epsilon_{+i}\pi_i\Phi\left(\frac{\Delta}{2} + \frac{(-1)^i}{\Delta}\ln\frac{\pi_1}{\pi_2}\right). \tag{5.30}$$

Theorem 5.2 gives the robust decision rule, which is also linear:

$$d = d^*(x) = \mathbf{1}(b^T x - \gamma_*), \; \gamma_* = \gamma_0 + \ln\frac{1-\epsilon_{+1}}{1-\epsilon_{+2}}.$$

It differs from the Bayesian decision rule (5.28) by threshold γ_* only. By Corollary 5.6,

$$r_+(d^*) = r_+(d_0) - \frac{\sqrt{\pi_1(1-\pi_1)}(\epsilon_{+2}-\epsilon_{+1})^2}{2\sqrt{2\pi}\Delta\exp(\Delta^2/8)} + \mathcal{O}(\epsilon_+^3). \tag{5.31}$$

The characteristics κ_0, κ_*, κ of relative robustness for decision rules $d_0(\cdot)$, $d^*(\cdot)$ are easily evaluated now by formulas (5.24), (5.29)—(5.31). For example:

$$\kappa = \frac{\sqrt{\pi_1(1-\pi_1)}(\epsilon_{+2}-\epsilon_{+1})^2}{2\sqrt{2\pi}\Delta\exp(\Delta^2/8)r_0}. \tag{5.32}$$

Taking into account (5.32), we transform the inequality (5.26) for critical distortion levels $\epsilon_{+1}, \epsilon_{+2}$ to the form:

$$|\epsilon_{+2} - \epsilon_{+1}| > \left(2\sqrt{2\pi}\Delta\delta r_0\exp(\Delta^2/8)/\sqrt{\pi_1(1-\pi_1)}\right)^{1/2}.$$

Figure 5.3 plots the dependence of κ_* on Δ for $\pi_1 = 0.1; 0.3; 0.5$ with $\epsilon_{+1} = 0.6$, $\epsilon_{+2} = 0$. It can be seen that κ_* increases monotonously as the values Δ, π_1 increase. The Tukey–Huber distortions influence on Bayesian DR risk most in the case of distant classes (Δ is large).

Figure 5.4 shows similar plots for the value of κ. It can be seen that κ decreases as the values Δ, π_1 increase. Therefore, the maximal gain value of the robust decision rule with respect to Bayesian decision rule is achieved for the situations where the classes $\{\Omega_i^0\}$ are nondistant (Δ is small) and large distortions (ϵ_{+i}) are present in the classes with small prior probabilities.

5.2 Distortions in L_2–metric

Consider now another type of nonparametric distortions of probability distributions of observations to be classified, namely, case D.2.2.2 of distortions in L_2–metric, described in Section 2.2.

Suppose that each class Ω_i, $i \in S$, is assigned with a hypothetical probability distribution density $p_i^0(x)$, but the true distribution density $p_i(x)$ is unknown and belongs to the ϵ_{+i}–neighborhood of $p_i^0(x)$ in L_2–metric with weight function $\Psi_i(x)$ normed to one:

$$\mathcal{P}_i(\epsilon_{+i}) = \left\{p_i(x), x \in R^N : p_i(x) \geq 0, \int_{R^N} p_i(x)dx = 1,\right.$$

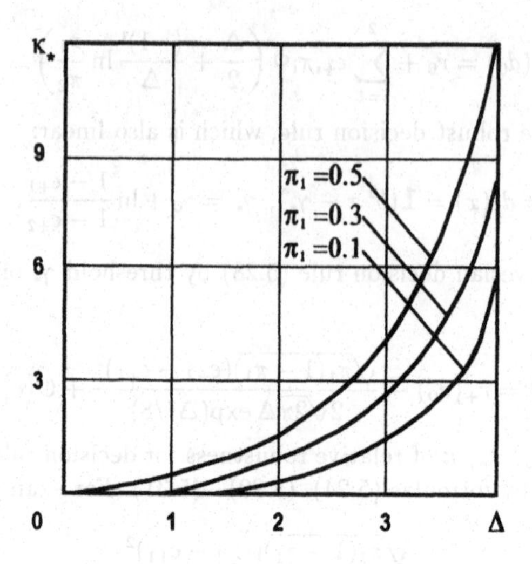

Figure 5.3: Plots of robustness factor dependence for RDR

$$\int_{R^N} (p_i(x) - p_i^0(x))^2 / \Psi_i(x) dx = \epsilon_i^2, 0 \le \epsilon_i \le \epsilon_{+i}\}. \tag{5.33}$$

As it was mentioned in Section 2.2, if $\Psi_i(x) = p_i^0(x)$, $x \in R^N$, then we obtain the ϵ_{+i}–neighborhood (5.33) in χ^2–metric (2.16).

Assume the following notations : $\mathbf{E}_{\Psi_i}\{\cdot\}$ denotes the expectation with respect to the distribution $\Psi_i(\cdot)$;

$$\langle z \rangle = \{ | z |, \text{ if } z < 0; 0, \text{ if } z \ge 0\};$$

$$\epsilon_i^* = \left(\mathbf{E}_{\Psi_i} \left\{ \left(\sum_{k=1}^{L} w_{ik}(\chi_k(x) - q(\Psi_i; \chi_k)) \right)^2 \right\} \right)^{1/2} \times$$

$$\times \left(\left\langle \inf_x \left(\frac{\Psi_i(x)}{p_i^0(x)} \sum_{l=1}^{L} w_{il}(\chi_l(x) - q(\Psi_i; \chi_l)) \right) \right\rangle \right)^{-1} \ge 0, \ i \in S. \tag{5.34}$$

Theorem 5.4 *If distortions of densities in L_2–metric (5.33) take place and the neighborhood radius ϵ_{+i} does not exceed the critical value $\epsilon_i^*(i \in S)$ determined by*

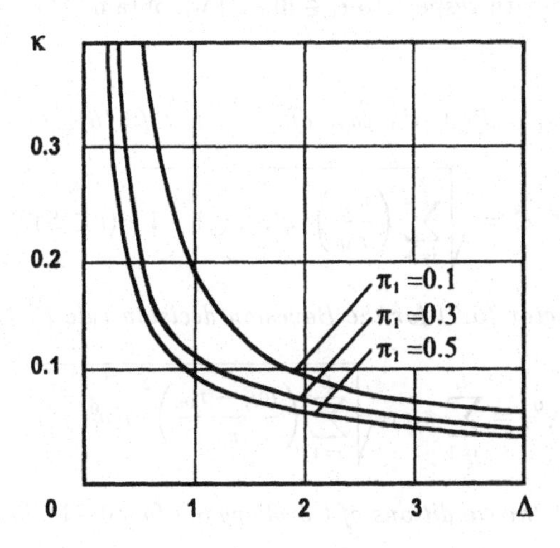

Figure 5.4: Plots of $\kappa(\Delta, \pi_1)$

(5.34), then the guaranteed risk for the decision rule χ is

$$r_+(\chi) = r(\chi; \{p_i^0\}) + \sum_{i=1}^{L} \pi_i \epsilon_{+i} \sqrt{\mathbf{E}_{\Psi_i} \left\{ \left(\sum_{k=1}^{L} w_{ik}(\chi_k(x) - q(\Psi_i; \chi_k)) \right)^2 \right\}}. \qquad (5.35)$$

Proof. Solving the problem (5.3), (5.5) of variational calculus under the constraints (5.33) (at fixed ϵ_i) for $p_i(\cdot)$ by the method of indefinite Lagrange multipliers, we obtain that the supremum of risk is attained for the function

$$p_i^*(x) = p_i^0(x) + \epsilon_i \Psi_i(x) \times$$

$$\times \frac{\sum_{k=1}^{L} w_{ik}(\chi_k(x) - q(\Psi_i; \chi_k))}{\sqrt{\mathbf{E}_{\Psi_i} \left\{ \left(\sum_{k=1}^{L} w_{ik}(\chi_k(x) - q(\Psi_i; \chi_k)) \right)^2 \right\}}} \quad (i \in S). \qquad (5.36)$$

Here, if $\epsilon_i \leq \epsilon_i^*$, where ϵ_i^* is defined by (5.34), then the function $p_i^*(x)$ is nonnegative and is some probability density. Substituting (5.36) into the risk functional (5.3) we obtain

$$\sup r(\chi; \{p_i\}) = r(\chi; \{p_i^0\}) + \sum_{i=1}^{L} \pi_i \epsilon_{+i} \sqrt{\mathbf{E}_{\Psi_i} \left\{ \left(\sum_{k=1}^{L} w_{ik}(\chi_k(x) - q(\Psi_i; \chi_k)) \right)^2 \right\}}.$$

Maximizing this value with respect to $\epsilon_i \in [0, \epsilon_{+i}]$ we obtain (5.35).

■

Corollary 5.7 *If $\Psi_i(\cdot) = p_i^0(\cdot)$ (the case of χ^2-metric (2.16)) and*

$$\epsilon_{+i} \le \epsilon_i^* = \sqrt{\sum_{j=1}^{L} \left(\frac{w_{ij}}{r_{0i}}\right)^2 q(p_i^0; \chi_j) - 1} \quad (i \in S),$$

then the robustness factor (5.7) for the Bayesian decision rule (5.4) is

$$\kappa_+(\chi^0) = \sum_{i=1}^{L} \pi_i \epsilon_{+i} \sqrt{\sum_{j=1}^{L} \left(\frac{w_{ij} - r_{0i}}{r_0}\right)^2 q(p_i^0; \chi_j^0)}.$$

Corollary 5.8 *Under the conditions of Corollary 5.7 for $(0-1)$-loss matrix (1.18),*

$$\epsilon_i^* = \sqrt{q(p_i^0; \chi_i^0)/(1 - q(p_i^0; \chi_i^0))}, \quad i \in S,$$

the robustness conditions (2.30), (2.31) assume the form:

$$\sum_{i=1}^{L} \pi_i \epsilon_{+i} \sqrt{q(p_i^0; \chi_i^0)(1 - q(p_i^0; \chi_i^0))} \le \delta,$$

$$\frac{1}{r_0} \sum_{i=1}^{L} \pi_i \epsilon_{+i} \sqrt{q(p_i^0; \chi_i^0)(1 - q(p_i^0; \chi_i^0))} \le \delta',$$

and for the breakdown point (2.29) the formula

$$\epsilon_+^* = (1 - L^{-1} - r_0) \left(\sum_{i=1}^{L} \pi_i \sqrt{q(p_i^0; \chi_i^0)(1 - q(p_i^0; \chi_i^0))} \right)^{-1}$$

is valid.

Proof of Corollaries 5.7, 5.8 is performed by substitution of $\{\Psi_i(\cdot) = p_i^0(\cdot)\}$ and $\{\chi_i^0(\cdot)\}$ into (5.34), (5.35).

■

Note that the constraints for the neighborhood radius ϵ_{+i} indicated in Theorem 5.4 and its Corollaries are not too restrictive. For example, if we set $\epsilon_{+i} = \epsilon_i^*$ $(i \in S)$ under the conditions of Corollary 5.8, then the guaranteed error probability $r_+(\chi^0)$ for the Bayesian decision rule attains its maximal value $r_+(\chi^0) = 1$.

The Gaussian model of Fisher type defined in Section 1.4:

$$p_i^0(x) = (2\pi)^{-N/2} \mid \Sigma \mid^{-1/2} \exp\{-(x - \mu_i)^T \Sigma^{-1} (x - \mu_i)/2\}, \qquad (5.37)$$

$$\pi_i = 1/2,\ i = 1,2,\ L = 2,$$

is often used in discriminant analysis as the hypothetical model. Here $\mu_i \in R^N$ is the mean vector and Σ is the common covariance matrix for observations from Ω_i in absence of distortions. In this case, as it was noted in Section 1.4, for $(0-1)$–loss matrix (1.18) the Bayesian decision rule (5.4) is linear:

$$d = d_0(x) = \begin{cases} 1, & l(x) \leq 0, \\ 2, & l(x) > 0,\ x \in R^N,\ d \in \{1,2\}, \end{cases}$$

$$l(x) = b^T x - \gamma,\ b = \Sigma^{-1}(\mu_2 - \mu_1),\ \gamma = \frac{1}{2}(\mu_2^T \Sigma^{-1} \mu_2 - \mu_1 \Sigma^{-1} \mu_1),$$

and the Bayesian error probability is

$$r_0 = \Phi(-\Delta/2),$$

where

$$\Delta = \left((\mu_2 - \mu_1)^T \Sigma^{-1} (\mu_2 - \mu_1)\right)^{1/2}$$

is the interclass Mahalanobis distance and $\Phi(\cdot)$ is the standard normal probability distribution function.

Corollary 5.9 *If conditions of Corollary 5.8 hold, the Gaussian model (5.37) is assumed, and the classes Ω_1, Ω_2 are equidistorted ($\epsilon_{+1} = \epsilon_{+2} = \epsilon_+$), then the "breakdown point" is*

$$\epsilon_+^* = \left(\Phi\left(\frac{\Delta}{2}\right) - \frac{1}{2}\right)\left(\Phi\left(\frac{\Delta}{2}\right)\left(1 - \Phi\left(\frac{\Delta}{2}\right)\right)\right)^{-1/2}.$$

If

$$\epsilon_+ \leq \epsilon_+(\delta', \Delta) = ((\Phi(\Delta/2))^{-1} - 1)^{1/2}\delta',$$

then δ'-robustness condition (2.31) holds.

Proof. It follows from (5.37) that

$$q(p_i^0; \chi_i^0) = \Phi\left(\frac{\Delta}{2}\right), \quad i = 1,2,$$

therefore, from Corollary 5.8 we obtain the required expression.

■

At Figure 5.5 two plots are given: the solid line plots the dependence of ϵ_+^* on Δ and the dashed line is for the critical value $\epsilon^* = (\Phi(\Delta/2)/(1 - \Phi(\Delta/2)))^{1/2}$.

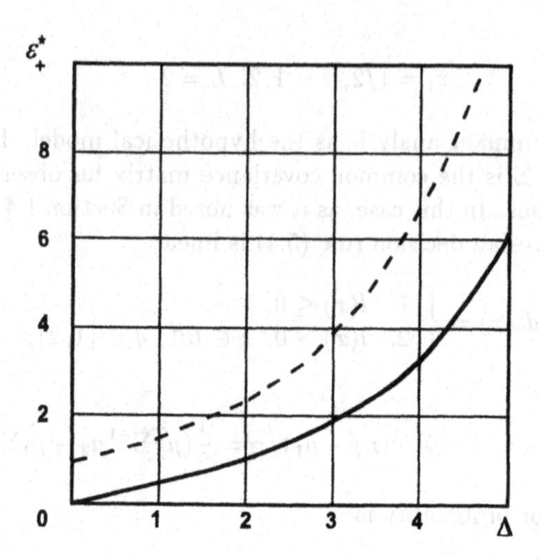

Figure 5.5: Breakdown point $\epsilon_+^*(\Delta)$ dependence

Figure 5.6 plots the dependence $\epsilon_+(\delta', \Delta)$. These figures allow to find maximal allowable distortion levels for the Fisher model. For example, if $\Delta = 2.56$ ($r_0 = 0.1$), then the "breakdown point" is $\epsilon_+^* = 1.29$ (here the maximal error probability for the Bayesian decision rule is 0.5). If $\epsilon_{+i} \leq \epsilon_+(1, \Delta) = 0.34$, then it is guaranteed that the robustness factor κ will not exceed $\delta' = 1$.

Consider now the problem of construction of robust decision rule under distortions (5.33). Assume the notations:

$$a_{kj}(x; \chi) = \sum_{i=1}^{L} \pi_i \epsilon_{+i} w_{ik} w_{ij} c_i(\chi) \Psi_i(x),$$

$$b_j(x; \chi) = \sum_{i=1}^{L} \pi_i w_{ij} \left(p_i^0(x) - \epsilon_{+i} c_i(\chi) \Psi_i(x) \sum_{k=1}^{L} w_{ik} q(\Psi_i; \chi_k) \right),$$

$$c_i(\chi) = \left(\mathbf{E}_{\Psi_i} \left\{ \left(\sum_{k=1}^{L} w_{ik}(\chi_k(x) - q(\Psi_i; \chi_k)) \right)^2 \right\} \right)^{1/2} \quad (i, j, k \in S). \qquad (5.38)$$

Theorem 5.5 *Under the conditions of Theorem 5.4 the robust critical functions* $\chi^* = (\chi_j^*(x))$, $j \in S$, *are the solution of the optimization problem with constraints (5.2):*

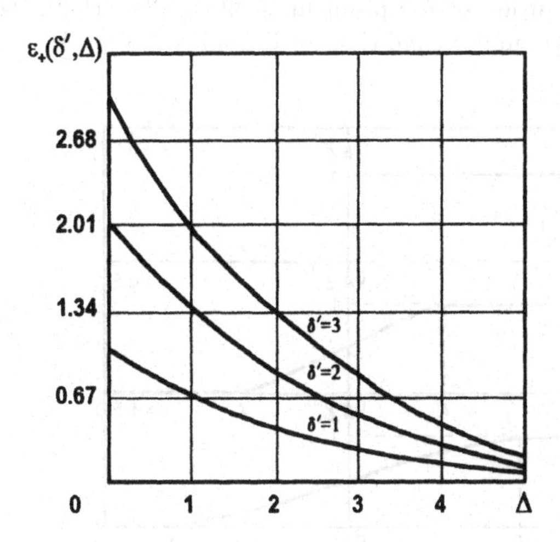

Figure 5.6: Critical distortion level $\epsilon_+(\delta', \Delta)$ dependence

$$\int_{R^N} \left(\sum_{k,j=1}^{L} a_{kj}(x;\chi)\chi_j(x)\chi_k(x) + \sum_{j=1}^{L} b_j(x;\chi)\chi_j(x) \right) dx \to \min_{\chi}. \tag{5.39}$$

Proof. Substituting the "least favorable density" (5.36) at $\epsilon_i = \epsilon_{+i}$ ($i \in S$) into the guaranteed risk functional (5.5) and using the notations (5.38) we obtain the optimization problem in the equivalent form (5.39).

■

To find the solution of the (perturbed) optimization problem (5.39), (5.38), (5.2), it is advisable to use the method of successive approximations (Pervozvanskij *et al.*, 1979). In fact, if $\epsilon_{+1} = \ldots = \epsilon_{+L} = 0$, then the sequence of Bayesian critical functions χ^0 defined by (5.4) is the solution of this problem. Therefore we take χ^0 to be an initial approximation for $\chi^* : \chi^{(0)} = \chi^0$. Then by (5.38) we find initial approximations for the coefficients of the objective function from (5.39):

$$a_{kj}^{(0)}(x) = a_{kj}(x;\chi^{(0)}),\ b_j^{(0)}(x) = b_j(x;\chi^{(0)}),\ (k,j \in S)$$

and solve the minimization problem for the quadratic form (5.39) with these fixed coefficients under the linear constraints (5.2).

As a result, we obtain a new approximation for the sequence of critical functions $\chi^{(1)}$, which will be used for finding new approximations for the coefficients of the quadratic form in (5.39), and so on. These steps are iterated M times (M is given *a priori* or it is determined from the condition of given proximity of consecutive approximations $\chi^{(M)}$ and $\chi^{(M-1)}$). The final result $\chi^{(M)} = (\chi_j^{(M)}(x))$ is considered as an approximate solution of the problem (5.39), (5.38), (5.2). For example, for the Fisher model (5.37) under conditions of Corollary 5.9 for $M = 1$ we obtain:

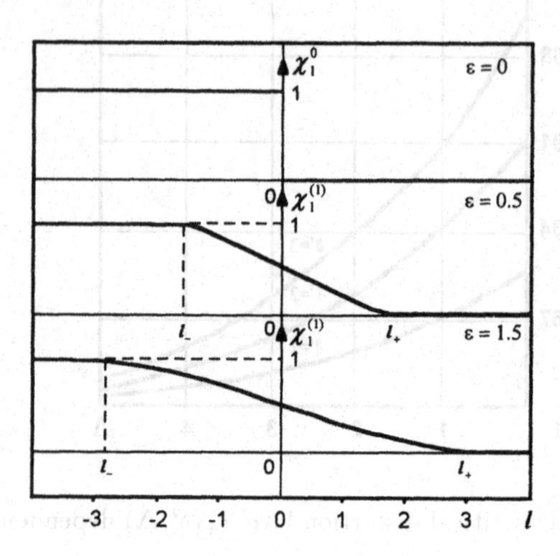

Figure 5.7: Critical function $\chi_1^{(1)}$ vs. l for different distortion levels

$$\chi_1^{(0)}(x) = \begin{cases} 1, & l(x) \le 0, \\ 0, & l(x) > 0, \end{cases}$$

$$\chi_2^{(0)}(x) = 1 - \chi_1^{(0)}(x), \quad x \in R^N;$$

$$\chi_1^{(1)}(x) = \begin{cases} 1, & l(x) \le l_-, \\ \left(1 + \frac{r_0}{2}(\epsilon_*/\epsilon - 1)(1 - e^{l(x)})\right)(1 + e^{l(x)}), & l < l(x) < l_+, \\ 0, & l(x) \ge l_+, \end{cases}$$

$$\chi_2^{(1)}(x) = 1 - \chi_1^{(1)}(x), \quad x \in R^N,$$

$$l_\pm = \pm \ln\left(1 + \left(\frac{r_0}{2}(\epsilon_*/\epsilon - 1)\right)^{-1}\right).$$

Figure 5.7 plots the dependence of $\chi_1^{(1)}$ on the values of the Fisher linear discriminant function $l = l(x)$ for $L = 2$, $\Delta = 2.56$ $(r_0 = 0.1)$, $\epsilon_{+1} = \epsilon_{+2} = \epsilon_+ = 0; 0.5; 1.5$.

In absence of distortions $(\epsilon_+ = 0)$ we have $\chi_1^{(1)} = \chi_1^0$ and randomization in decision making is absent. If $\epsilon_+ > 0$, then the "step" between the two extreme values of χ_1 disappears and randomization emerges. In particular, if $l = 0$, then the decisions $d = 1$ (class Ω_1) and $d = 2$ (class Ω_2) are equiprobable $(\chi_1^{(1)} = \chi_2^{(1)} = 0.5)$.

5.3 Robustness of Error Probability for Distortions in Variation Metric

Consider now the third kind of nonparametric distortions of observations to be classified (given in Section 2.2 and at Figure 2.1), specifically, the $D.2.2.3$ case of distortions defined by the probabilistic variation metric.

Suppose that in a given feature space R^N classes $\Omega_1, \ldots, \Omega_L$, $L \geq 2$, are recorded with prior probabilities $\pi_1, \ldots, \pi_L$. According to the hypothetical model M_0, an observation from class Ω_i ($i \in S = \{1, 2, \ldots, L\}$) is a random N-vector $X_i^0 \in R^N$ with ascribed hypothetical multivariate probability density function $p_i^0(x)$. However this main model assumption is often violated for real observed data: an observation from Ω_i is a random N-vector with actual probability density function $p_i(x)$ which is an element of ϵ_{+i}-neighborhood $\mathcal{P}_i(\epsilon_{+i})$ centered at $p_i^0(\cdot)$ and defined by the probabilistic variation metric:

$$
\begin{aligned}
\mathcal{P}_i(\epsilon_{+i}) = \{p_i(x), x \in R^N : p_i(x) \geq 0, \textstyle\int_{R^N} p_i(x)dx = 1, \\
\rho(p_i, p_i^0) = \epsilon_i, 0 \leq \epsilon_i \leq \epsilon_{+i}\}.
\end{aligned}
\tag{5.40}
$$

Here ρ is the variation distance between two points $p_i(\cdot)$ and $p_i^0(\cdot)$ in the space of N-variate probability density functions:

$$
\rho = \rho(p_i, p_i^0) = \frac{1}{2} \int_{R^N} | p_i(x) - p_i^0(x) | \, dx,
$$

where $\epsilon_{+i} \in [0, 1/2)$ is an *a priori* given maximal distortion level for Ω_i (radius of neighborhood $\mathcal{P}_i(\epsilon_{+i})$).

Let us consider a nonrandomized decision rule defined by the Borel function

$$
d = d(x), x \in R^N, d \in S,
$$

where d is the class number to which the observation x will be ascribed. The performance of decision making will be characterized by the functional of error probability

$$
r = r(d(\cdot); \{p_i(\cdot)\}) = 1 - \sum_{i \in S} \pi_i \int_{d(x)=i} p_i(x)dx.
$$

As it was shown in Section 1.3, the Bayesian decision rule $d = d_0(x)$ for the hypothetical model M_0 minimizes the functional $r(d(\cdot); \{p_i^0(\cdot)\})$ and up to zero Lebesgue measure

$$
\text{mes}\{x \in R^N : p_i^0(x) = p_j^0(x)\} = 0 \, (i \neq j)
$$

has the form

$$
d = d_0(x) = arg \max_{j \in S} \pi_j p_j^0(x), x \in R^N.
$$

The minimal error probability value attained by this Bayesian decision rule for the hypothetical model M_0 assumes the form

$$r_0(\{p_i^0(\cdot)\}) = r(d_0(\cdot); \{p_i^0(\cdot)\}) = \inf_{d(\cdot)} r(d(\cdot); \{p_i^0(\cdot)\}) =$$

$$= 1 - \int_{R^N} \max_{i \in S}(\pi_i p_i(x))dx.$$

It is called the *hypothetical Bayesian error probability for M_0*.

Let us consider the problem of guaranteed error probability estimation under distortions (5.40):

$$r_+(d(\cdot)) = \sup_{\{p_i(\cdot) \in \mathcal{P}_i(\epsilon_{+i})\}} r(d(\cdot); \{p_i(\cdot)\}).$$

First of all we shall list the main properties of probabilistic variation metric $\rho(\cdot)$, which can be easily proved.

1. For any probability density functions $p(x)$, $p'(x)$, $x \in R^N$,

$$\rho(p, p') \in [0, 1].$$

2. If $p'(\cdot) = p(\cdot)$, then $\rho(p, p) = 0$. If $p(x)p'(x) = 0$, $x \in R^N$, then $\rho(p, p') = 1$.

3. If B is the Borel σ-algebra on the real line R^1 and $\mathbf{P}(B), \mathbf{P}\prime(B)$ are the probability measures induced by probability density functions $p(x), p'(x)$ respectively, then

$$\sup_{B \in \mathrm{B}} \mid \mathbf{P}(B) - \mathbf{P}\prime(B) \mid = \rho(p, p').$$

4. If $r_0(p(\cdot), p'(\cdot))$ is the Bayesian error probability for discrimination of two equiprobable classes specified by probability density functions $p(\cdot)$ and $p'(\cdot)$ respectively, then the following functional dependence of Bayesian error probability and probabilistic variation metric $\rho(\cdot)$ takes place:

$$r_0(p(\cdot), p'(\cdot)) = (1 - \rho(p(\cdot), p'(\cdot)))/2,$$

$$\rho(p(\cdot), p'(\cdot)) = 1 - 2r_0(p(\cdot), p'(\cdot)).$$

Note that these four properties make variation distance ρ the most important performance characteristic for statistical classification.

For the hypothetical model M_0 let us define the family of decision rules to be investigated:

$$D_0 = \{d(\cdot) : \int_{d(x)=i} p_i^0(x)dx \geq \epsilon_{+i}, \, i \in S\}.$$

This means that we shall investigate only the decision rules for which the conditional probability of correct decision for each class is not smaller than a given distortion level.

Theorem 5.6 *If probability density functions $\{p_i(\cdot)\}$ are subjected to the distortions defined by (5.40), then the guaranteed error probability for the family D_0 of decision rules is represented as follows:*

$$r_+(d(\cdot)) = r(d(\cdot); \{p_i^0(\cdot)\}) + \sum_{i\in S} \pi_i \epsilon_{+i}, \ d(\cdot) \in D_0. \tag{5.41}$$

Proof. An arbitrary probability density function $p_i(\cdot) \in \mathcal{P}_i(\epsilon_{+i})$ $(i \in S)$ may be represented in the form convenient for the computation of the upper bound $r_+(\cdot)$:

$$p_i(x) = p_i^0(x) - \epsilon_i h_i(x), \ x \in R^N,$$

where $h_i(x) : R^N \to R^1$ is an integrable function such that

$$h_i(x) \leq \frac{1}{\epsilon_i} p_i^0(x), \ \int_{R^N} h_i(x)dx = 0, \ \int_{R^N} | \, h_i(x) \, | \, dx = 2. \tag{5.42}$$

Substituting this representation of $p_i(\cdot)$ into the functional $r(\cdot)$ we have:

$$r(d(\cdot); \{p_i(\cdot)\}) = r(d(\cdot); \{p_i^0(\cdot)\}) + \sum_{i\in S} \pi_i \epsilon_i \int_{d(x)=i} h_i(x)dx.$$

This expansion shows that finding the supremum $r_+(d(\cdot))$ comes to the solution of L problems in calculus of variations $(i = 1, \ldots, L)$ of the same type:

$$J(h_i(\cdot)) = \int_{d(x)=i} h_i(x)dx \to \max_{h_i(\cdot)}$$

under restrictions (5.42).

Denote:

$$\int_{d(x)=i, h_i(x)>0} h_i(x)dx = a_{i+}, \ \int_{d(x)=i, h_i(x)\leq 0} h_i(x)dx = -a_{i-},$$

$$\int_{d(x)\neq i, h_i(x)>0} h_i(x)dx = b_{i+}, \ \int_{d(x)\neq i, h_i(x)\leq 0} h_i(x)dx = -b_{i-},$$

where a_{i-}, b_{i-}, a_{i+}, $b_{i+} \geq 0$ are appropriate nonnegative real numbers. Then the i-th maximization problem assumes the form:

$$J(h_i(\cdot)) = a_{i+} - a_{i-} \to \max_{h_i(\cdot)},$$

$$h_i(x) \leq \frac{1}{\epsilon_i} p_i^0(x), \ a_{i+} - a_{i-} + b_{i+} - b_{i-} = 0, \ a_{i+} + a_{i-} + b_{i+} + b_{i-} = 2.$$

Resolving the last two linear equations for b_{i+}, b_{i-}, we obtain

$$b_{i+} = 1 - a_{i+} \geq 0,$$

$$b_{i-} = 1 - a_{i-} \geq 0.$$

Therefore, $0 \leq a_{i-} \leq 1, 0 \leq a_{i+} \leq 1$.

Integrating the inequality restriction for $h_i(\cdot)$ in (5.42) w.r.t. x over the region $\{x : d(x) = i\}$ we obtain:

$$a_{i+} - a_{i-} \leq \frac{1}{\epsilon_i} \int_{d(x)=i} p_i^0(x)dx.$$

According to the condition of Theorem 5.6, the investigated decision rule $d(\cdot)$ is in D_0, therefore,

$$\min\{1, \frac{1}{\epsilon_i} \int_{d(x)=i} p_i^0(x)\} = 1,$$

and

$$a_{i+} - a_{i-} \leq 1,$$

Consequently, the maximal value of the functional $J(h_i(\cdot))$ is attained at the functions $h_i^*(x)$ such that

$$\int_{d(x)=i, h_i^*(x)>0} h_i^*(x)dx = a_{i+}^* = 1,$$

$$\int_{d(x)=i, h_i^*(x)\leq 0} h_i^*(x)dx = a_{i-}^* = 0.$$

Thus we have:

$$\sup_{h_i(\cdot)} J(h_i(\cdot)) = a_{i+} - a_{i-} = 1.$$

Then the supremum of the functional $r(d(\cdot); \{p_i(\cdot)\})$ is

$$\sup_{\{h_i(\cdot)\}} r(d(\cdot); \{p_i(\cdot)\}) = r(d(\cdot); \{p_i^0(\cdot)\}) + \sum_{i \in S} \pi_i \epsilon_i.$$

The maximization of this last expression w.r.t. $\epsilon_i \in [0, \epsilon_{i+}]$ $(i \in S)$ results in (5.41). ■

Corollary 5.10 *If the Bayesian decision rule $d_0(\cdot)$ for the hypothetical model M_0 belongs to the family D_0, then the robust decision rule $d_*(\cdot)$ coincides with the Bayesian decision rule $d_0(\cdot)$.*

Proof. According to the definition (5.6) for the robust decision rule,

$$d = d_*(x) : R^N \to S$$

minimizes the guaranteed error probability:

$$r_+(d_*(\cdot)) = \inf_{d(\cdot) \in D_0} r_+(d(\cdot)).$$

But by (5.41) this minimization problem assumes the form

$$r(d_*(\cdot); \{p_i^0(\cdot)\}) = \inf_{d \in D_0} r(d(\cdot); \{p_i^0(\cdot)\})$$

equivalent to the form of the minimization problem determining the Bayesian decision rule $d_0(\cdot)$. Hence,

$$d_*(\cdot) = d_0(\cdot).$$

■

Corollary 5.11 *If there are two ($L = 2$) equiprobable ($\pi_1 = \pi_2 = 1/2$) classes, the hypothetical model of classes is Gaussian:*

$$p_i^0(x) = n_N(x \mid \theta_i^0, \Sigma),\ i \in S = \{1, 2\},$$

and distortions (5.40) are determined by the inequalities:

$$\frac{1}{2} \int_{R^N} \mid p_i(x) - n_N(x \mid \theta_i^0, \Sigma) \mid dx \leq \epsilon_{+i},$$

then the guaranteed error probability for the Fisher linear decision rule (1.55) (which is in fact the Bayesian decision rule for the hypothetical model M_0 and the robust decision rule for the distorted model) has the form:

$$r = \Phi(-\frac{\Delta}{2}) + \frac{1}{2}(\epsilon_{+1} + \epsilon_{+2}),$$

where $\Phi(\cdot)$ is the standard normal distribution function and

$$\Delta = ((\theta_2^0 - \theta_1^0)^T \Sigma^{-1} (\theta_2^0 - \theta_1^0))^{1/2}$$

is the hypothetical Mahalanobis interclass distance.

Proof is performed using (5.41) and previous results from Section 1.4.

■

Let us analyze the results given by Theorem 5.6 and its corollaries and compare them with the results from Sections 5.1, 5.2. It can be noted that the Bayesian decision rule $d_0(\cdot)$ constructed for the hypothetical model M_0 is the robust decision rule for the distortions (5.40) defined by variation distance, but at the same time it is not robust for Tukey-Huber contaminations or for the distortions defined by L_2-metric (5.33). This strange, at the first glance, combination takes place only if the classifier performance criterion is error probability (it means that loss matrix is of (0-1)-type), and can be explained by two reasons.

The first reason is that the neighborhood (5.40) specifies a reacher family of probability density functions than the neighborhoods (5.1), (5.33) do. Let us illustrate this fact by four properties.

1) If $p_i(x)$ is from the Tukey-Huber neighborhood (5.1), then

$$\rho(p_i, p_i^0) = \frac{1}{2}\int_{R^N} \epsilon_i \mid h_i(x) - p_i^0(x) \mid dx \leq \epsilon_i.$$

Hence, $p_i(\cdot)$ is from the neighborhood (5.40).

2) If $p_i(\cdot)$ is from the L_2-metric neighborhood (5.33), then by Cauchy–Schwartz–Bunyakovsky inequality (see, e.g., (Gnedenko and Kolmogorov, 1954)),

$$\rho(p_i, p_i^0) = \frac{1}{2}\int_{R^N} \frac{\mid p_i(x) - p_i^0(x) \mid}{\sqrt{\Psi_i(x)}}\sqrt{\Psi_i(x)}dx \leq$$

$$\leq \frac{1}{2}\sqrt{\int_{R^N} \frac{(p_i(x) - p_i^0(x))^2}{\Psi_i(x)}dx \int_{R^N} \Psi_i(x)dx} = \frac{\epsilon_i}{2},$$

and $p_i(\cdot)$ is also from the neighborhood (5.40).

3) Let $A_i \subset \mathcal{B}$ be a measurable subset in R^N,

$$P_i^0(A_i) = \int_{A_i} p_i^0(x)dx = \epsilon_i \in (0, \epsilon_{+i}],$$

and

$$p_i(x) = \begin{cases} 0, & if\; x \in A_i, \\ \frac{1}{1-P_i^0(A_i)}p_i^0(x), & if\; x \notin A_i. \end{cases}$$

This density belongs to the neighborhood (5.40):

$$\rho(p_i, p_i^0) = \frac{1}{2}(\int_{A_i} p_i^0(x)dx + \int_{R^N \setminus A_i} P_i^0(A_i)(1 - P_i^0(A_i))^{-1}p_i^0(x)dx) = \epsilon_i,$$

but does not belong to the Tukey-Huber neighborhood (5.1) since the inequality

$$p_i(x) \geq (1 - \epsilon_i)p_i^0(x),\; x \in R^N$$

fails for $x \in A_i : p_i(x)/p_i^0(x) = 0$.

4) Suppose that

$$p_i(x) = (1 - \epsilon_i)p_i^0 + \epsilon_i\delta(x - z_i),$$

where $\delta(\cdot)$ is the generalized Dirac function, $z_i \in R^N$. Then

$$\rho(p_i, p_i^0) = \frac{\epsilon_i}{2}\int_{R^N} \mid \delta(x - z_i) - p_i^0(x) \mid dx = \epsilon_i,$$

$$\int_{R^N} \frac{(p_i(x) - p_i^0(x))^2}{\Psi_i(x)}dx = +\infty.$$

Therefore, $p_i(\cdot)$ is from the variation metric neighborhood (5.40) but not from the L_2-metric neighborhood (5.33).

The second reason is that by the fourth property of $\rho(\cdot)$, the probabilistic variation metric is well suited to the classifier performance characteristic, i.e., to error probability. If we use risk functional from Section 1.3 with general loss matrix W, then the robust decision rule $d_*(\cdot)$ will be different from the Bayesian decision rule $d_0(\cdot)$.

5.4 Decision Rule Robustness to Random Distortions of Densities

In Sections 5.1, 5.2 we considered only deterministic distortions of probability distributions. However in applications the probability model $\{\Omega_i\}$ is often constructed from experimental data, and therefore D.2.2.4 type of random distortions of model is observed:

$$p_i^0(x) - p_i(x) = \xi_i(x), \qquad x \in R^N \quad (i \in S). \tag{5.43}$$

Here $\xi_i(x)$ is a random field on R^N describing random errors in the assignment of probability density for class Ω_i and satisfying the constraints:

$$\xi_i(x) \overset{a.s.}{\leq} p_i^0(x), \qquad \int_{R^N} \xi_i(x)dx = 0.$$

Assume that this random field has moments of the first and second order:

$$\mathbf{E}\{\xi_i(x)\} = -\epsilon_i^{\alpha_i} G_{1i}(x) + o(\epsilon_i^{\alpha_i}),$$

$$\text{Cov}\{\xi_i(x), \xi_i(x')\} = \epsilon_i G_{2i}(x, x') + o(\epsilon_i), \qquad x, x' \in R^N, \tag{5.44}$$

where $0 \leq \epsilon_i \leq 1$ is the parameter characterizing the distortion level in mean-square sense (if $\epsilon_i \to 0$, then $\xi_i(x) \overset{m.s.}{\to} 0$); α_i, $0 < \alpha_i < 1$, is the parameter defining the decrease rate for $\xi_i(\cdot)$; $G_{2i}(\cdot) : R^N \times R^N \to R^1$ is a nonnegative definite function; $G_{1i}(\cdot)$ and $G_{2i}(\cdot)$ are absolutely integrable functions.

The distortion model (5.43) is appropriate, for example, if the hypothetical density $p_i^0(x)$ is in fact the statistical estimator $\hat{p}_i(x)$ of the true unknown distribution density $p_i^*(x)$ based on a random sample $\{z_{it} : t = 1, \ldots, n_i\} \subset R^N$ of size n_i. In particular, if the nonparametric Rosenblatt–Parzen estimator (see Section 4.5) is used:

$$\hat{p}_i(x) = \frac{1}{n_i \mid H_i \mid} \sum_{t=1}^{n_i} \mathcal{K}\left(H_i^{-1}(x - z_{it})\right),$$

where $\mathcal{K}(z)$, $z \in R^N$ is a "kernel", $H_i = \text{diag}\{h_{i1}, \ldots, h_{iN}\}$ is a diagonal $(N \times N)$-matrix, $h_{ij} = c_{ij} n_i^{-1/(N+4)}$ are "smoothness parameters" $(c_{ij} > 0, j = 1, \ldots, N)$, then, according to (Epanechnikov, 1969),

$$\epsilon_i = n_i^{-4/(N+4)}, \quad \alpha_i = 1/2,$$

$$G_{1i}(x) = \frac{1}{2}\sum_{j=1}^{N} c_{ij}^2 \frac{\partial^2 p_i^*(x)}{\partial x_j^2},$$

$$G_{2i}(x, x') = p_i^*(x) \prod_{j=1}^{N} c_{ij}^{-1} \int_{R^N} \mathcal{K}(y)\mathcal{K}\left(y + H_i^{-1}(x' - x)\right) dy.$$

Since the distortions (5.43) are random, the robustness characteristics

$$\beta(\chi^0) = r(\chi^0) - r_0, \kappa(\chi^0) = (r(\chi^0) - r_0)/r_0$$

defined in Section 2.3 for the Bayesian decision rule become random values. Therefore we will characterize the robustness of the Bayesian decision rule χ^0 by the expected values of risk increments:

$$\overline{\beta}(\chi^0) = \mathbf{E}\{r(\chi^0) - r_0\}, \quad \overline{\kappa^2}(\chi^0) = \mathbf{E}\left\{\left((r(\chi^0) - r_0)/r_0\right)^2\right\}.$$

Assume the notations: $\epsilon_+ = \max \epsilon_i$,

$$g_{1ik} = \int_{d_0(x)=k} G_{1i}(x)dx, \quad g_{2ikl} = \int_{d_0(x)=k, d_0(x')=l} G_{2i}(x, x')dxdx',$$

$$g_{1i} = \sum_{k=1}^{L} w_{ik} g_{1ik}, \quad g_{2i} = \sum_{k=1,l=1}^{L} w_{ik} w_{il} g_{2ikl} \quad (i, k \in S). \tag{5.45}$$

Theorem 5.7 *If $\{\xi_i(\cdot)\}$ are mutually independent random fields satisfying the conditions (5.44) for $\alpha_- = \min \alpha_i \geq 1/2$ $(i \in S)$, then the following asymptotic expansions for the expected robustness characteristics for the Bayesian decision rule χ^0 hold:*

$$\overline{\beta}(\chi^0) = \sum_{i=1}^{L} \epsilon_i^{\alpha_i} \pi_i g_{1i} + o(\epsilon_+^{\alpha_-}),$$

$$\overline{\kappa^2}(\chi^0) = \left(\sum_{i=1}^{L} \epsilon_i \pi_i g_{2i} + \overline{\beta}^2(\chi^0)\right) / r_0^2 + o(\epsilon_+). \tag{5.46}$$

Proof. From (5.3), (5.4) and (5.43) the expression for risk follows:

$$r(\chi^0; \{p_i\}) = r_0 - \sum_{i=1}^{L} \pi_i \int_{R^N} w_{i,d_0(x)} \xi_i(x) dx.$$

Taking into account the notations (5.45) and conditions (5.44) for moments $\{\xi_i(\cdot)\}$, we obtain (5.46).

■

Corollary 5.12 *If nonparametric Rosenblatt–Parzen estimators are used, then the following expressions take place for robustness characteristics (where* $(n_0 = \min n_i)$*):*

$$\overline{\beta}(\chi^0) = \mathcal{O}(n_0^{-2/(N+4)}), \qquad \overline{\kappa^2}(\chi^0) = \mathcal{O}(n_0^{-4/(N+4)}).$$

In a similar way it is possible to evaluate the robustness for the k-NN adaptive decision rule described in Section 4.2.

5.5 Additive Distortions of Observations

Consider the situation where the observations to be classified are subjected to additive random distortions of type D.2.1.3:

$$X_i = X_i^0 + \epsilon_{+i} Y_i, \qquad i \in S. \tag{5.47}$$

Here $X_i^0 \in R^N$ is an unobservable nondistorted random vector of feature variables with a given triply continuously differentiable probability density $p_i^0(x)$ describing the class Ω_i in absence of distortions; $Y_i \in R^N$ is the unobservable random vector of distortions with unknown probability density $h_i(\cdot) \in H_i$ (here Y_i and X_i^0 are independent); $X_i \in R^N$ is the observable vector of feature variables from the class Ω_i. The set H_i is a family of probability densities with finite third order moments and fixed first and second order moments:

$$H_i = \Big\{ h_i(\cdot) : h_i(y) \geq 0, \int_{R^N} h_i(y) dy = 1,$$

$$\mathbf{E}\{Y_i\} = \mu_i; \; \mathbf{Cov}\{Y_i, Y_i\} = \Sigma_i \Big\}, \tag{5.48}$$

where $\mu_i \in R^N$ is a given mean vector and Σ_i is a given nonsingular covariance matrix. The distortion model (5.47), (5.48) can be interpreted in an intuitive way as follows. Investigation of the classes $\{\Omega_i\}$, including estimation of densities $\{p_i^0(\cdot)\}$, was performed in ideal conditions (without noise), but real observations are recorded with noise background whose statistical properties are known only partially (only first and second order moments for noise are known). This situation is typical for applied problems of random signals recognition (Omelchenko, 1979).

First, let us evaluate the risk for the Bayesian decision rule (1.34):

$$d = d_0(x) = \sum_{i=1}^{L} i \prod_{\substack{k=1, \\ k \neq i}}^{L} \mathbf{1}(f_{ki}^0(x)) = \sum_{i=1}^{L} i \mathbf{I}_{V_i^0}(x), \; x \in R^N,$$

where $f_{ki}^0(x) = f_k^0(x) - f_i^0(x)$ and $\{f_k^0(x)\}$ are defined by (5.4), in presence of the distortions (5.47), (5.48). By additivity of model (5.47), the family of admissible distorted densities assumes the form:

$$\mathcal{P}_i(\epsilon_{+i}) = \left\{ p_i(\cdot) : p_i(x) = \int_{R^N} p_i^0(x - \epsilon_{+i} y) h_i(y) dy, \; h_i(\cdot) \in H_i \right\}. \tag{5.49}$$

Denote:

$$\alpha_i = -\sum_{j=1}^{L} w_{ij} \int_{R^N} \mathbf{I}_{V_j^0}(x) \nabla p_i^0(x) dx$$

is an N-vector-column,

$$\beta_i = \sum_{j=1}^{L} w_{ij} \int_{R^N} \mathbf{I}_{V_j^0}(x) \nabla^2 p_i^0(x) dx \tag{5.50}$$

is an $(N \times N)$-matrix,

$$q_i(x) = \epsilon_{+i} \left(-\mu_i^T \nabla p_i^0(x) + \epsilon_{+i} \left(-\mu_i^T \nabla^2 p_i^0(x) \mu_i + \mathrm{tr}(\Sigma_i \nabla^2 p_i^0(x)) \right) / 2 \right).$$

Theorem 5.8 *The asymptotic expansion of the guaranteed risk for the Bayesian decision rule $d_0(\cdot)$ under additive distortions (5.47)—(5.49) assumes the form:*

$$r_+(d_0) = r_0 + \sum_{i=1}^{L} \epsilon_{+i} \pi_i \left(\mu_i^T \alpha_i + \epsilon_{+i} (\mu_i^T \beta_i \mu_i + \mathrm{tr}(\Sigma_i \beta_i))/2 \right) + O(\epsilon_+^3). \tag{5.51}$$

Proof. Applying the Taylor formula to the density $p_i(\cdot)$ defined by the convolution integral in (5.49) and taking into account the notations (5.50), we find:

$$p_i(x) = p_i^0(x) + q_i(x) + O(\epsilon_+^3), \quad i \in S. \tag{5.52}$$

Substitute (5.52) into the expression for risk:

$$r = r(d_0; \{p_i\}) = \sum_{i=1}^{L} \pi_i r_i(d_0; p_i),$$

where $r_i(d_0; p_i) = \mathbf{E}\left\{ w_{i, d_0(X_i)} \right\}$ is the conditional classification risk for the observations from Ω_i that have density $p_i(\cdot)$. Use (5.50) and the expression for the Bayesian risk:

$$r_0 = r(d_0; \{p_i^0\}) = \sum_{i=1}^{L} \pi_i r_{0i}, \; r_{0i} = \sum_{j=1}^{L} w_{ij} \int_{R^N} \mathbf{I}_{V_j^0}(x) p_i^0(x) dx.$$

Grouping the terms in $r(d_0; \{p_i^0\})$ by powers of $\{\epsilon_{+i}\}$ we obtain:

$$r(d_0; \{p_i\}) = r_0 + \sum_{i=1}^{L} \epsilon_{+i} \pi_i \left(\mu_i^T \alpha_i + \epsilon_{+i} (\mu_i^T \beta_i \mu_i + \mathrm{tr}(\Sigma_i \beta_i))/2 \right) + O(\epsilon_+^3). \tag{5.53}$$

The main term of this expression is independent of density $h_i(\cdot) \in H_i$ $(i \in S)$; only the remainder term in (5.53) depends on $\{h_i(\cdot)\}$. Therefore (5.53) implies (5.51). ■

Construction and investigation of the robust decision rule for the model (5.47), (5.48) are conducted in the same way as in Section 5.1 for the Tukey–Huber distortion type. The only peculiarity is that in the minimization problem (5.6) we use the main terms of the asymptotic expansion of risk $r_+(\cdot)$ in powers of $\{\epsilon_{+i}\}$.

Theorem 5.9 *Under additive distortions of observations, the robust decision rule up to $\mathcal{O}(\epsilon_+^3)$ has the form (5.13), where*

$$f_i(x) = f_i^0(x) + \sum_{l=1}^{L} \pi_l w_{li} q_l(x), \quad f_{ki}(x) = f_k(x) - f_i(x). \tag{5.54}$$

Moreover, the following asymptotic expansion of the guaranteed risk for the robust decision rule holds:

$$r_+(d^*) = r_+(d_0) - \sum_{l,m=1}^{L} \rho_{lm}\pi_l\pi_m\epsilon_{+l}\epsilon_{+m} + \mathcal{O}(\epsilon_+^3), \tag{5.55}$$

where the expansion coefficients are determined by the formulas:

$$\rho_{lm} = \mu_l^T u_{lm}\mu_m,$$

$$u_{lm} = \sum_{i=1}^{L-1}\sum_{j=i+1}^{L}\Big(\frac{(w_{lj}-w_{li})(w_{mj}-w_{mi})}{2}\int_{\Gamma'_{ij}} \nabla p_l^0(x)(\nabla p_m^0(x))^T|\nabla_x f_{ij}^0(x)|^{-1}ds_{N-1}-$$

$$-\sum_{k=j+1}^{L}((w_{lj}-w_{li})(w_{mj}-w_{mi})+(w_{lk}-w_{li})(w_{mk}-w_{mj}))\times$$

$$\times\int_{\Gamma'_{ij}\cap\Gamma'_{kj}} f_i^0(x)\nabla p_l^0(x)(\nabla p_m^0(x))^T[f_{ij}^0(x), f_{kj}^0(x)]^{-1}ds_{N-2}\Big). \tag{5.56}$$

Proof. Substituting (5.52) into the risk functional gives

$$r_+(d) = \int_{R^N}\sum_{i=1}^{L}\pi_i w_{id(x)}(p_i^0(x)+q_i(x))dx + \mathcal{O}(\epsilon_+^3).$$

From (5.6) we find the robust decision rule

$$d = d_*(x) = \arg\min_j f_j(x),$$

which has an equivalent form (5.13) similar to the robust decision rule under Tukey–Huber distortions. Hence, risk expansion (5.55), (5.56) is constructed similarly to the risk expansion from Theorem 5.3.

■

As in Section 5.1, by neglecting the remainder term $\mathcal{O}(\epsilon_+^3)$ and by Theorems 5.8, 5.9 it is possible to obtain the formulas for approximate computation of the guaranteed risk values $r_+(d_0), r_+(d^*)$ and robustness factors $\kappa_0, \kappa_*, \kappa$ introduced by the formulas (5.24). In particular,

$$\kappa_0 = \sum_{i=1}^{L} \epsilon_{+i}\pi_i \left(\mu_i^T \alpha_i + \epsilon_{+i}(\mu_i^T \beta_i \mu_i + \operatorname{tr}(\Sigma_i \beta_i))/2\right)/r_0,$$

$$\kappa = \sum_{l,m=1}^{L} \pi_l \pi_m \mu_l^T u_{lm} \mu_m \epsilon_{+l}\epsilon_{+m}/r_0. \tag{5.57}$$

As for the case of Tukey–Huber distortions (see Section 5.1), the critical distortion levels $\{\epsilon_{+i}\}$ assuring significant gain in guaranteed risk for the robust decision rule as compared with the Bayesian decision rule are determined by inequality (5.27) where the coefficients $\{\rho_{ij}\}$ of the quadratic form are computed by (5.56).

Consider now the situation where the hypothetical nondistorted model of the classes $\{\Omega_i^0\}$ is assumed to be the Fisher type Gaussian model described in Section 1.4:

$$p_i^0(x) = n_N(x \mid a_i, B).$$

This situation appears in practice (Omelchenko, 1979) in classification of random signals observed over random noise background. In this case X_i^0 in (5.47) is a vector of N spectral samples of the signal (its distribution is assumed to be Gaussian, by the well known asymptotic normality of the spectral density statistical estimator (Anderson, 1971)), and Y_i is a vector of N "noise background" spectral density estimator samples for the class Ω_i.

Evaluating (5.50), (5.51) by the properties of the Gaussian distribution (Anderson, 1958), we find the guaranteed risk for the Bayesian decision rule:

$$r_+(d_0) = r_0 + \sum_{i=1}^{2} \pi_i \left((-1)^{i+1}\epsilon_{+i} b^T \mu_i + \epsilon_{+i}^2 \left((b^T\mu_i)^2 + b^T \Sigma_i b\right)/4\right) \times$$

$$\times \exp\left(-\left(\frac{(-1)^i \Delta^2}{2} - \ln\frac{\pi_1}{\pi_2}\right)^2 /(2\Delta^2)\right) / \left(\sqrt{2\pi}\Delta\right) + O(\epsilon_+^3), \tag{5.58}$$

where $\Delta = \sqrt{(a_2 - a_1)^T B^{-1}(a_2 - a_1)}$ is the Mahalanobis interclass distance.

According to Theorem 5.9, the robust decision rule is

$$d = d_*(x) = \mathbf{1}(b^T x - \gamma_0 + \lambda_2(x) - \lambda_1(x)) + 1, \tag{5.59}$$

where b, γ_0 are defined by (3.56), and

$$\lambda_i(x) = \ln\left(1 + \epsilon_{+i}\mu_i^T B^{-1}(x - a_i) + \epsilon_{+i}^2 \left(\left(\mu_i^T B^{-1}(x - a_i)\right)^2 - \mu_i^T B^{-1}\mu_i - \right.\right.$$

$$\left.\left. - \operatorname{tr}(\Sigma_i B^{-1}) + (x - a_i)^T B^{-1}\Sigma_i B^{-1}(x - a_i)\right)/2\right),\ i \in S = \{1, 2\}.$$

Since the decision rule (5.59) is constructed using the expansion with the remainder term $\mathcal{O}(\epsilon_+^3)$, it is possible to simplify the expressions for $\{\lambda_i(x)\}$ using Taylor formulas keeping only the terms of $\{\epsilon_{+i}\}, \{\epsilon_{+i}^2\}$ orders of magnitude:

$$\lambda_i(x) = \epsilon_{+i}\mu_i^T B^{-1}(x - a_i) + \frac{\epsilon_{+i}^2}{2}((x - a_i)^T B^{-1}\Sigma_i B^{-1}(x - a_i) - \mu_i^T B^{-1}\mu_i -$$

$$- \operatorname{tr}(\Sigma_i B^{-1})) + \mathcal{O}(\epsilon_+^3), \quad i \in S. \tag{5.60}$$

Let us analyze in more detail the situation

$$\pi_1 = \pi_2 = 0.5, \epsilon_{+1} = \epsilon_{+2} = \epsilon_+, \mu_1 = \mu_2 = \mu, \Sigma_1 = \Sigma_2 = \Sigma,$$

that corresponds to equiprobable and equidistorted classes. Here the Bayesian risk r_0 is $\Phi(-\Delta/2)$; the guaranteed risk values for the Bayesian and robust decision rules are defined by (5.58) and by (5.55), (5.56) respectively:

$$r_+(d_0) = r_0 + \epsilon_+^2\left((b^T\mu)^2 + b^T\Sigma b\right)\exp(-\Delta^2/8)/\left(4\sqrt{2\pi}\Delta\right) + \mathcal{O}(\epsilon_+^3),$$

$$r_+(d_*) = r_0 + \epsilon_+^2 b^T\Sigma b \exp(-\Delta^2/8)/\left(4\sqrt{2\pi}\Delta\right) + \mathcal{O}(\epsilon_+^3). \tag{5.61}$$

The robustness factors $\kappa_0, \kappa_*, \kappa$ are easily computed now by (5.24), (5.61). In particular,

$$\kappa = \epsilon_+^2 \frac{(b^T\mu)^2\exp(-\Delta^2/8)}{4\sqrt{2\pi}\Delta\Phi(-\Delta/2)}. \tag{5.62}$$

Figure 5.8 plots the dependencies of κ_0/ϵ_+^2 and κ_*/ϵ_+^2 on interclass distance Δ for $\mu = a_2 - a_1, \Sigma = B$. It can be seen that distortions are most influential for "strongly" distinct classes. But as Δ increases, κ_*/ϵ_+^2 increases significantly more slowly than κ_0/ϵ_+^2 does. Moreover, their difference $\kappa/\epsilon_+^2 = \kappa_0/\epsilon_+^2 - \kappa_*/\epsilon_+^2$ increases infinitely. The last fact means that as Δ increases, the robust decision rule gains more in stability over the Bayesian decision rule.

At Figure 5.9, solid lines plot the dependencies of $r_+(d_0)/r_0$ (curve 1) and $r_+(d_*)/r_0$ (curve 2) on distortion level ϵ_+ for two-dimensional observation space ($N = 2$), $\mu = a_2 - a_1$, $\Sigma = B, \Delta = 2.56$ ($r_0 = 0.1$). Small circles and triangles indicate point statistical estimates $\hat{r}(d_0)$ and $\hat{r}(d_*)$ respectively. These estimates are produced by computer modeling of decision rules $d_0(\cdot)$ and $d_*(\cdot)$ (400 computer experiments were performed). The corresponding dashed lines denote hypothetical 95%–confidence intervals for the decision rules $d_0(\cdot)$ and $d_*(\cdot)$ (the hypothesis is that the values $r(d_0), r(d^*)$ coincide with the main terms in (5.61)).

It can be seen from Figure 5.9 that the computer results are in good accordance with the approximation formulas derived from (5.61) for $\epsilon_+ \leq 0.45$. Hence these approximation formulas can be recommended for robustness analysis of decision rules.

Let us use (5.62) to find the critical level of distortions $\epsilon_+(\delta)$ at which $\kappa = \delta$. For the case $\mu = a_2 - a_1$, $\Sigma = B$,

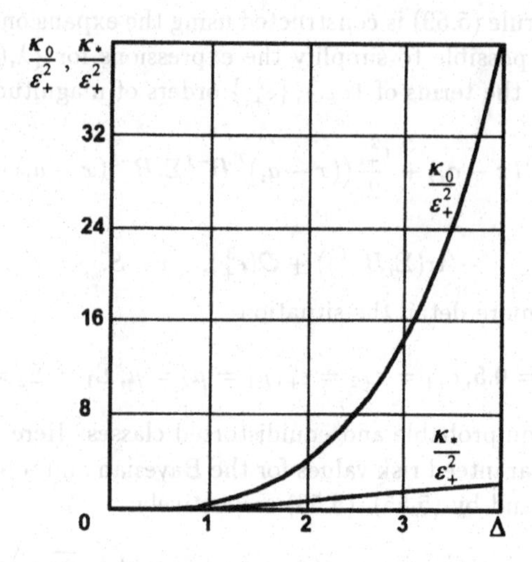

Figure 5.8: Robustness factors vs. interclass distance

$$\epsilon_+(\delta) = \frac{2}{\Delta^2}\left(\sqrt{2\pi}\Delta\Phi(-\Delta/2)\exp(\Delta^2/8)\delta\right)^{1/2}.$$

Table 5.1 lists critical levels $\epsilon_+(\delta)$ for different values of Δ, δ.

Table 5.1: Critical values $\epsilon_+(\delta)$

	Δ							
δ	2.0	2.5	3.0	3.5	4.0	4.5	5.0	6.0
0.25	0.29	0.20	0.14	0.10	0.08	0.07	0.06	0.04
0.50	0.41	0.28	0.20	0.15	0.11	0.09	0.08	0.06
1.00	0.58	0.39	0.28	0.21	0.15	0.13	0.11	0.08

It can be seen from Table 5.1 that, for example, in order to achieve relative gain of 100% ($\delta = 1$) in the guaranteed risk for the robust decision rule in comparison with the Bayesian decision rule for $\Delta = 2.5$ $(r_0 = 0.1)$, the distortion level ϵ_+ must exceed the critical value $\epsilon_+(1) = 0.39$, and for $\Delta = 4.5$ $(r_0 = 0.01)$ it must exceed $\epsilon_+(1) = 0.13$.

Let us now apply the results of this section to the investigation of robustness of pattern recognition algorithms in the situation where the observed features are subjected to round-off errors. This is a common practical situation, because all features $x_1, x_2, \ldots, x_N \in R$ are recorded by real devices with finite measurement precision.

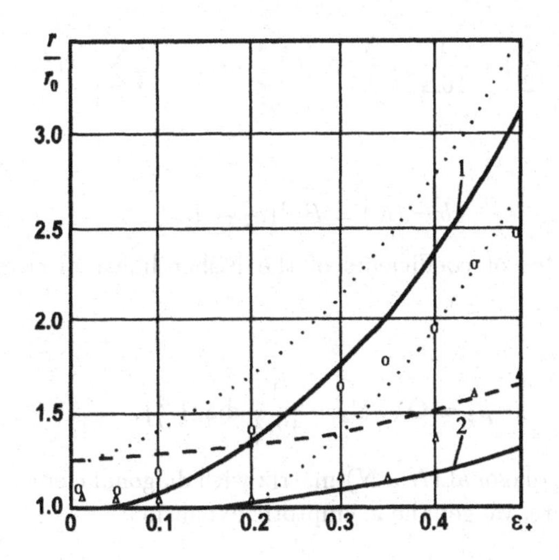

Figure 5.9: Analytical and computer results for BDR and RDR

For mathematical description of round-off errors let us use the well-known probability model (see, e.g., (Hamming, 1962) and (Feller, 1971)). According to this model, determined by the vector equation (5.47), the unobservable random vector of distortions $Y_i = (Y_{ij}) \in R^N$ for the i-th class consists of N independent random components $Y_{i1}, \ldots, Y_{iN}$ with uniform (rectangular) probability distributions. Parameters of these probability distributions are determined by type of round-off.

Two main types of round-off are distinguished. The first type (R_1) of rounding off for the j-th feature value ($j = 1, \ldots, N$) is truncation of least significant digits, i.e., t_j fixed-point digits are left. Here, the round-off error does not exceed $\epsilon_j > 0$, which equals to the value of the t_j-th digit. Therefore, in (5.47) the distortion level $\epsilon_{+i} = \epsilon_+ = \max \epsilon_j$ does not depend on the class number i, and the probability distribution is

$$\mathcal{L}\{Y_{ij}\} = \begin{cases} R(-\epsilon_j/\epsilon_+, 0] & \text{if } X^0_{ij} \geq 0, \\ R[0, \epsilon_j/\epsilon_+) & \text{if } X^0_{ij} < 0. \end{cases}$$

According to the second type (R_2), the round-off for the j-th feature value is the nearest t_j-digit number. Here the absolute value of round-off error is less than $\epsilon_j/2$:

$$\mathcal{L}\{Y_{ij}\} = R(-\frac{\epsilon_j}{2\epsilon_+}, \frac{\epsilon_j}{2\epsilon_+}).$$

Using the Fisher model (1.52) for $\{X^0_i\}$ with two ($L = 2$) equiprobable classes ($\pi_1 = \pi_2$) and formula (5.58), one can express the guaranteed risk for the Bayesian decision rule $d_0(\cdot)$ in the case R_1:

$$r_+(d_0) = \Phi(-\frac{\Delta}{2}) + \frac{1}{16\Delta}\varphi(\frac{\Delta}{2}) \times ((\sum_{j=1}^{N} b_j\epsilon_j)^2 + \frac{1}{3}\sum_{j=1}^{N} b_j^2\epsilon_j^2) + O(\epsilon_+^3),$$

where

$$b = (b_j) = B^{-1}(a_2 - a_1)$$

is the N-column-vector of coefficients of the Fisher linear discriminant function (1.56).

For the case R_2,

$$\mu_2 = \mathbf{O}_N,\ \Sigma_2 = \frac{1}{12\epsilon_+^2}\,\mathrm{diag}\{\epsilon_j^2\},$$

where $\mathrm{diag}\{\epsilon_j^2\}$ is the diagonal $(N \times N)$-matrix with diagonal elements $\epsilon_1^2, \epsilon_2^2, \ldots, \epsilon_N^2$. Using (5.58) once more, we get the asymptotic expansion:

$$r_+(d_0) = \Phi(-\frac{\Delta}{2}) + \frac{1}{48\Delta}\varphi(\frac{\Delta}{2})\sum_{j=1}^{N} b_j^2\epsilon_j^2 + O(\epsilon_+^3).$$

One can see that rounding off the features with maximal values of the squared coefficients in the Fisher linear discriminant function influence on instability of the Bayesian decision rule $d_0(\cdot)$ most significantly.

For practice, it is useful to compare the increments of error probability for the Bayesian decision rule $d_0(\cdot)$ derived by feature rounding off (distortion of type D.2.1.3) and by finiteness of training samples (distortion of type D.1.1, or small-sample effects). We make this comparison for the Fisher hypothetical model of data (1.52) with R_2-type rounding off. Let us assume, for simplicity, that the maximal round-off errors are the same for all features:

$$\epsilon_j = \epsilon_+ = \text{const}\ ,\ j = 1, \ldots, N,$$

and $B = \mathbf{I}_N$. Then R_2-round-off distortions produce an increment of error probability:

$$\delta_1 = r_+(d_0) - \Phi(-\frac{\Delta}{2}) = \epsilon_+^2\frac{\Delta}{48}\varphi(\frac{\Delta}{2}) + O(\epsilon_+^3).$$

Finiteness of sample sizes $n_1 = n_2 = n_0$ produces an increment of error probability according to (3.65):

$$\delta_2 = r_+(d_0) - \Phi(-\frac{\Delta}{2}) =$$

$$= \frac{1}{2\Delta}\varphi(\frac{\Delta}{2})(N - 1 + \frac{\Delta^2}{4})\frac{1}{n_0} + O(n_0^{-3/2}).$$

From the balance condition for main terms of these increments: $\delta_1 = \delta_2$,

Table 5.2: Values of "equivalent sample size" $\tilde{n}_0$

N	Δ 1	2	3	4	5
1	7	7	7	7	7
3	55	19	12	10	8
5	103	31	17	13	10
10	223	61	31	20	15

we find the "equivalent sample size" $\tilde{n}_0$, which produces the same instability level as ϵ_+-rounding off for N features:

$$\tilde{n}_0 = \left\lfloor \frac{6}{\epsilon_+^2}(1 + \frac{4(N-1)}{\Delta^2}) \right\rfloor + 1.$$

Table 5.2 presents values of "equivalent sample size" $\tilde{n}_0$ for $\epsilon_+ = 1$, different dimension values N of the feature space and different Mahalanobis distances Δ.

5.6 Distortions Induced by Finite Mixtures

Consider the situation where the observations to be classified are subjected to parametric distortions of type D.2.1.2 (see Section 2.2). Suppose that for the class Ω_i ($i \in S$) in the space R^N $M_i \geq 2$ probability distribution densities are defined:

$$\{p_{i1}(x), p_{i2}(x), \ldots, p_{iM_i}(x)\}, \tag{5.63}$$

where $p_{i1}(x) \equiv p_i^0(x)$ is the hypothetical probability distribution density of the feature vector and $p_{i2}(x), \ldots, p_{iM_i}(x)$ is a set of $M_i - 1$ distribution densities ϵ_i–close to $p_{i1}(x)$ under a given probabilistic metric $\rho(\cdot)$:

$$\rho(p_{ij}(\cdot), p_{i1}(\cdot)) = \epsilon_{ij}, \quad 0 \leq \epsilon_{ij} \leq \epsilon_{+i}, j = 2, \ldots, M_i. \tag{5.64}$$

For $\rho(\cdot)$ one can use L_2–metric, χ^2–metric from Section 5.2, or Hellinger metric (Ibragimov *et al.*, 1979). The observed random vector of feature variables $X_i \in R^N$ with unknown probability a_{ij} is described by the j-th probability distribution density from the given set (5.63), so that the true probability density of the random vector X_i is a mixture of M_i distributions (5.63) and belongs to the family of distorted densities:

$$\mathcal{P}_i(\epsilon_{+i}) = \Big\{ p_i(x), x \in R^N : p_i(x) = \sum_{j=1}^{M_i} a_{ij} p_{ij}(x), \tag{5.65}$$

$$0 \leq a_{ij} \leq 1, \sum_{j=1}^{M_i} a_{ij} = 1 \Big\}, \quad i \in S.$$

An interpretation for these distortions of data was presented in Section 2.2. In practice, the type D.2.1.2 distortions are generated by nonstationarity of training samples (Aivazyan, 1981) used for estimation of density $p_i(x)$: there are M_i samples, and the j-th sample is a random sample from the distribution $p_{ij}(\cdot), j = 1, \ldots, M_i$. The other source of the distortion (5.65) is that the investigated phenomenon (of class Ω_i) may admit M_i different "flow regimes" (Kharin, 1985) and switch randomly among them, according to an unknown discrete distribution $\{a_{ij}\}$.

Note that another interpretation of the set of distributions (5.63) is possible, when the hypothetical distribution $p_{i1}(\cdot)$ is not distinguished and all distributions (5.63) have "equal rights":

$$\rho\left(p_{ij}(\cdot), p_{ik}(\cdot)\right) = \epsilon_{jk}, \quad 0 \leq \epsilon_{jk} \leq \epsilon_{+i}, j \neq k \in \{1, \ldots, M_i\}.$$

As in Sections 5.1, 5.2, we shall consider the generalized randomized decision rule $\chi(x) = (\chi_i(x))$ satisfying the constraints (5.2). At first we shall evaluate the guaranteed risk $r_+(\chi)$ for an arbitrary decision rule $\chi(\cdot)$. Denote:

$$P_{ij} = P_{ij}(\chi) = \int_{R^N} p_{ij}(x) \sum_{k=1}^{L} w_{ik} \chi_k(x) dx \geq 0, \ j = 1, \ldots, M_i; \tag{5.66}$$

$P_i^T = (P_{i1}, \ldots, P_{iM_i})$ is a row-vector.

Theorem 5.10 *If probability distribution density distortions induced by the mixtures of ϵ-close distributions (5.65) occur, then the guaranteed risk for the randomized decision rule χ is*

$$r_+(\chi) = \sum_{i=1}^{L} \pi_i \max_{1 \leq j \leq M_i} P_{ij}(\chi). \tag{5.67}$$

Proof. By (5.3), (5.65), (5.66), the risk functional for the decision rule $\chi(\cdot)$ has the form:

$$r(\chi; \{a_i\}) = \sum_{i=1}^{L} \pi_i a_i^T P_i, \tag{5.68}$$

where $a_i^T = (a_{i1}, \ldots, a_{iM_i})$ is the row-vector of unknown elementary probabilities. It can be seen from (5.65), (5.66), (5.68) that the computation of the guaranteed risk

$$r_+(\chi) = \sup_{\{a_i\}} r(\chi; \{a_i\})$$

is reduced to the solution of L independent conditional maximization problems ($i = 1, \ldots, L$):

$$a_i^T P_i \rightarrow \max_{a_i},$$

$$a_{ij} \geq 0,\ j = 1, \ldots, M_i; \quad \sum_{j=1}^{M_i} a_{ij} = 1.$$

The solution of the i-th maximization problem is the vector

$$a_i^{0T} = (a_{i1}^0, \ldots, a_{iM_i}^0),\ a_{ij}^0 = \delta_{j,j^0},\ j^0 = \arg \max_{1 \leq k \leq M_i} P_{ik}. \tag{5.69}$$

Substituting (5.69) into (5.68), we obtain (5.67).

■

Corollary 5.13 *The guaranteed risk for the Bayesian decision rule (5.4) is*

$$r_+(\chi^0) = \sum_{i=1}^{L} \pi_i \max_{1 \leq j \leq M_i} \sum_{k=1}^{L} w_{ik} p_{ijk}, \tag{5.70}$$

where

$$p_{ijk} = \int_{d_0(x)=k} p_{ij}(x) dx \tag{5.71}$$

is the probability of the Bayesian decision "$d_0(x) = k$" making under the condition that the random vector of feature variables to be classified has probability density function $p_{ij}(\cdot)$.

Proof. From (5.66), (5.4), (5.71) we find

$$P_{ij}(\chi^0) = \sum_{k=1}^{L} w_{ik} p_{ijk},$$

therefore (5.67) assumes the form (5.70).

■

Corollary 5.14 *If $W = (w_{ij})$ is a $(0-1)$-loss matrix (1.18), then the guaranteed error probability for the Bayesian decision rule χ^0 equals to*

$$r_+(\chi^0) = 1 - \sum_{i=1}^{L} \pi_i \min_{1 \leq j \leq M_i} p_{iji}. \tag{5.72}$$

Proof is based on substituting of (1.18) into (5.70) and using the norming condition:

$$\sum_{k=1}^{L} p_{ijk} = 1,$$

which follows from (5.71).

■

Consider an assumption, common in practice, that (5.63) is a Gaussian family of probability distributions ($i \in S$):

$$p_{ij}(x) = \begin{cases} p_i^0(x) = n_N(x \mid \mu_i, \Sigma), & j = 1, \\ n_N(x \mid \mu_{ij}, \Sigma_{ij}), & j \in \{2, 3, \ldots, M_i\}, \end{cases} \tag{5.73}$$

where mean vectors $\mu_i = \mu_{i1}, \mu_{ij} \in R^N$ and covariance matrices $\Sigma = \Sigma_{i1}, \Sigma_{ij}$ satisfy the ϵ-closeness condition (5.64). As in Section 1.4, let

$$\Delta = \left((\mu_2 - \mu_1)^T \Sigma^{-1} (\mu_2 - \mu_1)\right)^{1/2} > 0$$

denote the Mahalanobis interclass distance for hypothetical distributions $p_1^0(\cdot), p_2^0(\cdot)$.

Corollary 5.15 *If in the case of two equiprobable classes ($L = 2$) ($\pi_1 = \pi_2 = 0.5$) the Gaussian family (5.73) and ($0-1$)-loss matrix (1.18) are assumed, then the guaranteed error probability for the Bayesian decision rule (constructed for the hypothetical distributions $p_1^0(\cdot), p_2^0(\cdot)$) is*

$$r_+(\chi^0) = 1 - \frac{1}{2} \sum_{i=1}^{2} \Phi\left(\min_{1 \le j \le M_i} \frac{\frac{\Delta}{2} + \frac{(-1)^i}{\Delta}(\mu_2 - \mu_1)^T \Sigma^{-1} (\mu_{ij} - \mu_i)}{\sqrt{1 + \frac{1}{\Delta^2}(\mu_2 - \mu_1)^T \Sigma^{-1} (\Sigma_{ij} - \Sigma) \Sigma^{-1} (\mu_2 - \mu_1)}} \right). \tag{5.74}$$

Proof. Under the conditions of Corollary 5.15 the Bayesian decision rule (5.4) is the linear decision rule (1.55), (1.56). Moreover, by (5.73), the linear discriminant function $f(x)$ has Gaussian distribution:

$$\mathcal{L}\{x\} = N_N(\mu_{ij}, \Sigma_{ij}) \Rightarrow \mathcal{L}\{f(x)\} = N_1(b^T \mu_{ij} + \beta, b^T \Sigma_{ij} b).$$

Substituting the identities

$$\mu_{ij} = \mu_i + (\mu_{ij} - \mu_i), \quad \Sigma_{ij} = \Sigma + (\Sigma_{ij} - \Sigma)$$

and transforming the parameters using (1.56) we obtain

$$\mathcal{L}\{f(x)\} = N_1\Big((-1)^i \frac{\Delta^2}{2} + (\mu_2 - \mu_1)^T \Sigma^{-1} (\mu_{ij} - \mu_i),$$

$$\Delta^2 + (\mu_2 - \mu_1)^T \Sigma^{-1} (\Sigma_{ij} - \Sigma) \Sigma^{-1} (\mu_2 - \mu_1)\Big). \tag{5.75}$$

Further, from (5.71), (1.55) we have:

$$p_{iji} = P_{ij}\left\{(-1)^i f(X_i) > 0\right\},$$

where $P_{ij}\{\cdot\}$ denotes the probability of a random event under the assumption that the random vector X_i has density $p_{ij}(\cdot)$. By (5.75) and the properties of the function $\Phi(\cdot)$, we come to the expression

$$p_{iji} = \Phi\left(\frac{\frac{\Delta^2}{2} + (-1)^i(\mu_2 - \mu_1)^T\Sigma^{-1}(\mu_{ij} - \mu_i)}{\sqrt{\Delta^2 + (\mu_2 - \mu_1)^T\Sigma^{-1}(\Sigma_{ij} - \Sigma)\Sigma^{-1}(\mu_2 - \mu_1)}}\right).$$

Substituting it into (5.72) we obtain(5.74).

■

Note that if Gaussian distributions (5.73) differ only in mean vectors $\{\mu_{ij}\}$ and have equal covariance matrices ($\Sigma_{ij} = \Sigma$, $j = 1, \ldots, M_i, i \in S$), then the ϵ–closeness condition can be transformed to the form

$$(\mu_{ij} - \mu_i)^T\Sigma^{-1}(\mu_{ij} - \mu_i) \leq \epsilon_{+i}, \quad j = 2, \ldots, M_i, \tag{5.76}$$

and the guaranteed error probability for the linear Bayesian decision rule according to Corollary 5.15 is

$$r_+(\chi^0) = 1 - \frac{1}{2}\sum_{i=1}^{2}\Phi\left(\frac{\Delta}{2} + \min_{1\leq j\leq M_i}\frac{(-1)^i}{\Delta}(\mu_2 - \mu_1)^T\Sigma^{-1}(\mu_{ij} - \mu_i)\right).$$

Applying the linear Taylor formula to the last expression under the conditions (5.76) we obtain

$$r_+(\chi^0) = \Phi\left(-\frac{\Delta}{2}\right) - \frac{e^{-\Delta^2/8}}{2\sqrt{2\pi}\Delta}\sum_{i=1}^{2}\min_{1\leq j\leq M_i}\left((-1)^i(\mu_2 - \mu_1)^T\Sigma^{-1}(\mu_{ij} - \mu_i)\right) + \mathcal{O}(\epsilon_+^2),$$

where $\epsilon_+ = \max \epsilon_{+i}$ is the maximal distortion level for all classes. The Bayesian error probability, as it follows from Section 1.4, is $r_0 = \Phi(-\Delta/2)$, therefore we obtain the asymptotic expansion for the robustness factor $\kappa_+(\chi^0)$ of the Bayesian decision rule:

$$\kappa_+(\chi^0) = \frac{e^{-\Delta^2/8}}{2\sqrt{2\pi}\Delta\Phi(-\Delta/2)}\times$$

$$\times\sum_{i=1}^{2}\max_{1\leq j\leq M_i}\left((-1)^i(\mu_2 - \mu_1)^T\Sigma^{-1}(\mu_{ij} - \mu_i)\right) + \mathcal{O}(\epsilon_+^2).$$

Note that $\kappa_+(\chi^0) \geq 0$, since $\mu_{i1} = \mu_i$, and therefore,

$$\max_{1\leq j\leq M_i}\left((-1)^i(\mu_2 - \mu_1)^T\Sigma^{-1}(\mu_{ij} - \mu_i)\right) \geq 0, \quad i \in S.$$

Let us turn now to the problem of robust decision rule synthesis.

Theorem 5.11 *Under distortions (5.64), (5.65) the robust decision rule $\chi^*(x)$ has the form*

$$\chi^*(x) = (\chi_1^*(x), \ldots, \chi_L^*(x)), \quad \chi_k^*(x) = \delta_{k,k^0},$$

$$k^0 = \arg\min_{k \in S} \sum_{i=1}^{L} \pi_i w_{ik} p_{ij_i^0}(x), \tag{5.77}$$

and $j_i^0 \in \{1, 2, \ldots, M_i\}$ $(i \in S)$ *is the solution of the optimization problem:*

$$\int_{R^N} \min_{k \in S} \left(\sum_{i=1}^{L} \pi_i w_{ik} p_{ij_i}(x) \right) dx \to \max_{j_1, \ldots, j_L}. \tag{5.78}$$

Proof. By Theorem 5.10 and the definition (5.6), in order to find the robust decision rule we have to solve the following optimization problem:

$$\sum_{i=1}^{L} \pi_i \max_{1 \le j \le M_i} P_{ij}(\chi) \to \min_{\chi}$$

under the constraints (5.2).Taking (5.66) into account, rewrite this problem to an equivalent form:

$$\max_{j_1, \ldots, j_L} \int_{R^N} \sum_{k=1}^{L} \chi_k(x) \left(\sum_{i=1}^{L} \pi_i w_{ik} p_{ij_i}(x) \right) dx \to \min_{\chi}.$$

Change the order of optimization operations in this problem:

$$\min_{\chi} \int_{R^N} \sum_{k=1}^{L} \chi_k(x) \left(\sum_{i=1}^{L} \pi_i w_{ik} p_{ij_i}(x) \right) dx \to \max_{j_1, \ldots, j_L}.$$

Minimization with respect to $\chi(\cdot)$ is performed in the same way as in Theorem 1.1, therefore we obtain (5.77), (5.78).

■

Corollary 5.16 *For* $(0-1)$*–loss matrix (1.18) the robust decision rule has the form:*

$$\chi_k^*(x) = \delta_{k,k^0}, \quad k^0 = \arg\max_{k \in S} \pi_k p_{kj_k^0}(x),$$

where $j_k^0 \in \{1, 2, \ldots, M_k\}$ $(k \in S)$ *is the solution of the optimization problem:*

$$F(j_1, \ldots, j_L) = \int_{R^N} \max_{k \in S} (\pi_k p_{kj_k}(x)) dx \to \min_{j_1, \ldots, j_L}. \tag{5.79}$$

Proof is conducted by substituting (1.18) into (5.77), (5.78) and by using the norming condition.

■

It is computationally difficult to solve the combinatorial optimization problems (5.78), (5.79), therefore we give a method of approximate robust decision rule synthesis. For this purpose we construct the upper bound for the functional (5.67):

$$r_+(\chi) \leq \overline{r}_+(\chi) = \sum_{i=1}^{L} \pi_i \int_{R^N} \max_{1 \leq j \leq M_i} p_{ij}(x) \sum_{k=1}^{L} w_{ik} \chi_k(x) dx. \qquad (5.80)$$

From the minimality condition for the functional (5.80) we obtain an approximation of the robust decision rule:

$$\overline{\chi_k^*}(x) = \delta_{k,k^0}, \qquad k^0 = \arg\min_{k \in S} \sum_{i=1}^{L} \pi_i w_{ik} \max_{1 \leq j \leq M_i} p_{ij}(x),$$

and, in particular, for $(0-1)$–loss matrix,

$$k^0 = \arg\max_{k \in S} (\pi_k \max_{1 \leq j \leq M_k} p_{kj}(x)). \qquad (5.81)$$

Let us evaluate the gain of the robust decision rule (5.81) with respect to the Bayesian decision rule:

$$\chi_k^0(x) = \delta_{k,k^0}, \qquad k^0 = \arg\max_{1 \leq i \leq L} \pi_i p_{i1}(x) \qquad (5.82)$$

basing on the guaranteed error probability $\overline{r_+}(\cdot)$ defined by (5.80). It follows from (5.80)–(5.82), (1.18) that

$$\overline{r}_+(\chi^0) = 1 - \int_{R^N} \pi_{d_0(x)} \max_{1 \leq j \leq M_{d_0(x)}} p_{d_0(x),j}(x) dx,$$

$$\overline{r}_+(\chi^*) = 1 - \int_{R^N} \max_{1 \leq k \leq L} (\pi_k \max_{1 \leq j \leq M_k} p_{kj}(x)) dx.$$

Then we obtain

$$\overline{r}_+(\chi^0) - \overline{r}_+(\chi^*) =$$

$$\int_{R^N} (\max_{1 \leq k \leq L} (\pi_k \max_{1 \leq j \leq M_k} p_{kj}(x)) - \pi_{d_0(x)} \max_{1 \leq j \leq M_{d_0(x)}} p_{d_0(x),j}(x)) dx.$$

Note that the integrand in this expression is nonnegative, therefore

$$\overline{r}_+(\chi^*) \leq \overline{r}_+(\chi^0).$$

5.7 Errors in Assignment of Distribution Parameters

This section considers the type D.2.1.1 (according to Figure 2.1) of distortions of data model often met in applied pattern recognition problems. It is induced by errors in assignment of hypothetical distribution parameters when the Bayesian decision rule is built.

Suppose that a separable m–parametric family of probability distribution densities

$$Q = \{q(x;\theta), x \in R^N : \theta \in R^m\}$$

is defined in R^N and L different hypothetical densities $q(\cdot;\theta_1^0), \ldots, q(\cdot;\theta_L^0)$ are fixed. The true probability distribution density of observations from class $\Omega_i (i \in S)$ differs from the hypothetical density $q(\cdot;\theta_i^0)$ by parametric ϵ–distortions:

$$p_i(x) = q(x;\theta_i), \quad (\theta_i - \theta_i^0)^T B_i(\theta_i - \theta_i^0) = \epsilon_i^2, \quad 0 \leq \epsilon_i \leq \epsilon_{+i}, \tag{5.83}$$

where $B_i = (b_{ijk})$ is a given symmetric positive definite matrix determining the shape of the ϵ_{+i}–neighborhood of the point θ_i^0 in the parametric space (for example, $B_i = \mathbf{I}_m$ determines a hyperspherical neighborhood), and ϵ_{+i} is the distortion level for the class Ω_i.

The distortions (5.83) often appear in practice, since a parametric probability model is often used (Aivazyan *et al.*, 1989) that has either *a priori* unknown parameters to be estimated or known ones, but with some errors. In the first case, finiteness of the samples leads to estimation errors, and it is possible to represent them in the form (5.83), for example, using confidence regions. In the second case, if the "uncertainty region" for parameter values is of ellipsoidal shape in R^m, then we come to the expression (5.83) again.

Note that type (5.83) distortions can be generalized to the case of neighborhoods of point θ_i^0 more complex than ellipsoidal. Moreover, the distortions represented by *"fuzzy parametric models"* (Das Gupta *et al.*, 1981) are close to (5.83).

Let us construct the guaranteed risk functional for the family of generalized randomized decision rules $\chi(x) = (\chi_i(x))$ defined by the set of L critical functions (5.2). Assume the notations:

$$\epsilon_+ = \max_i \epsilon_{+i};$$

$$\alpha_i^{(k)} = \sum_{j=1}^{L} w_{ij} \int_{R^N} \chi_j(x) \nabla_{\theta_i^0}^k q(x;\theta_i^0)dx, \quad k \in \{1,2\} \tag{5.84}$$

is an m-column-vector for $k = 1$ and an $(m \times m)$-matrix for $k = 2$;

$$r(\chi;\{\theta_i\}) = \sum_{i,j=1}^{L} \pi_i w_{ij} \int_{R^N} q(x;\theta_i)\chi_j(x)dx \tag{5.85}$$

is risk for the situation where observations to be classified by the decision rule $\chi(\cdot)$ are specified by densities $\{q(\cdot;\theta_i)\}$.

Theorem 5.12 *Let Q be the family of triply continuously differentiable densities such that*

$$\int_{R^N} \left| \frac{\partial^3 q(x;\theta)}{\partial\theta_j \partial\theta_k \partial\theta_l} \right| dx \le c < \infty,\ \theta = (\theta_j) \in R^m,\ j,k,l \in \{1,2,\ldots,m\}. \tag{5.86}$$

Then in presence of distortions (5.83) the guaranteed risk for the randomized decision rule $\chi(\cdot)$ admits the following asymptotic expansion:

$$r_+(\chi) = r(\chi;\{\theta_i^0\}) + \sum_{i=1}^{L} \pi_i \epsilon_{+i} \sqrt{\alpha_i^T B_i^{-1} \alpha_i} + \mathcal{O}(\epsilon_+^2). \tag{5.87}$$

Proof. Let us use the second-order Taylor formula for density $q(\cdot) \in Q$:

$$q(x;\theta_i) = q(x;\theta_i^0) + (\theta_i - \theta_i^0)^T \nabla_{\theta_i^0} q(x;\theta_i^0) + \frac{1}{2}(\theta_i - \theta_i^0)^T \nabla_{\theta_i^0}^2 q(x;\theta_i^0) \times$$

$$\times (\theta_i - \theta_i^0) + \frac{1}{6} \sum_{j,k,l=1}^{m} \frac{\partial^3 q(x;\overline{\theta}_i)}{\partial\overline{\theta}_{ij} \partial\overline{\theta}_{ik} \partial\overline{\theta}_{il}} (\theta_{ij} - \theta_{ij}^0)(\theta_{ik} - \theta_{ik}^0)(\theta_{il} - \theta_{il}^0),$$

where $\overline{\theta} = (\overline{\theta}_{ij}) \in R^m$ satisfies the inequality $|\overline{\theta}_i - \theta_i^0| < |\theta_i - \theta_i^0|$. Substitute this expansion into the risk functional (5.85), taking (5.84), (5.86) into account:

$$r(\chi;\{\theta_i\}) = r(\chi;\{\theta_i^0\}) + \sum_{i=1}^{L} \pi_i {\alpha_i^{(1)}}^T (\theta_i - \theta_i^0) +$$

$$+ \frac{1}{2} \sum_{i=1}^{L} \pi_i (\theta_i - \theta_i^0)^T \alpha_i^{(2)} (\theta_i - \theta_i^0) + \mathcal{O}(\epsilon_+^3). \tag{5.88}$$

In order to find the guaranteed risk (5.5), we solve the conditional maximization problem:

$$r(\chi;\{\theta_i\}) \to \max_{\{\theta_i\}}, \tag{5.89}$$

$$(\theta_i - \theta_i^0)^T B_i (\theta_i - \theta_i^0) = \epsilon_i^2, \quad i \in S.$$

Using the method of indefinite Lagrange multipliers and the main term of the asymptotic expansion (5.88) we obtain the Lagrange function in the form:

$$L(\{\theta_i\},\{\lambda_i\}) = r(\chi;\{\theta_i^0\}) + \sum_{i=1}^{L} \pi_i {\alpha_i^{(1)}}^T (\theta_i - \theta_i^0) +$$

$$+\frac{1}{2}\sum_{i=1}^{L}\pi_i(\theta_i-\theta_i^0)^T\alpha_i^{(2)}(\theta_i-\theta_i^0)+\frac{1}{2}\sum_{i=1}^{L}\frac{\pi_i}{\lambda_i}\times$$

$$\times\left(\epsilon_i^2-(\theta_i-\theta_i^0)^T B_i(\theta_i-\theta_i^0)\right),$$

where $\{\lambda_i\}$ are indefinite Lagrange multipliers. Differentiating this function with respect to $\{\theta_i\}$ we obtain the expression for the stationary point θ_i^*:

$$\theta_i^*-\theta_i^0=\lambda_i(\mathbf{I}_m-\lambda_i B_i^{-1}\alpha_i^{(2)})^{-1}B_i^{-1}\alpha_i^{(1)},\quad i\in S. \tag{5.90}$$

Applying the constraint in (5.89) we obtain the equation for λ_i:

$$\lambda_i^2=\frac{\epsilon_i^2}{\alpha_i^{(1)^T}B_i^{-1}(\mathbf{I}_m-\lambda_i B_i^{-1}\alpha_i^{(2)})^{-1}B_i(\mathbf{I}_m-\lambda_i B_i^{-1}\alpha_i^{(2)})^{-1}B_i^{-1}\alpha_i^{(1)}}.$$

Here we see that $\lambda_i=\mathcal{O}(\epsilon_i)$. Use the asymptotic expansion for the inverse matrix (Bellman, 1960):

$$(\mathbf{I}_m-\lambda_i B_i^{-1}\alpha_i^{(2)})^{-1}=\mathbf{I}_m+\lambda_i B_i^{-1}\alpha_i^{(2)}+\mathcal{O}(\epsilon_i^2)\mathbf{1}_{m\times m}.$$

Then we find the asymptotic expansion for λ_i:

$$\lambda_i=\pm\epsilon_i/\sqrt{\alpha_i^{(1)^T}B_i^{-1}\alpha_i^{(1)}}+\mathcal{O}(\epsilon_i^2),\quad i\in S.$$

Substituting this expansion into (5.90) we obtain

$$\theta_i^*=\theta_i^0\pm\frac{\epsilon_i}{\sqrt{\alpha_i^{(1)^T}B_i^{-1}\alpha_i^{(1)}}}B_i^{-1}\alpha_i^{(1)},\quad i\in S. \tag{5.91}$$

From (5.88), (5.91) it follows that

$$r(\chi;\{\theta_i^*\})=r(\chi;\{\theta_i^0\})\pm\sum_{i=1}^{L}\pi_i\epsilon_i\sqrt{\alpha_i^{(1)^T}B_i^{-1}\alpha_i^{(1)}}+\mathcal{O}(\epsilon_i^2).$$

Choosing the " + " sign in this expression, which corresponds to the maximal risk value, and maximizing this expression with respect to $\epsilon_i\in[0,\epsilon_{+i}]$ according to (5.83) we come to (5.87).

■

Note that similarly to Theorem 5.12, using the " − " sign in (5.91), one can obtain the expression for the greatest lower bound of the risk for the decision rule $\chi(\cdot)$:

$$r_-(\chi)=\inf r(\chi;\{\theta_i\})=r(\chi;\{\theta_i^0\})-\sum_{i=1}^{L}\pi_i\epsilon_{+i}\sqrt{\alpha_i^{(1)^T}B_i^{-1}\alpha_i^{(1)}}+\mathcal{O}(\epsilon_+^2).$$

The asymptotic expansion (5.87) has an obvious interpretation. The main term $r(\chi;\{\theta_i^0\})$ of the asymptotic expansion corresponds to the value of risk for the decision rule χ in absence of distortions (absence of any errors in assignment of distribution parameters). The second set of nonnegative summands characterizes the risk bias induced by the errors (5.83) in assignment of distribution parameters. The remainder term is of order $\mathcal{O}(\epsilon_+^2)$.

Notice that by the same method as was used in Theorem 5.12 one can construct asymptotic expansions with remainders of smaller order.

Corollary 5.17 *For the* $(0-1)$*–loss matrix (1.18) the guaranteed error probability for the Bayesian decision rule (5.4) satisfies the asymptotic expansion*

$$r_+(\chi^0) = r_0 + \sum_{i=1}^{L} \pi_i \epsilon_{+i} \sqrt{\alpha_i^{(1)T} B_i^{-1} \alpha_i^{(1)}} + \mathcal{O}(\epsilon_+^2), \tag{5.92}$$

where $r_0 = r(\chi^0;\{\theta_i^0\})$ *is the Bayesian risk for the hypothetical (nondistorted) model;*

$$\alpha_i^{(1)} = -\int_{V_i^0} q(x;\theta_i^0)\nabla_{\theta_i^0} \ln q(x;\theta_i^0)dx; \tag{5.93}$$

$V_i^0 \subset R^N$ *is the region of the Bayesian decision* $d_0(x) = i$ *making.*

Proof is by substitution of (1.18) and (5.4) into (5.84), (5.85).

■

Note that for a regular family Q the well known identity (Ibragimov *et al.*, 1979)

$$\int_{R^N} q(x;\theta_i^0)\nabla_{\theta_i^0} \ln q(x;\theta_i^0)dx \equiv \mathbf{O}_m$$

holds. However in this case the integration region in (5.93) does not coincide with R^N, therefore in general case, $\alpha_i^{(1)} \neq \mathbf{O}_m$.

Corollary 5.18 *The* δ*–robustness condition (see Section 2.3) for the Bayesian decision rule assumes the form:*

$$\sum_{i=1}^{L} \pi_i \epsilon_{+i} \sqrt{\alpha_i^{(1)T} B_i^{-1} \alpha_i^{(1)}} \leq \delta.$$

Corollary 5.19 *If* Q *is an* N*-variate Gaussian family of probability distributions with shift parameter* $\theta \in R^N$ *and nonsingular covariance matrix* Σ*, then in the case of two equiprobable classes* ($L = 2$, $\pi_1 = \pi_2 = 0.5$) *and* $(0-1)$*–loss matrix (1.18) at* $B_i = \Sigma$ *the Bayesian DR robustness factor admits the asymptotic expansion:*

$$\kappa_+(\chi^0) = \frac{\phi(\Delta/2)}{2\Phi(-\Delta/2)}(\epsilon_{+1} + \epsilon_{+2}) + \mathcal{O}(\epsilon_+^2), \tag{5.94}$$

where

$$\Delta = ((\theta_2^0 - \theta_1^0)^T \Sigma^{-1} (\theta_2^0 - \theta_1^0))^{1/2}$$

is the Mahalanobis interclass distance; $\phi(\cdot), \Phi(\cdot)$ *are the density and the distribution function of the standard normal distribution respectively.*

Proof. Note first of all that the Gaussian family Q

$$q(x;\theta) = (2\pi)^{-N/2} \mid \Sigma \mid^{-1/2} \exp(-(x-\theta)^T\Sigma^{-1}(x-\theta)/2) \qquad (5.95)$$

satisfies the regularity conditions (5.86), therefore we use (5.92). According to (5.93), (5.95)

$$\alpha_i^{(1)} = -\mathbf{E}_{\theta_i^0}\left\{\mathbf{1}((-1)^i f(X_i))\Sigma^{-1}(X_i - \theta_i^0)\right\},$$

where $f = f(x) = b^T x + \beta$ is the Bayesian linear discriminant function (1.55), (1.56). Compute the j-th component ($j = 1, \ldots, N$) of the vector $\alpha_i^{(1)}$ using the notation $\Sigma^{-1} = (\overline{\sigma}_{jk})$ and the formula for total expectation:

$$\alpha_{ij}^{(1)} = -\sum_{k=1}^{N} \overline{\sigma}_{jk}\mathbf{E}_{\theta_i^0}\left\{\mathbf{1}((-1)^i f(X_i))\mathbf{E}\{(X_{ik} - \theta_{ik}^0) \mid f(X_i) = f\right\}.$$

Applying the theorem about conditional Gaussian distributions (Anderson, 1958) and formula (1.56) we find:

$$\alpha_{ij}^{(1)} = -\sum_{k=1}^{N} \overline{\sigma}_{jk}\mathbf{E}_{\theta_i^0}\left\{\mathbf{1}((-1)^i f(X_i))\frac{\mathbf{Cov}\{X_{ik}, f(X_i)\}}{\mathbf{D}\{f(X_i)\}}(f(X_i) - \mathbf{E}_{\theta_i^0}\{f(X_i)\})\right\} =$$

$$= -\frac{b_j}{\Delta^2}\mathbf{E}_{\theta_i^0}\left\{\mathbf{1}((-1)^i f(X_i))(f(X_i) - \mathbf{E}_{\theta_i^0}\{f(X_i)\})\right\}.$$

The probability distribution of $f = f(X_i)$, according to (1.61), (1.64), is Gaussian, therefore

$$\alpha_i^{(1)} = \Delta^{-1}\phi(\Delta/2)b, \quad i \in \{1,2\}.$$

Substituting this expression into (5.92) and using the equalities $r_0 = \Phi(-\Delta/2)$, $b^T\Sigma b = \Delta^2$, and the definition (5.7) we obtain (5.94).

■

Corollary 5.19 allows to find the "breakdown point" (2.29) for the Bayesian decision rule:

$$\epsilon_+^* = (\frac{1}{2} - \Phi(-\frac{\Delta}{2}))/\phi(\frac{\Delta}{2}).$$

Figure 5.10 plots the dependence of the "breakdown point" ϵ_+^* on Mahalanobis interclass distance Δ (between the hypothetical models of classes Ω_1, Ω_2), important for applied pattern recognition problems.

It can be seen from Figure 5.10 that if the interclass distance is $\Delta = 2.56$ (the hypothetical error probability is $r_0 = 0.1$), then the "breakdown point" ϵ_+^* is 2.2. Consequently, if the error level ϵ_{+i}, $i \in S$ in parameter assignment exceeds 2.2, then the Bayesian decision rule may turn out to be equivalent to the equiprobable coin tossing.

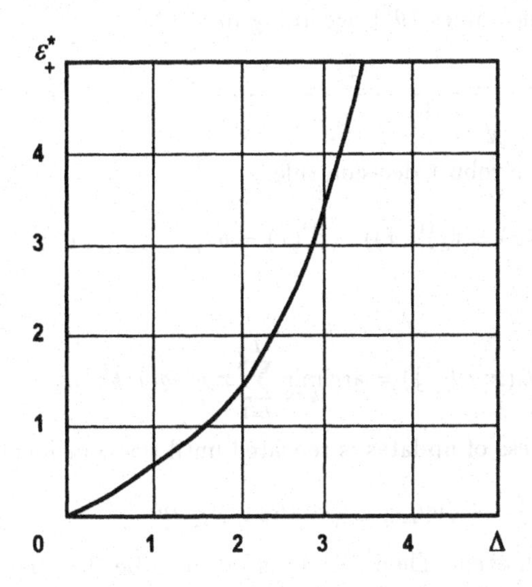

Figure 5.10: Breakdown point vs. interclass distance

Consider now the robust decision rule synthesis problem in the presence of distortions (5.83). Using Theorem 5.11 and definition (5.6) to find the robust decision rule, we obtain the following nonlinear multidimensional variational problem:

$$r(\chi; \{\theta_i^0\}) + \sum_{i=1}^{L} \pi_i \epsilon_{+i} \sqrt{\alpha_i^{(1)^T} B_i^{-1} \alpha_i^{(1)}} \to \min_{\chi} \tag{5.96}$$

with constraints (5.2), where risk functional $r(\cdot)$ is defined by (5.85) and $\{\alpha_i^{(1)}\}$ depend on χ according to (5.84).

Let us describe the method of approximate solution of the problem (5.96) based on its equivalent representation:

$$r(\chi; \{\theta_i^*\}) \to \min_{\chi}, \tag{5.97}$$

where $\theta_i^* = \theta_i^*(\chi)$ is the maximum point, which depends on χ according to (5.91) (the sign "+" is chosen in (5.91)). For an initial approximation of the solution of the problem (5.97) we take the Bayesian decision rule $d_0(x)$ constructed for the hypothetical models of the classes:

$$\chi^{(0)}(x) = (\chi_i^{(0)}(x)), \ \chi_i^{(0)}(x) = \delta_{i, d_0(x; \{\theta_i^0\})}, i \in S.$$

Further, the algorithm of approximate solution of the problem (5.97) consists in iteration of the following two stages.

<u>*Stage 1.*</u> Update the values $\{\theta_i^*\}$ according to (5.91):

$$\theta_i^{(1)} = \theta_i^0 + \frac{\epsilon_{+i}}{\sqrt{\alpha_i^{(1)^T}(\chi^{(0)}) B_i^{-1} \alpha_i^{(1)}(\chi^{(0)})}} B_i^{-1} \alpha_i^{(1)}(\chi^{(0)}), \; i \in S.$$

<u>*Stage 2.*</u> Update the robust decision rule:

$$\chi^{(1)}(x) = (\chi_i^{(1)}(x)), \; \chi_i^{(1)}(x) = \delta_{i, d_0(x; \{\theta_i^{(1)}\})}, i \in S,$$

where

$$d_0(x; \{\theta_i^{(1)}\}) = \arg\min_{k \in S} \sum_{i=1}^{L} \pi_i w_{ik} q(x; \theta_i^{(1)}).$$

This iterative process of updates is repeated until the coincidence

$$\chi^{(k)}(x) = \chi^{(k-1)}(x), x \in R^N$$

at k-th iteration is achieved. Then $\chi^{(k)}(x)$ is taken to be the final estimate of the robust decision rule $\chi^*(x)$. Note that for practical usage the "one–step estimate" (Orlov, 1986) $\chi^{(1)}(x)$ may also be recommended.

Chapter 6

Decision Rule Robustness under Distortions of Training Samples

In this chapter we investigate pattern recognition problems for which the hypothetical assumptions about training samples are disturbed in various ways: the class-i training sample is contaminated by elements from alien classes, or samples contain outliers, or elements of the training sample are statistically dependent. We estimate the robustness factor and analyze its dependence on sample sizes, distortion levels, and other factors. We construct new decision rules with higher order of robustness and illustrate their stability by computer results.

6.1 Risk Robustness under Statistical Dependence of Sample Elements

We shall consider the model of adaptive pattern recognition under parametric prior uncertainty. It is often encountered in applications; we described it in Section 2.1 and used in Chapter 3.

Assume that in the Euclidean feature space R^N a parametric family of probability densities

$$Q = \{q(x;\theta), x \in R^N : \theta \in \Theta \subseteq R^m\}$$

is given, $L \geq 2$ different densities $q(\cdot;\theta_1^0), \ldots, q(\cdot;\theta_L^0) \in Q$ are fixed, and feature vectors $x \in R^N$ from L classes $\Omega_1, \ldots, \Omega_L$ are observed with given prior probabilities $\pi_1, \ldots, \pi_L$ $(\pi_1 + \ldots + \pi_L = 1)$. The mathematical model of an observed feature vector from $\Omega_i (i \in S)$ is a random N-vector $X_i \in R^N$ with probability distribution density $q(x;\theta_i^0)$, and the true value of the parameter vector $\theta_i^0 = (\theta_{ij}^0)$ is unknown. For tuning, a pattern recognition algorithm uses a classified training random sample $A = \bigcup_{i=1}^{L} A_i$ consisting of L subsamples;

$$A_i = \{z_{i1}, z_{i2}, \ldots, z_{in_i}\} \subset R^N$$

is a subsample of n_i random observations from Ω_i. A random observation $X \in R^N$ to be classified belongs to one of L classes $\{\Omega_i\}$ and does not depend on A.

In pattern recognition theory and its applications, it is traditionally assumed that the elements $z_{i1}, z_{i2}, \ldots, z_{in_i} \in R^N$ of each sample $A_i (i \in S)$ are mutually independent. Under this assumption a family of plug-in decision rules (PDR) (3.4)—(3.8) is constructed using minimum contrast estimators (Pfanzagl, 1969), (Chibisov, 1973):

$$\hat{\theta}_i = \arg\min_{\theta \in \overline{\Theta}} L_i(\theta), \quad L_i(\theta) = n_i^{-1} \sum_{t=1}^{n_i} g(z_{it}; \theta) \quad (i = 1, \ldots, L), \tag{6.1}$$

where $\overline{\Theta} \subseteq R^m$ is the closure of Θ; $g(z;\theta)$ is a contrast function (in particular, if $g(z;\theta) = -\ln q(z;\theta)$, then $\hat{\theta}_i = (\hat{\theta}_{ik})$ is the maximum likelihood estimator). The family of plug-in decision rules is derived by the substitution of estimates $\{\hat{\theta}_i\}$ for $\{\theta_i^0\}$ in the Bayesian decision rule (3.1), (3.2):

$$d = d(x; A) = \arg\min_{i \in S} f_i(x; \{\hat{\theta}_j\}), x \in R^N, d \in S,$$

$$f_i(x; \{\theta_j\}) = \sum_{j=1}^{L} \pi_j \omega_{ji} q(x; \theta_j), i \in S; \tag{6.2}$$

here d is the number of the class to which the observed object with feature vector x will be assigned. The perfomance of the plug-in decision rule (6.2) will be characterized by classification risk (expected losses):

$$r = r(d) = \sum_{i=1}^{L} \pi_i \mathbf{E}\{\omega_{i,d(\mathbf{X}_i, \mathbf{A})}\}; \tag{6.3}$$

The classical assumption of independence of sample elements is often violated for application data (Aivazyan, 1981), (Aivazyan *et al.*, 1989), (Huber, 1981), (Lawoko *et al.*, 1986), (McLachlan, 1992). Therefore the problem of robustness evaluation for the plug-in decision rule (6.2) considered here is very topical for the situation where sample elements are dependent (for the distortions of type D.3.2). It is convenient to apply the asymptotic expansion method (Kharin, 1981, 1983) to solve this problem.

In contrast to traditional assumptions, we shall assume that A_i is a sample of n_i dependent random observations from Ω_i $(i \in S)$ with joint probability density function $p(z_{i1}, \ldots, z_{in_i}; \theta_i^0, \epsilon_i)$, where $\epsilon_i \in (-1, +1)$ is the parameter characterizing the level of dependence: if $\epsilon_i = 0$, then $z_{i1}, \ldots, z_{in_i}$ are independent; the more $|\epsilon_i|$, the "stronger" the dependence. The following concordance relations of this density with the family Q take place:

$$\int_{R^{N(n_i-1)}} p(x_1, \ldots, x_{n_i}; \theta_i^0, \epsilon_i) dx_1 \ldots dx_{t-1} dx_{t+1} \ldots dx_{n_i} = q(x_t; \theta_i^0);$$

$$p(z_{i1}, \ldots, z_{in_i}; \theta_i^0, 0) = \prod_{t=1}^{n_i} q(z_{it}; \theta_i^0).$$

Let us consider four main particular cases of the investigated type D.3.2 of distortions. For brevity we shall omit the class index $i \in S$, where this leads to no confusion.

D.3.2.1. Stationary time series

It is the most general kind of dependence: $z_1, \ldots, z_n \in R^N$ is a strictly stationary time series (vector random process in discrete time) (Anderson, 1971).

D.3.2.2. Dependence induced by contamination:

$$p(x_1, \ldots, x_{n_i}; \theta, \epsilon) = q(x_1; \theta) \prod_{t=1}^{n-1} ((1-\epsilon) q(x_{t+1}; \theta) + \epsilon h(x_{t+1} \mid x_t, \ldots x_1, \theta)),$$

where $0 \leq \epsilon < 1$ is the probability of occurence (at moment $t+1$) of a contamination generating the dependence of z_{t+1} on previous observations $z_1, \ldots, z_t$; $h(x_{t+1} \mid x_t, \ldots x_1, \theta) \geq 0$ is the probability density of the observation z_{t+1} conditional on that the observations $z_t, \ldots, z_1$ are fixed;

$$\int_{R^N} h(x_{t+1} \mid x_t, \ldots x_1, \theta) dx_{t+1} \equiv 1.$$

The probability density function $h(\cdot)$ is in concordance with the family Q by the relations:

$$\int_{R^{Nt}} h(x_{t+1} \mid x_t, \ldots x_1, \theta) \prod_{i=1}^{t} q(x_i; \theta) dx_i = q(x_{t+1}; \theta), x_{t+1} \in R^N.$$

D.3.2.3. Markov dependence of observations:

$$p(x_1, \ldots, x_n; \theta, \epsilon) = q(x_1; \theta) \prod_{t=1}^{n-1} h(x_{t+1} \mid x_t, \theta),$$

where $h(x_{t+1} \mid x_t, \theta) \geq 0$ is a transition probability density function concorded with the family Q:

$$\int_{R^N} h(x_{t+1} \mid x_t, \theta) q(x_t; \theta) dx_t = q(x_{t+1}; \theta), x_{t+1} \in R^N.$$

In particular, if Q is the Gaussian family with shift parameter θ, then

$$q(x_t; \theta) = n_N(x_t \mid \theta, \Sigma),$$

$$h(x_{t+1} \mid x_t, \theta) = n_N(x_{t+1} \mid \theta + \epsilon \Sigma_{12} \Sigma^{-1}(x_t - \theta), \Sigma - \epsilon^2 \Sigma_{12} \Sigma^{-1} \Sigma_{12}^T),$$

where $n_N(x_t \mid \theta, \Sigma)$ is N-variate marginal normal density and $\epsilon\Sigma_{12} = \mathbf{Cov}\{x_t, x_{t+1}\}$ is an arbitrary covariance matrix of neighboring sample elements, for which the block matrix

$$\begin{pmatrix} \Sigma & \vdots & \epsilon\Sigma_{12} \\ \cdots & \vdots & \cdots \\ \epsilon\Sigma_{12}^T & \vdots & \Sigma \end{pmatrix}$$

is positive definite. For example, if $\Sigma_{12} = \Sigma$, $-1 < \epsilon < +1$, then

$$h(x_{t+1} \mid x_t, \theta) = n_N(x_{t+1} \mid \theta + \epsilon(x_t - \theta), (1 - \epsilon^2)\Sigma).$$

D.3.2.4. Autoregressive dependence of order s :

•.

$$x_{t+1} = B_1 x_t + B_2 x_{t-1} + \ldots + B_s x_{t-s+1} + \xi_{t+1} \quad (t = s, s+1, \ldots, n-1),$$

where $\xi_{s+1}, \ldots, \xi_n \in R^N$ is a sequence of mutually independent identically distributed random vectors; $x_1, \ldots, x_s \in R^N$ are "initial" random vectors. Thus, the observed sample is a vector time series of autoregressive type of order s, i.e., $AR(s)$ (Anderson, 1971). The matrix coefficients $B_1, \ldots, B_s$ and the probability distribution of ξ_{t+1} depend on ϵ in such a way that as $\epsilon \to 0$, $B_1, \ldots, B_s \to \mathbf{0}_{N \times N}$, and the probability density function of ξ_{t+1} tends to $q(\cdot; \theta)$.

At the begining, let us investigate statistical properties of the minimum contrast estimators $\{\hat{\theta}_i\}$ under the distortions of type D.3.2. Assume the notations:

$$g^{(k)}(z; \theta') = (g^{(k)}_{i_1, \ldots, i_k}(z; \theta')) = \nabla^k_{\theta'} g(z; \theta')$$

is the set of m^k k-th order partial derivatives of the function $g(z; \theta)$ with respect to θ at point $\theta = \theta'$;

$$a_i(\theta) = \mathbf{E}_{\theta_i^0}\{g^{(1)}(z_{it}; \theta)\}, \tag{6.4}$$

$$b_{i\tau}(\theta) = \mathbf{Cov}_{\theta_i^0}\{g^{(1)}(z_{it}; \theta), g^{(1)}(z_{i,t+\tau}; \theta)\} \quad (i \in S, \tau = 0, \pm 1, \ldots),$$

where $\mathbf{E}_{\theta_i^0}\{\cdot\}$, $\mathbf{Cov}_{\theta_i^0}\{\cdot\}$ are the mathematical expectation and the covariance with respect to joint distribution $p(\cdot; \theta_i^0, \epsilon_i)$.

Theorem 6.1 *Let the following regularity conditions* C1 – C4 $(i \in S)$ *be satisfied:*

C1: *Θ is a compact set, θ_i^0 is an internal point of Θ, and*

$$\int_{R^N} \mid q(x; \theta) - q(x; \theta_i^0) \mid dx = 0 \Leftrightarrow \theta = \theta_i^0;$$

C2: *A density $q(x; \theta)$ is triply continuously differentiable with respect to $\theta \in \Theta, \forall x \in R^N$;*

C3: *$a_i(\theta_i^0) = \mathbf{0}_m$, the matrix $J(\theta, \theta_i^0) = \mathbf{E}_{\theta_i^0}\{g^{(2)}(X_i;\theta)\}$ is nonsingular and continuous in $\theta \in \Theta$; $b_0(\theta_i^0) = J(\theta_i^0, \theta_i^0)$ is a positive definite matrix;*

C4: *$z_{it}(t = 1, 2, \ldots,)$ is a strictly stationary random sequence such that for $j = 1, 2$, $k = (k_1, \ldots, k_j)$, $k' = (k'_1, \ldots, k'_j)$, $k_l, k'_l \in \{1, \ldots, m\}$, $\theta_i \in \Theta$,*

$$\frac{1}{n_i}\sum_{\tau=0}^{n_i-1}(1-\tau/n_i)\mathbf{Cov}_{\theta_i^0}\{g_k^{(j)}(z_{i1};\theta_i), g_{k'}^{(j)}(z_{i,1+\tau};\theta_i)\} \to 0$$

as $n_i \to \infty$.

Then as $n_i \to \infty$ the MC-estimator (6.1) is consistent: $\hat{\theta}_i \xrightarrow{\mathbf{P}} \theta_i^0$.

Proof. By C1, C2, and (6.1), the estimator $\hat{\theta}_i$ is the root of the equation $\nabla_\theta L_i(\theta) = \mathbf{0}_m$. Applying the linear Taylor formula to the left side of this equation we get

$$B_{0n_i} + B_{1n_i}(\overline{\theta_i})(\hat{\theta}_i - \theta_i^0) = \mathbf{0}_m, \tag{6.5}$$

$$B_{0n_i} = n_i^{-1}\sum_{t=1}^{n_i} g^{(1)}(z_{it};\theta_i^0), B_{1n_i}(\overline{\theta_i}) = n_i^{-1}\sum_{t=1}^{n_i} g^{(2)}(z_{it};\overline{\theta_i}),$$

and moreover, $|\, \overline{\theta_i} - \theta_i^0 \,|\leq|\, \hat{\theta}_i - \theta_i^0 \,|$. Because of the regularity conditions C3, C4, the Markov condition (see, e.g., (Koroljuk, 1984)) is fulfilled and the law of large numbers holds: $B_{0n_i} \xrightarrow{\mathbf{P}} \mathbf{0}_m$.

Further, from (6.5) it may be proved by *reductio ad absurdum* that the estimator is consistent: $\hat{\theta}_i \xrightarrow{\mathbf{P}} \theta_i^0$.

■

In the sequel it will be necessary to have an asymptotic expansion of the mutual covariance matrix of the MC-estimator (6.1):

$$V_{ik} = \mathbf{E}\{(\hat{\theta}_i - \theta_i^0)(\hat{\theta}_k - \theta_k^0)^T\} \quad (i, k \in S).$$

Theorem 6.2 *Let the regularity conditions* C1 – C4 *and the following additional condition* C5 *be fulfilled:*

C5:

$$\sup_{n_i>0}\sum_{\tau=1}^{n_i-1}(1-\tau/n_i)\,\mathrm{tr}(b_{i\tau}(\theta_i^0)) < \infty \quad (i \in S; j, k, l = \overline{1, m}),$$

$$n_i^{-3}\sum_{t_1,t_2,t_3=1}^{n_i}\mathbf{E}_{\theta_i^0}\{g_j^{(1)}(z_{it_1};\theta_i^0)g_k^{(1)}(z_{it_2};\theta_i^0)g_l^{(1)}(z_{it_3};\theta_i^0)\} = o(n_i^{-1}).$$

Then the following asymptotic expansions hold:

$$V_{ii} = \frac{1}{n_i}(b_0(\theta_i^0) +$$

$$+ \sum_{\tau=1}^{n_i-1}(1-\tau/n_i)(b_0^{-1}(\theta_i^0)(b_{i\tau}(\theta_i^0) + b_{i\tau}^T(\theta_i^0))b_0^{-1}(\theta_i^0)) + o(n_i^{-1})\mathbf{1}_{m\times m},$$

$$V_{ik} = o(n_i^{-1})\mathbf{1}_{m\times m} \quad (i \neq k \in S); \tag{6.6}$$

$$\mathbf{E}\{|(\hat{\theta}_{ij} - \theta_{ij}^0)(\hat{\theta}_{ik} - \theta_{ik}^0)(\hat{\theta}_{il} - \theta_{il}^0)|\} = o(n_i^{-1}) \quad (j,l,k = \overline{1,m}).$$

Proof. Denote $\alpha_{n_i} = n_i B_{on_i} B_{on_i}^T$. Then by Theorem 6.1 as $n_i \to \infty$ we have from (6.5):

$$\alpha_{n_i} - n_i b_0(\theta_i^0)(\hat{\theta}_i - \theta_i^0)(\hat{\theta}_i - \theta_i^0)^T b_0(\theta_i^0) \xrightarrow{\mathbf{P}} \mathbf{0}_m.$$

Condition C5 and the Schwartz inequality imply the finiteness of the absolute moments:

$$\sup_{n_i} \mathbf{E}\{| (\alpha_{n_i})_{jk} |\} < \infty \quad (j,k = \overline{1,m}).$$

Therefore, by the properties of convergence in mean (see, e.g., (Koroljuk, 1984)) we get:

$$\mathbf{E}\{n_i B_{0n_i} B_{0n_i}^T\} - n_i b_0(\theta_i^0) V_{ii} b_0(\theta_i^0) = \mathbf{1}_{m\times m} o(1),$$

from which the first expansion in (6.6) follows. The second and third expansions in (6.6) can be found in the same way.

■

Let us construct now the asymptotic expansion of risk for the plug-in decision rule (6.2) in the case of distortions D.3.2 using the properties of MC-estimators formulated in Theorems 6.1, 6.2.

Denote:

$$n_0 = \min_i n_i, c_{ikj} = \pi_i(w_{ik} - w_{ij}),$$

$$f_{kj}(x;\{\theta_i\}) = f_k(x;\{\theta_i\}) - f_j(x;\{\theta_i\}), \quad G_{kj}(x) = | \nabla_x f_{kj}(x;\{\theta_i^0\}) |^{-1};$$

$$\Gamma_{kj} = \{x : f_{kj}(x;\{\theta_i^0\}) = 0\} \subset R^N$$

is the Bayesian discriminant surface for the pair of classes $\Omega_k, \Omega_j (k \neq j \in S)$; r_0 is the risk (3.3) of the Bayesian decision rule if $\{\theta_i^0\}$ are known and distortions are absent. Let us agree to mark a function by a small zero on top of it, if it is computed for true parameter values $\{\theta_i^0\}$; for example: $\overset{\circ}{q}{}_i^{(1)} = \nabla_{\theta_i^0} q(x;\theta_i^0)$. Moreover, let us denote the linear combinations of surface integrals over $\{\Gamma_{kj}\}$ $(i \in S; \tau = 0,1,\ldots)$ as follows:

$$\rho_{i\tau} = \frac{1}{2}(1 - \delta_{0\tau}/2) \sum_{j=2}^{L} \sum_{k=1}^{j-1} c_{ikj}^2 \int_{\Gamma_{kj}} \overset{\circ}{q}_i^{(1)T} b_0^{-1}(\theta_i^0)(b_{i\tau}(\theta_i^0) +$$

$$+ b_{i\tau}^T(\theta_i^0)) b_0^{-1}(\theta_i^0) \overset{\circ}{q}_i^{(1)} G_{kj}(x) ds_{N-1}. \tag{6.7}$$

Theorem 6.3 *Suppose that for each of the classes $\{\Omega_i : i \in S\}$ the conditions C1 - C5 are fulfilled, the integrals in (6.7) are finite and there are neighborhoods $U_1, \ldots, U_L$ for the points $\theta_1^0, \ldots, \theta_L^0$ respectively such that for any $\theta' = (\theta_1', \ldots, \theta_L') \in U_1 \times \ldots \times U_L$*

$$\int_{f_{kj}(x;\theta')=0} | q_{pst}^{(3)}(x;\theta') | \, q(x;\theta_j^0) G_{kj}(x) ds_{N-1} < \infty$$

$$(p, s, t = \overline{1, m}; \quad i, j, k, l, \theta = \overline{1, L}),$$

$$| \frac{d}{dv} \int_{f_{kj}(x;\theta')=v} q_{ps}^{(2)}(x;\theta_l') q_t^{(1)}(x;\theta_i') \, \overset{\circ}{q}_j \, G_{kj}(x) ds_{N-1} \, |_{v=0} < \infty,$$

$$| \frac{d^2}{dv^2} \int_{f_{kj}(x;\theta')=v} q_p^{(1)}(x;\theta_i') q_s^{(1)}(x;\theta_l') q_s^{(1)}(x;\theta_\theta') \, \overset{\circ}{q}_j \, G_{kj}(x) ds_{N-1} \, |_{v=0} < \infty.$$

Then the plug-in decision rule (6.2) admits the expansion

$$r(d) = r_0 + \sum_{i=1}^{L} \frac{1}{n_i} (\rho_{i0} + \sum_{\tau=1}^{n_i - 1} (1 - \tau/n_i) \rho_{i\tau}) + o(n_0^{-1}). \tag{6.8}$$

Proof is conducted by the same method as for Theorem 3.3, by applying Theorems 6.1, 6.2.

■

Corollary 6.1 *If there is no dependence in the samples $A_1, \ldots, A_L$, i.e., $\epsilon_i = 0$, $i = 1, \ldots, L$, then*

$$r(d) = r_0 + \sum_{i=1}^{L} \rho_{i0}/n_i + o(n_0^{-1}), \tag{6.9}$$

and $\rho_{10}, \ldots, \rho_{L0} \geq 0$.

Corollary 6.2 *If dependence of sample elements in the sample A_i ($i \in S$) is of Markov type :*

$$\frac{1}{2}(b_{i\tau}(\theta_i^0) + b_{i\tau}^T(\theta_i^0)) = \epsilon_i^{|\tau|} b_0(\theta_i^0), \tau = 0, \pm 1, \ldots,$$

where $-1 < \epsilon_i < 1$, then the risk expansion assumes the form

$$r(d) = r_0 + \sum_{i=1}^{L} \frac{1 + \epsilon_i}{1 - \epsilon_i} \frac{\rho_{i0}}{n_i} + o(n_0^{-1}).$$

Proof of Corollaries 6.1, 6.2 is conducted by direct insertion of $\{b_{i\tau}(\theta_i^0)\}$ into (6.7), (6.8).

■

Note that the expansion (6.9) coincides with the expansion (3.33) constructed in Theorem 3.3 under the classical independence condition for sample elements.

Define the robustness factor for the plug-in decision rule (6.2) as the relative bias of risk $r(d)$ with respect to the Bayesian risk $r_0 > 0$:

$$\kappa = \kappa(d) = (r(d) - r_0)/r_0 \geq 0.$$

The less κ, the more stable the decision rule.

Let us investigate the dependence of the robustness factor κ on sample sizes $\{n_i\}$ and distortion levels $\{\epsilon_i\}$.

Corollary 6.3 *If* $\epsilon_1 = \ldots = \epsilon_L = \epsilon \in (-1, 1)$ *under the conditions of Corollary 6.2, then the robustness factor of the plug-in decision rule (6.2) admits the expansion:*

$$\kappa = \kappa(d; \epsilon, \{n_i\}) = K(\epsilon)\kappa_0(d; \{n_i\}) + o(n_0^{-1}),$$

$$K(\epsilon) = \frac{1+\epsilon}{1-\epsilon}, \quad \kappa_0(d; \{n_i\}) = \sum_{i=1}^{L} \frac{\rho_{i0}}{n_i r_0}. \tag{6.10}$$

In (6.10), dependencies of κ on ϵ and on $\{n_i\}$ are factorized (split). The factor $\kappa_0(d; \{n_i\})$ characterizes the decision rule robustness in dependence on training sample sizes $\{n_i\}$ under type D.1 distortions: the more n_0, the more stable the plug-in decision rule (6.2). The factor $K(\epsilon)$ in (6.10) describes the dependence of the robustness factor κ on "dependence level" ϵ. It is seen that dependence of sample elements may lead to decreasing as well as to increasing of the factor κ. Robustness decreases if $\epsilon > 0$ and the decrease becomes more significant as ϵ grows. Robustness increases if $\epsilon < 0$ and the increase becomes more significant as $|\epsilon|$ grows.

Let us introduce, according to Section 2.3, two characteristics important in applications. The first one is *δ-admissible dependence level* ϵ_+: if $\epsilon > \epsilon_+$ the robustness factor exceeds the given value $\delta (0 < \delta < \infty) : \kappa > \delta$. The second one is *$\delta$-admissible sample size* n_0^- (for $n_1 = \ldots = n_L = n_0$), which guarantees the given level δ for the κ factor. Using the main expansion term in (6.10) we find:

$$\epsilon_+ = \epsilon_+(\delta, \{n_i\}) = \left(\delta - \sum_{i=1}^{L} \frac{\rho_{i0}}{n_i r_0}\right)\left(\delta + \sum_{i=1}^{L} \frac{\rho_{i0}}{n_i r_0}\right)^{-1},$$

$$n_0^- = n_0^-(\delta, \epsilon) = \left\lfloor K(\epsilon) \sum_{i=1}^{L} \frac{\rho_{i0}}{\delta r_0} \right\rfloor + 1, \tag{6.11}$$

where $\lfloor z \rfloor$ denotes the floor function.

Let us illustrate these results for the multivariate Gaussian model of classes described in Section 1.4, which is often used in practice. Assume that $L = 2$ equiprobable classes ($\pi_i = 0.5$) are given, X_i has N-variate normal distribution

$N_N(\mu_i^0, \Sigma_i^0)$ with mathematical expectation vector $\mu_i^0 = (\mu_{ik}^0) \in R^N$ and covariance matrix $\Sigma_i^0 (i = 1,2; k = \overline{1,N})$. The loss matrix W has the form (1.18), so that risk r is in fact the classification error probability. A sample A_i $(i = 1,2)$ is an autoregressive sequence in R^N:

$$z_{i,t+1} = (1-\epsilon_i)\mu_i^0 + \epsilon_i z_{it} + \xi_{i,t+1} \quad (t = 1,2,\ldots), 0 \leq |\epsilon_i| < 1,$$

where z_{i1} has normal distribution $N_N(\mu_i^0, \Sigma_i^0)$ and does not depend on independent Gaussian random vectors $\xi_{i2}, \xi_{i3}, \ldots$ with distribution $N_N(\mathbf{0}, (1-\epsilon_i^2)\Sigma_i^0)$. Let us consider two situations.

1) $\Sigma_1^0 = \Sigma_2^0 = \Sigma$ is known, $\{\theta_i^0 = \mu_i^0 : i = 1,2\}$ are unknown.

In this situation the plug-in decision rule (6.2) using ML-estimators is linear:

$$d = d_1(x; A) = \mathbf{1}(b^T x + \beta) + 1, \quad x \in R^N; \quad b = \Sigma^{-1}(\hat{\mu}_2 - \hat{\mu}_1),$$

$$\beta = (\hat{\mu}_2 + \hat{\mu}_1)^T \Sigma^{-1}(\hat{\mu}_1 - \hat{\mu}_2)/2, \quad \hat{\mu}_i = n_i^{-1} \sum_{t=1}^{n_i} z_{it} \quad (i = 1,2).$$

Theorem 6.3 leads to a simple approximation formula for the robustness factor (for $\epsilon_1 = \epsilon_2 = \epsilon$):

$$\kappa(d_1; \epsilon, \{n_i\}) = \frac{(1+\epsilon)(N - 1 + \Delta^2/4)}{(1-\epsilon)4\Delta\Psi(\Delta/2)}\left(\frac{1}{n_1} + \frac{1}{n_2}\right), \tag{6.12}$$

where

$$\Psi(a) = (1 - \Phi(a))/\varphi(a)$$

is the Mills ratio for the standard normal distribution (Bolshev and Smirnov, 1983), $\Delta = \sqrt{(\theta_2^0 - \theta_1^0)^T \Sigma^{-1} (\theta_2^0 - \theta_1^0)}$ is the interclass Mahalanobis distance.

Table 6.1: Analytical and experimental values of κ

ϵ	Robustness factors	n_0					
		10	15	20	30	40	50
0	κ	0.17	0.11	0.08	0.05	0.04	0.03
	$\hat{\kappa}$	0.20	0.08	0.06	0.04	0.02	0.02
0.8	κ	1.10	0.75	0.55	0.37	0.30	0.22
	$\hat{\kappa}$	0.99	0.66	0.42	0.30	0.36	0.19

In Table 6.1 the values of the robustness factor κ evaluated by (6.12) and the estimates $\hat{\kappa}$ found by computer modeling of the plug-in decision rule $d_1(\cdot)$ are presented for $N = 2, \Delta = 2.56$ $(r_0 = 0.1)$, $n_1 = n_2 = n_0$ and for different values of ϵ, n_0.

It is evident that the accuracy of the approximation (6.12) derived with the help of asymptotic expansions is sufficiently high, and this formula can be recommended for practical usage. It follows from the table that introduction of dependence for sample elements significantly affects the risk of the plug-in decision rule (6.2).

Formula (6.11) leads to convenient formulas for δ-admissible dependence level ϵ_+ and δ-admissible sample size n_0^-:

$$\epsilon_+ = \frac{\delta - (N - 1 + \Delta^2/4)/(2\Delta n_0 \Psi(\Delta/2))}{\delta + (N - 1 + \Delta^2/4)/(2\Delta n_0 \Psi(\Delta/2))},$$

$$n_0^- = [K(\epsilon)\frac{N - 1 + \Delta^2/4}{2\Delta\delta\Psi(\Delta/2)}] + 1.$$

Figure 6.1 plots the dependence of δ-admissible dependence level ϵ_+ on interclass distance Δ for $N = 2$, $n_o = 10$, $\delta = 0.1; 0.3$.

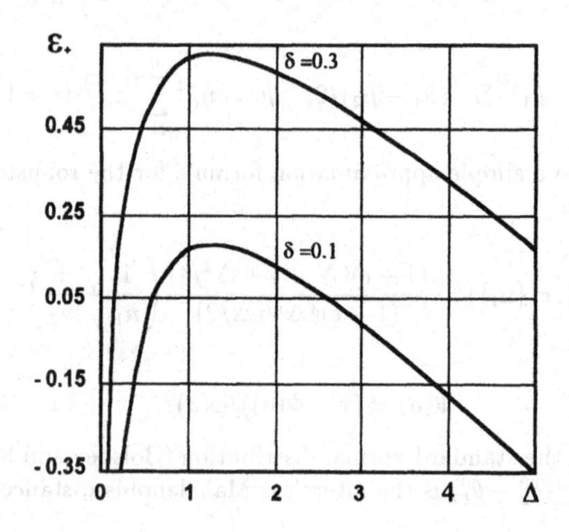

Figure 6.1: Plots of δ-admissible dependence level $\epsilon_+(\Delta, \delta)$

2) $\{\Sigma_i^0 = \mathrm{diag}\{\sigma_{i11}^0, \ldots \sigma_{iNN}^0\}, \mu_i^0 : i = 1, 2\}$ are unknown.

In this situation, the plug-in decision rule (6.2) using MC-estimators is quadratic:

$$d = d_2(x; A) = \mathbf{1}\left(\sum_{j=1}^{N}\left(\frac{(x_j - \hat{\mu}_{1j})^2}{\hat{\sigma}_{1jj}} - \frac{(x_j - \hat{\mu}_{2j})^2}{\hat{\sigma}_{2jj}} + \ln\frac{\hat{\sigma}_{1jj}}{\hat{\sigma}_{2jj}}\right)\right) + 1, x \in R^N;$$

$$\hat{\sigma}_{ijj} = n_i^{-1}\sum_{t=1}^{n_i}(z_{itj} - \hat{\mu}_{ij})^2.$$

If $\Sigma_1^0 = \Sigma_2^0 = \Sigma$, $\epsilon_1 = \epsilon_2 = \epsilon$, $n_1 = n_2 = n_0$, then Theorem 6.3 leads to the following approximation formula for the robustness factor of the plug-in decision rule $d = d_2(x; A)$:

$$\kappa(d_2; \epsilon, n_0) \approx \frac{1}{2\Delta n_0 \Psi(\Delta/2)} \left(2N - 3 + \frac{3\Delta^2}{4} + \frac{\nu}{2}(\frac{\Delta^4}{16} - \frac{3\Delta^2}{2} + 3) + \right.$$

$$\left. +2(N - 1 + \frac{\Delta^2}{4})G(\epsilon) + 2(N - 2 + \frac{\Delta^2}{2} + \frac{\nu}{2}(\frac{\Delta^4}{16} - \frac{3\Delta^2}{2} + 3))G(\epsilon^2) \right),$$

where

$$\nu = \Delta^{-4} \sum_{j=1}^{N} (\mu_{2j}^0 - \mu_{1j}^0)^4 / (\sigma_{jj}^0)^2, G(\epsilon) = \epsilon/(1-\epsilon).$$

6.2 Robustness under Misclassification Errors in Training Sample

In Chapter 3, the problem of adaptive pattern recognition with minimal risk was considered in the situations where conditional probability distributions of feature variables are known up to some parameters and a training sequence of observations A is given such that for each observation x_t the true class number $\nu_t^0 \in S$ is exactly known. Such sequence of observations is called *classified training sample.* Pattern recognition was performed by the plug-in decision rule (PDR), which is derived from the Bayesian decision rule by substitution of MC-estimators (in particular, ML-estimators) for unknown true values of parameters. For this plug-in decision rule, based on the classified training sample, consistency was proved, and asymptotic expansions for risk were constructed. These expansions allow to investigate the robustness of the plug-in decision rule with respect to small-sample effects.

The listed results were obtained under the assumption that A is classified error-free, i.e., the class numbers $\{\nu_t^0 : \quad t = 1, \ldots, n\}$ are known certainly. In practice this assumption is often violated for the following reasons:

a) "direct recording" of class number is not always available;

b) methods of registration of class numbers $\{\nu_t^0 : t = 1, \ldots, n\}$are not perfect;

c) during creation of data archives (and then of the training sample A) some errors of type "class numbers are confused" are introduced.

Therefore in practice a training sample is in fact "contaminated" by the observations from alternative classes: the sample of observations of i-th class (for which $\nu^0 = i$ is assumed) in fact contains $\epsilon_i 100\%$ observations from other classes $(0 \leq \epsilon_i < 1)$. In other words, the distortions of type D.3.3 described in Section 2.2 are often present in practice (Aitkin *et al.*, 1980).

For this reason the following three problems of robust pattern recognition are very topical:

1) to investigate the influence of training sample misclassifications on PDR risk;

2) to construct and investigate new decision rules robust with respect to small-rate misclassifications in training sample;

3) to find critical values of contamination levels $\{\epsilon_i\}$, under which significant gain in risk for the robust decision rule is achieved with respect to the traditionally used plug-in Bayesian decision rule.

Note that the problem 1) is considered by (Lachenbruch, 1966) for the simple case when there are two classes only and the decision rule is linear. In (Pugachev, 1967), (Agrawala, 1970), (Shanmugam, 1972) the problem of "Bayesian learning with imperfect (probabilistic) teacher", close to the problem 2), is considered under the assumption that the parameters (of conditional probability distributions of observations) are random and have known density. Practical applications of these methods (Pugachev, 1967), (Agrawala, 1970), (Shanmugam, 1972) are limited by well-known computer difficulties of Bayesian estimation of parameters (Hartigan, 1983). In (Greblicki, 1980) nonparametric Rosenblatt–Parzen density estimators are used to solve the problem 2). This approach, unfortunately, loses significant information about the parametric form of the densities.

In this section, in order to solve the listed problems 1)–3), we use the method of asymptotic expansion of risk in powers of $\{\epsilon_i\}$.

Thus, assume that in the feature space R^N random observations from L classes $\Omega_1, \ldots, \Omega_L$ are recorded with given probabilities $\pi_1, \ldots, \pi_L$ $(\pi_1 + \ldots + \pi_L = 1,$ $\pi_i > 0, i \in S)$. An observation from Ω_i is a random vector $X_i \in R^N$ with probability density function $p(x; \theta_i^0)$, $\quad x \in R^N, \quad \theta_i^0 \in \Theta \subseteq R^m \quad (i \in S)$. True values $\{\theta_i^0\}$ of the parameters are unknown. A training sample A that consists of L independent subsamples $A_1, \ldots, A_L$ is recorded:

$$A^T = (A_1^T \vdots \ldots \vdots A_L^T), \quad A_i^T = (z_{i1}^T \vdots \ldots \vdots z_{in_i}^T)$$

is a sample of n_i independent observations such that an observation $z_{ij} \in R^N$ belongs (in fact) to class Ω_k with probability $\epsilon_{ik} \geq 0 \quad (j = 1, \ldots, n_i); \quad \epsilon_{i1} + \ldots + \epsilon_{iL} = 1 \quad (i, k \in S)$. Here the value $\epsilon_i = 1 - \epsilon_{ii}$ is the misclassification error, which characterizes the level of contamination of the subsample A_i by the observations from "alien" classes $\{\Omega_k : k \neq i\}$. If the matrix $E = (\epsilon_{ik}) = \mathbf{I}_L$ is the identity matrix, then contaminations in the sample A are absent. The classification loss matrix $W = (w_{ij})$ is given.

For the model formulated in this way, let us consider the problems 1)-3). It is known from Section 3.1 that for given true parameters $\{\theta_i^0\}$ the Bayesian decision rule (3.1), (3.2) ensures minimal risk r_0 defined by (3.3). As $\{\theta_i^0\}$ are unknown, we shall consider the plug-in decision rule (3.4), (3.5) using MC-estimators (3.8) for $\{\theta_i^0\}$ with contrast functions $\{g(x; \theta_i)\}$.

Let us introduce the notations: $g^{(k)}(x; \theta_i')$ is the set of k-th order partial derivatives of $g(x; \theta_i)$ with respect to θ_i at point $\theta_i' \quad (g^{(0)}(\cdot) \equiv g(\cdot))$; the notation $p^{(k)}(x; \theta_i')$ is similar;

$$G_{ij}^{(k)} = G_{ij}^{(k)}(\theta_i; \theta_j^0) = \mathbf{E}_{\theta_j^0}\{g^{(k)}(X_j; \theta_i)\} \quad (k = 0, 1, 2, 3);$$

$$G_i^{(k)}(\theta_i) = \sum_{j=1}^{L} \epsilon_{ij} G_{ij}^{(k)}; \quad G_i^{(0)}(\theta_i) = G_i(\theta_i); \tag{6.13}$$

$$J_{ij} = J_{ij}(\theta_i, \theta_j^0) = \mathbf{E}_{\theta_j^0}\{g^{(1)}(X_j; \theta_i) g^{(1)T}(X_j; \theta_i)\} \quad (i, j =\in S).$$

For $\theta_i = \theta_i^0$ the values in (6.13) will be denoted by $\overset{\circ}{G}{}_{ij}^{(k)}, \overset{\circ}{G}{}_i^{(k)}, \overset{\circ}{g}{}_i^{(k)}, \overset{\circ}{P}{}_i^{(k)}, \overset{\circ}{J}_{ij}$. Note that if the Chibisov regularity conditions (Section 1.6) are fulfilled, then

$$\overset{\circ}{G}{}_{ii}^{(1)} \equiv 0_m, \quad \overset{\circ}{G}{}_{ii}^{(2)} \succ 0 \tag{6.14}$$

is a positive definite $m \times m$-matrix. Moreover, if

$$g(x; \theta_i) = -\ln p_i(x; \theta_i), \tag{6.15}$$

then $\overset{\circ}{G}{}_{ii}^{(2)} = \overset{\circ}{J}_{ii}$ is the Fisher information matrix for class Ω_i (such choice of the contrast function means that $\hat{\theta}_i$ is the ML-estimator).

Theorem 6.4 *If probability densities $\{p(x; \theta_i)\}$ are triply continuously differentiable with respect to $\{\theta_i\}$, satisfy the Chibisov regularity conditions, the mathematical expectations in (6.13) are finite and the minimum point*

$$\theta_i^* = \arg \min_{\theta_i \in \bar{\Theta}_i} G_i(\theta_i) \tag{6.16}$$

is unique, then under type D.3.3 distortions the following almost sure convergence takes place:

$$\hat{\theta}_i \xrightarrow{a.s.} \theta_i^*; \tag{6.17}$$

as $n_i \to \infty$ and θ_i^* admits the following asymptotic expansion ($i \in S$):

$$\theta_i^* = \theta_i^0 - (\overset{\circ}{G}{}_{ii}^{(2)})^{-1} \sum_{j=1, j \neq i}^{L} \epsilon_{ij} \overset{\circ}{G}{}_{ij}^{(1)} + o(\epsilon_i^2) \mathbf{1}_m. \tag{6.18}$$

Proof. By conditions of the theorem, the probability density function of an observation z_{ij} is the mixture of the densities $\{p(x; \theta_l^0)\}$ with discrete mixing probability distribution $(\epsilon_{i1}, \ldots, \epsilon_{iL})$:

$$q_i(x; \{\theta_i^0\}) = \sum_{l=1}^{L} \epsilon_{il} p(x; \theta_l^0) \quad (j = 1, \ldots, n_i, i \in S). \tag{6.19}$$

Therefore under the notations (6.13),

$$\mathbf{E}\{g(z_{ij}; \theta_i)\} = G_i(\theta_i).$$

The strong law of large numbers implies the following asymptotic behavior of the objective function for the estimator $\hat{\theta}_i$:

$$L_i(\theta_i) \stackrel{a.s.}{\to} G_i(\theta_i), \theta_i \in \Theta.$$

Further, (6.17) is proved by this result in the same way as the consistency of ML-estimators (see, for example, (Borovkov, 1987), (Ibragimov, 1979), (Le Cam, 1990)).

Let us prove now the correctness of the asymptotic expansion (6.18). A necessary condition for (6.16) is

$$G_i^{(1)}(\theta_i^*) = \mathbf{0}_m. \tag{6.20}$$

Apply the Taylor formula to the left side of (6.20) in the neighborhood of θ_i^0 under the notations (6.13):

$$\theta_i^* - \theta_i^0 = -(\overset{\circ}{G}{}_i^{(2)})^{-1} \overset{\circ}{G}{}_i^{(1)} + \mathcal{O}(|\theta_i^* - \theta_i^0|^2)\mathbf{1}_m. \tag{6.21}$$

From (6.13), (6.14) and the equality $\epsilon_{ii} = 1 - \epsilon_i$, we find:

$$\overset{\circ}{G}{}_i^{(1)} = \sum_{j=1, j\neq i}^{L} \epsilon_{ij} \overset{\circ}{G}{}_{ij}^{(1)} = o(\epsilon_i)\mathbf{1}_m, \tag{6.22}$$

$$(\overset{\circ}{G}{}_i^{(2)})^{-1} = (\overset{\circ}{G}{}_{ii}^{(2)} + (\sum_{j\neq i} \epsilon_{ij} \overset{\circ}{G}{}_{ij}^{(2)} - \epsilon_i \overset{\circ}{G}{}_{ii}^{(2)}))^{-1} =$$

$$= (\overset{\circ}{G}{}_{ii}^{(2)})^{-1} - \sum_{j\neq i} \epsilon_{ij} (\overset{\circ}{G}{}_{ii}^{(2)})^{-1} \overset{\circ}{G}{}_{ij}^{(2)} (\overset{\circ}{G}{}_{ii}^{(2)})^{-1} + \epsilon_i (\overset{\circ}{G}{}_{ii}^{(2)})^{-1} + \mathcal{O}(\epsilon_i^2)\mathbf{1}_{m\times m}. \tag{6.23}$$

Then (6.21) implies: $\theta_i^* - \theta_i^0 = \mathcal{O}(\epsilon_i)\mathbf{1}_m$. Substituting (6.22) into (6.21) and keeping the main terms only up to the order $\mathcal{O}(\epsilon_i)$ we obtain (6.18).

■

Theorem 6.4 implies that if the sample A contains misclassifications, then the estimators $\{\theta_i\}$ defined by (3.7), (3.8) lose their consistency; a systematic error is introduced (see (6.18)). In addition, the finiteness of sample sizes $\{n_i\}$ leads to random error. These two components determine the risk of the plug-in decision rule (3.4) (it is convenient to denote it by $d = d_1(x; A)$):

$$r(d_1) = \sum_{i=1}^{L} \pi_i \mathbf{E}\{w_{i,d_1(X_i;A)}\}. \tag{6.24}$$

Denote:

$$\Delta\theta_i = (\Delta\theta_{is}) = \hat{\theta}_i - \theta_i^0$$

is the deviation of the estimator $(s = 1, \ldots, m)$;

$$\tau_i^2 = \max\{\epsilon_i^2, (1+\epsilon_i)/n_i\}, \tau_* = \max_{i \in S} \tau_i; \tag{6.25}$$

$$b_{ij}(x) = 1/ \mid \nabla_x \overset{\circ}{f}_{ij}(x) \mid; \quad \overset{\circ}{\Gamma}_{ij} = \{x : \overset{\circ}{f}_{ij}(x) = 0\}$$

is the $(N-1)$-dimensional Bayesian discriminant surface for pair of the classes $\{\Omega_i, \Omega_j\}$; $\overset{\circ}{\Gamma}{}'_{ij} \subset \overset{\circ}{\Gamma}_{ij}$ is the "part" of $\overset{\circ}{\Gamma}_{ij}$ which is the boundary of the Bayesian region V_j^0; $\Gamma_{tj} = \{x : f_{tj}(x; \{\theta_l\}) = 0\}$ is a hypersurface depending on parameters $\{\theta_l\}$;

$$f_{tj}(x; \{\theta_l\}) = \sum_{l=1}^{L} c_{ltj} p(x; \{\theta_l\}), \; c_{ltj} = \pi_l(w_{lt} - w_{lj});$$

$p_{f_{tjl}, f_{t'jl}, f_{t''jl}}(z_1, z_2, z_3)$ is the joint probability density of the statistics

$$f_{tjl} = f_{tj}(X_l; \{\theta_l\}), \quad f_{t'jl} = f_{t'j}(X_l; \{\theta_l\}), \quad f_{t''jl} = f_{t''j}(X_l; \{\theta_l\}),$$

where $j \neq t \neq t' \neq t'' \quad (j, l, t, t', t'' \in S)$.

Corollary 6.4 *Under the conditions of Theorem 6.4, if*

$$\tau_* \to 0 \;\; (\epsilon_i \to 0, n_i \to \infty : i \in S),$$

then the estimators $\{\hat{\theta}_i\}$ *are strongly consistent.*

Let us construct an asymptotic expansion of the risk $r(d_1)$.

Theorem 6.5 *Assume that the Chibisov regularity conditions are fulfilled and for some neighborhoods* $U_1, \ldots, U_L \subset \Theta$ *of the points* $\theta_1^0, \ldots, \theta_L^0$ *respectively the partial derivatives with respect to* $\{\theta_i\}$ $p'(x; \theta_i), p''(x; \theta_i), p'''(x; \theta_i)$ *are uniformly bounded on surfaces* $\Gamma_{tj} \subset R^N \quad (\theta_i \in U_i, x \in \Gamma_{tj}; i, t, j \in S)$, *so that* $\{p(x; \theta_i)\}$ *are triply continuously differentiable,* $\{p'(x; \theta_i)\}$ *are twice continuously differentiable and* $\{p'''(x; \theta_i)\}$ *are differentiable with respect to* x. *Suppose that for any* $\{\theta_i \in U_i : i \in S\}$, $j, l, t, t', t'' \in S, \quad (j \neq t \neq t' \neq t'')$ *the following values are bounded:*

$$p_{f_{tjl}}(0) < \infty, \quad p_{f_{tjl}, f_{t'jl}}(0,0) < \infty, \tag{6.26}$$

$$p_{f_{tjl}, f_{t'jl}, f_{t''jl}}(0,0,0) < \infty.$$

If $\{\hat{\theta}_i\}$ *are strongly consistent (as* $\tau_* \to 0$*) estimators such that*

$$\mathbf{E}\{|\Delta\theta_{ij}\Delta\theta_{kl}\Delta\theta_{st}|\} = o(\tau_*^2) \quad (i, k, s \in S, \; j, l, t \in \{1, \ldots, m\}), \tag{6.27}$$

$$\mathbf{E}\{\Delta\theta_i(\Delta\theta_k^T)\} = V_{ik} + o(\tau_*^2)\mathbf{1}_{m \times m}, \; V_{ik} = (v_{ik\mu\nu}), \; \mu = 1, \ldots, m, \; \nu = 1, \ldots, m,$$

then the risk functional (6.24) of the plug-in decision rule using the estimators $\{\hat{\theta}_i\}$ *admits the following asymptotic expansion:*

$$r(d_1) = r_0 + \sum_{i,k=1}^{L} \sum_{j=2}^{L} \sum_{l=1}^{j-1} c_{ilj} c_{klj} \int_{\overset{\circ}{\Gamma}{}'_{lj}} \overset{\circ}{p}{}_i^{(1)T} V_{ik} \overset{\circ}{p}{}_k^{(1)} b_{lj}(x) ds_{N-1}/2 + o(\tau_*^2). \quad (6.28)$$

Proof. Let us express the unconditional risk $r(d_1)$ in terms of the conditional one:

$$r(d_1) = \mathbf{E}\{\mathrm{rc}(A)\}, \quad (6.29)$$

where $\mathrm{rc}(A)$ is the conditional risk (3.10) of the plug-in decision rule $d_1(\cdot)$ for fixed sample A; $\mathbf{E}\{\cdot\}$ denotes mathematical expectation with respect to the distribution of A. Let us use the stochastic expansion of the conditional risk given by Theorem 3.1. This result can be used under the conditions of Theorem 6.5 if instead of the asymptotics $n_* \to 0$ (in Theorem 3.1) we consider the asymptotics $\tau_* \to \infty$ (in Theorem 6.5). Therefore, using (3.17), (3.19) we may write the stochastic expansion of the conditional risk as follows:

$$\mathrm{rc}(A) = r_0 + \sum_{i,k=1}^{L} (\Delta\theta_i)^T \alpha_{ik} \Delta\theta_k + \zeta_{\tau_0}, \quad (6.30)$$

where the $(m \times m)$-matrix α_{ik} is defined by the expression:

$$\alpha_{ik} = \frac{1}{2} \sum_{j=2}^{L} \sum_{l=1}^{j-1} c_{ilj} c_{klj} \int_{\overset{\circ}{\Gamma}{}'_{lj}} \overset{\circ}{p}{}_i^{(1)} (\overset{\circ}{p}{}_k^{(1)})^T b_{lj}(x) ds_{N-1}; \quad (6.31)$$

$$\zeta_{\tau_0} = \sum_{i,\mu,p=1}^{L} \sum_{s=1}^{m} \sum_{t=1}^{m} \sum_{q=1}^{m} \zeta_{is\mu tpq} \Delta\theta_{is} \Delta\theta_{\mu t} \Delta\theta_{pq}; \quad (6.32)$$

$\zeta_{is\mu tpq}$ is a bounded random variable.

It follows from (6.32), (6.27) that $\mathbf{E}\{|\zeta_{\tau_0}|\} = o(\tau_*^2)$. From (6.31), (6.27) we have (by the properties of $\mathrm{tr}(\cdot)$ operation):

$$\mathbf{E}\{(\Delta\theta_i)^T \alpha_{ik} \Delta\theta_k\} = \mathrm{tr}\{\alpha_{ik} V_{ik}\} + o(\tau_*^2) =$$

$$= \frac{1}{2} \sum_{j=2}^{L} \sum_{l=1}^{j-1} c_{ilj} c_{klj} \int_{\overset{\circ}{\Gamma}{}'_{lj}} \overset{\circ}{p}{}_i^{(1)T} V_{ik} \overset{\circ}{p}{}_k^{(1)} b_{lj}(x) ds_{N-1} + o(\tau_*^2).$$

Substituting now (6.30) into (6.29) and using the found relations, we arrive at (6.28). ■

In order to apply Theorem 6.5 to MC-estimators $\{\hat{\theta}_i\}$, we have to obtain the expressions (6.27) for these estimators.

Define the following $(m \times m)$-matrices D_{ij}, T_{ij} $(i, j \in S)$:

$$D_{ij} = \mathbf{E}\{g^{(1)}(X_i; \theta_i^0) \overset{\circ}{G}{}_{ij}^{(2)} (\overset{\circ}{G}{}_{ii}^{(2)})^{-1} g^{(2)}(X_i; \theta_i^0)\}, \quad (6.33)$$

$$T_{ij} = (t_{ijpq}), \quad t_{ijpq} = \sum_{k=1}^{m} (\overset{\circ}{G}{}^{(3)}_{ii})_{kpq} ((\overset{\circ}{G}{}^{(2)}_{ii})^{-1} \overset{\circ}{G}{}^{(1)}_{ij})_k \quad (p, q = 1, \ldots, m). \tag{6.34}$$

Lemma 6.1 *Let the conditions of Theorem 6.4 be fulfilled and let the elements of the matrices $\{D_{ij} : i \neq j \in S\}$ be finite. Then:*

$$\mathbf{E}\{\hat{\theta}_i - \theta_i^*\} = \mathcal{O}(\tau_*^2)\mathbf{1}_m, \tag{6.35}$$

$$\Xi_i = \mathbf{E}\{(\hat{\theta}_i - \theta_i^*)(\hat{\theta}_i - \theta_i^*)^T\} = n_i^{-1} (\overset{\circ}{G}{}^{(2)}_{ii})^{-1} \overset{\circ}{J}_{ii} (\overset{\circ}{G}{}^{(2)}_{ii})^{-1} +$$

$$+ \sum_{j=1, j\neq i}^{L} \epsilon_{ij} (\overset{\circ}{G}{}^{(2)}_{ii})^{-1} (\overset{\circ}{J}_{ii} + \overset{\circ}{J}_{ij} + 2T_{ij} (\overset{\circ}{G}{}^{(2)}_{ii})^{-1} \overset{\circ}{J}_{ii} - 2 \overset{\circ}{G}{}^{(2)}_{ij} (\overset{\circ}{G}{}^{(2)}_{ii})^{-1} \overset{\circ}{J}_{ii} -$$

$$- D_{ij} - D_{ij}^T)(\overset{\circ}{G}{}^{(2)}_{ii})^{-1}/n_i + o(\tau_*^2)\mathbf{1}_{m\times m}; \tag{6.36}$$

$$\mathbf{E}\left\{\left|(\hat{\theta}_{is} - \theta_{is}^*)(\hat{\theta}_{it} - \theta_{it}^*)(\hat{\theta}_{iq} - \theta_{iq}^*)\right|\right\} = o(\tau_*^2). \tag{6.37}$$

Proof. By Theorem 6.4, $\hat{\theta}_i \xrightarrow{a.s.} \theta_i^*$. Therefore, for the left side of the equation

$$\nabla_{\hat{\theta}_i} L_i(\hat{\theta}_i) = \mathbf{0}_m,$$

that determines the MC-estimator $\hat{\theta}_i$ (see (3.8)), we may apply the Taylor formula in the neighborhood of the limit point θ_i^*:

$$\hat{\theta}_i - \theta_i^* = -\left(\nabla^2_{\bar{\theta}_i} L_i(\bar{\theta}_i)\right)^{-1} \nabla_{\theta_i^*} L_i(\theta_i^*), \tag{6.38}$$

$|\ \bar{\theta}_i - \theta_i^*\ | < |\ \hat{\theta}_i - \theta_i^*\ |$. Since $\hat{\theta}_i \xrightarrow{a.s.} \theta_i^*$, it follows from the last inequality that $\bar{\theta}_i \xrightarrow{a.s.} \theta_i^*$. Because of (3.7), (6.13), and the strong law of large numbers, the following convergence takes place:

$$\left(\nabla^2_{\bar{\theta}_i} L_i(\bar{\theta}_i)\right)^{-1} \xrightarrow{a.s.} (G_i^{(2)}(\theta_i^*))^{-1}. \tag{6.39}$$

Let us prove now the expansion (6.36). From (6.38), (6.39), (3.7) and (6.13) we have (as $n_i \to \infty$) :

$$\Xi_i = (G_i^{(2)}(\theta_i^*))^{-1} \mathbf{E}\{\nabla_{\theta_i^*} L_i(\theta_i^*)(\nabla_{\theta_i^*} L_i(\theta_i^*))^T\} (G_i^{(2)}(\theta_i^*))^{-1} =$$

$$= \frac{1}{n_i} (G_i^{(2)}(\theta_i^*))^{-1} \Big(\sum_{k=1}^{L} \epsilon_{ik} J_{ik}(\theta_i^*, \theta_k^0)\Big) (G_i^{(2)}(\theta_i^*))^{-1}. \tag{6.40}$$

Using (6.13), (6.18), and the Taylor formula:

$$g^{(1)}(x;\theta_i^*) = g^{(1)}(x;\theta_i^0) + g^{(2)}(x;\theta_i^0)(\theta_i^* - \theta_i^0) + \mathcal{O}(\epsilon_i^2)\mathbf{1}_m,$$

we obtain

$$J_{ik}(\theta_i^*, \theta_k^0) = \overset{\circ}{J}_{ik} + \mathcal{O}(\epsilon_i)\mathbf{1}_{m\times m} \quad (i,k \in S, k \neq i); \tag{6.41}$$

$$J_{ii}(\theta_i^*, \theta_i^0) = \overset{\circ}{J}_{ii} + \mathbf{E}\{g^{(2)}(X_i;\theta_i^0)(\theta_i^* - \theta_i^0)g^{(1)T}(X_i;\theta_i^0)\} +$$

$$\mathbf{E}\{g^{(1)}(X_i;\theta_i^0)(\theta_i^* - \theta_i^0)^T g^{(2)}(X_i;\theta_i^0)\} + \mathcal{O}(\epsilon_i^2)\mathbf{1}_{m\times m},$$

$$G_i^{(2)}(\theta_i^*) = \overset{\circ}{G}{}_i^{(2)} + (\sum_{t=1}^{m}(\overset{\circ}{G}{}_{ii}^{(3)})_{tpq}(\theta_{it}^* - \theta_{it}^0))_{pq} + \mathcal{O}(\epsilon_i^2)\mathbf{1}_{m\times m} \quad (p,q = 1,\dots,m).$$

Substituting $\theta_i^* - \theta_i^0 \in R^m$ from (6.18)into (6.41), we find by (6.13), (6.14), (6.33), (6.34):

$$J_{ik}(\theta_i^*, \theta_k^0) = \begin{cases} \overset{\circ}{J}_{ik} + \mathcal{O}(\epsilon_i)\mathbf{1}_{m\times m}, & k \neq i, \\ \overset{\circ}{J}_{ii} - \sum_{j=1, j\neq i}^{L} \epsilon_{ij}(D_{ij} + D_{ij}^T) + \mathcal{O}(\epsilon_i^2)\mathbf{1}_{m\times m}, & k = i, \end{cases}$$

$$G_i^{(2)}(\theta_i^*) = \overset{\circ}{G}{}_{ii}^{(2)} + \sum_{j=1, j\neq i}^{L} \epsilon_{ij}(\overset{\circ}{G}{}_{ij}^{(2)} - \overset{\circ}{G}{}_{ii}^{(2)} - T_{ij}) + \mathcal{O}(\epsilon_i^2)\mathbf{1}_{m\times m},$$

$$(G_i^{(2)}(\theta_i^*))^{-1} = (\overset{\circ}{G}{}_{ii}^{(2)})^{-1} -$$

$$-\sum_{j=1, j\neq i}^{L} \epsilon_{ij}(\overset{\circ}{G}{}_{ii}^{(2)})^{-1} - (\overset{\circ}{G}{}_{ij}^{(2)} - \overset{\circ}{G}{}_{ii}^{(2)} - T_{ij})(\overset{\circ}{G}{}_{ii}^{(2)})^{-1} + \mathcal{O}(\epsilon_i^2)\mathbf{1}_{m\times m}.$$

Inserting these expressions into (6.40) results in (6.36). The relations (6.35), (6.37) are proved in the same way.

■

Define the following linear combinations of surface integrals ($i, k \in S$):

$$\gamma(E) = \frac{1}{2}\sum_{j=2}^{L}\sum_{l=1}^{j-1}\int_{\overset{\circ}{\Gamma}{}'_{lj}} \left(\sum_{i=1}^{L} c_{ilj} \sum_{k=1, k\neq i}^{L} \epsilon_{ik} \overset{\circ}{P}{}_i^{(1)T} (\overset{\circ}{G}{}_{ii}^{(2)})^{-1} \overset{\circ}{G}{}_{ik}^{(1)}\right)^2 \times$$

$$\times b_{lj}(x) ds_{N-1} \geq 0,$$

$$\alpha_i = \frac{1}{2}\sum_{j=2}^{L}\sum_{l=1}^{j-1} c_{ilj}^2 \int_{\overset{\circ}{\Gamma}{}'_{lj}} \overset{\circ}{P}{}_i^{(1)T} (\overset{\circ}{G}{}_{ii}^{(2)})^{-1} \overset{\circ}{J}_{ii} (\overset{\circ}{G}{}_{ii}^{(2)})^{-1} \overset{\circ}{P}{}_i^{(1)} b_{lj}(x) ds_{N-1} \geq 0,$$

$$\beta_{ik} = \frac{1}{2}\sum_{j=2}^{L}\sum_{l=1}^{j-1} c_{ilj}^2 \int_{\overset{\circ}{\Gamma}_{lj}'} \overset{\circ}{p}_i^{(1)T} (\overset{\circ}{G}_{ii}^{(2)})^{-1} \times$$

$$\times \left(\overset{\circ}{J}_{ii} + \overset{\circ}{J}_{ik} + 2(T_{ik} - \overset{\circ}{G}_{ik}^{(2)})(\overset{\circ}{G}_{ii}^{(2)})^{-1} \overset{\circ}{J}_{ii} - D_{ik} - D_{ik}^T \right) \times \tag{6.42}$$

$$\times (\overset{\circ}{G}_{ii}^{(2)})^{-1} \overset{\circ}{p}_l^{(1)} b_{lj}(x) ds_{N-1} \geq 0.$$

Theorem 6.6 *Let the conditions of Theorem 6.5 be fulfilled, the mathematical expectations in (6.13) and the elements of matrices $\{D_{ij} : i \neq j \in S\}$ be finite, and for each $i \in S$ the point θ_i^* determined by (6.16) be unique. Then the risk of the plug-in decision rule using MC-estimators under the contaminations of the training sample by observations from "alien" classes admits the following expansion:*

$$r(d_1) = r_0 + \gamma(E) + \sum_{i=1}^{L} \alpha_i/n_i + \sum_{i=1}^{L} \sum_{k=1,k\neq i}^{L} \beta_{ik}\epsilon_{ik}/n_i + o(\tau_*^2). \tag{6.43}$$

Proof. Theorem 6.4, Lemma 6.1, the independence of random subsamples $A_1, \ldots, A_L$ (independence of $\hat{\theta}_1, \ldots, \hat{\theta}_L$), and (6.27) together imply:

$$V_{ik} = \delta_{ik}\Xi_i + (\theta_i^* - \theta_i^0)(\theta_k^* - \theta_k^0)^T =$$

$$= \delta_{ik}\Xi_i + \sum_{s=1,s\neq i}^{L} \sum_{t=1,t\neq k}^{L} \epsilon_{is}\epsilon_{kt}(\overset{\circ}{G}_{ii}^{(2)})^{-1} \overset{\circ}{G}_{is}^{(1)} \overset{\circ}{G}_{kt}^{(1)T} (\overset{\circ}{G}_{kk}^{(2)})^{-1} + o(\tau_*^2)\mathbf{1}_{m\times m}. \tag{6.44}$$

The conditions of Theorem 6.5 are fulfilled. Substituting the expressions (6.44), (6.36) into (6.28) and using the notations (6.42), we obtain (6.43).

■

Formula (6.43) determines the expansion of risk in powers of "small" parameters: $\{\epsilon_{ik}, 1/n_i\}$. The terms $\{\alpha_i n_i^{-1}\}$ are induced only by random errors of the estimators $\{\hat{\theta}_i\}$, the quadratic form $\gamma(E)$ in elements of matrix $E = (\epsilon_{ik})$ is induced only by systematic errors, and $\{\beta_{ik}\epsilon_{ik}/n_i\}$ is the result of their joint influence on risk. Note that if $E = \mathbf{I}_L$ is the identity matrix (i. e. if the samples are without any contaminations, $\epsilon_i \equiv 0$), then the expansion (6.43) coincides with the expansion (3.33) - (3.35), constructed earlier for this particular case.

Consider some other particular cases of the expansion (6.43).

Corollary 6.5 *If $\hat{\theta}_i$ is the ML-estimator based on the subsample A_i $(i \in S)$, then the coefficients of risk expansion (6.43) are determined by the following formulas $(i, k \in S)$:*

$$\gamma(E) = \frac{1}{2}\sum_{j=2}^{L}\sum_{l=1}^{j-1}\int_{\overset{\circ}{\Gamma}{}'_{lj}}\left(\sum_{i=1}^{L} c_{ilj}\sum_{k=1,k\neq i}^{L}\epsilon_{ik}\overset{\circ}{P}{}_i^{(1)T}\overset{\circ}{J}{}_{ii}^{-1}\overset{\circ}{G}{}_{ik}^{(1)}\right)^2 b_{lj}(x)ds_{N-1},$$

$$\alpha_i = \frac{1}{2}\sum_{j=2}^{L}\sum_{l=1}^{j-1} c_{ilj}^2\int_{\overset{\circ}{\Gamma}{}'_{lj}}\overset{\circ}{P}{}_i^{(1)T}\overset{\circ}{J}{}_{ii}^{-1}\overset{\circ}{P}{}_i^{(1)} b_{lj}(x)ds_{N-1},$$

$$\beta_{ik} = \frac{1}{2}\sum_{j=2}^{L}\sum_{l=1}^{j-1} c_{ilj}^2\int_{\overset{\circ}{\Gamma}{}'_{lj}}\overset{\circ}{P}{}_i^{(1)T}\overset{\circ}{J}{}_{ii}^{-1}\times$$

$$\times\left(\overset{\circ}{J}_{ii}+\overset{\circ}{J}_{ik}+2(T_{ik}-\overset{\circ}{G}{}_{ik}^{(2)})-D_{ik}-D_{ik}^T\right)\overset{\circ}{J}{}_{ii}^{-1}\overset{\circ}{P}{}_i^{(1)} b_{lj}(x)ds_{N-1}.$$

Proof. It is sufficient to choose the contrast function for ML-estimators $\{\hat{\theta}_i\}$ in the form (6.15) and to take this fact into account in (6.42). ■

Corollary 6.6 *If there are only two equiprobable classes ($L = 2, \pi_i = 0.5$) and the loss matrix is the $(0-1)$-matrix (1.18), then under conditions of Corollary 6.5 the expansion coefficients for error probability $r(d_1)$ are as follows:*

$$\gamma(E) = \gamma(\epsilon_1, \epsilon_2) =$$

$$= \frac{1}{8}\int_{\overset{\circ}{\Gamma}_{12}}(\epsilon_2\overset{\circ}{P}{}_2^{(1)T}\overset{\circ}{J}{}_{22}^{-1}\overset{\circ}{G}{}_{21}^{(1)} - \epsilon_1\overset{\circ}{P}{}_1^{(1)T}\overset{\circ}{J}{}_{11}^{-1}\overset{\circ}{G}{}_{12}^{(1)})^2 b_{12}(x)ds_{N-1},$$

$$\alpha_i = \frac{1}{8}\int_{\overset{\circ}{\Gamma}_{12}}\overset{\circ}{P}{}_i^{(1)T}\overset{\circ}{J}{}_{ii}^{-1}\overset{\circ}{P}{}_i^{(1)} b_{12}(x)ds_{N-1}, \tag{6.45}$$

$$\beta_{ik} = \alpha_i + \frac{1}{8}\int_{\overset{\circ}{\Gamma}_{12}}\overset{\circ}{P}{}_i^{(1)T}\overset{\circ}{J}{}_{ii}^{-1}\left(\overset{\circ}{J}_{ik}+2(T_{ik}-\overset{\circ}{G}{}_{ik}^{(2)})-D_{ik}-D_{ik}^T\right)\times$$

$$\times\overset{\circ}{J}{}_{ii}^{-1}\overset{\circ}{P}{}_i^{(1)} b_{12}(x)ds_{N-1}.$$

Corollary 6.7 *Under the conditions of Theorem 6.6 the following approximations for risk $r(d_1)$ and robustness factor $\kappa(d_1)$ of the plug-in decision rule are valid up to the remainder $o(\tau_*^2)$:*

$$r(d_1) \approx r_1 = r_0 + \gamma(E) + \sum_{i=1}^{L}(\alpha_i/n_i + \sum_{k=1,k\neq i}^{L}\beta_{ik}\epsilon_{ik}/n_i),$$

$$\kappa(d_1) \approx \kappa_1 = (\gamma(E) + \sum_{i}^{L}(\alpha_i/n_i + \sum_{k=1,k\neq i}^{L}\beta_{ik}\epsilon_{ik}/n_i))/r_0. \tag{6.46}$$

Only L^2+2 main terms are evaluated in the expansion (6.43). In the same way, one can obtain subsequent terms and increase the accuracy of the approximate formulas (6.46).

Consider now the problem of robust decision rule synthesis. Here we shall use the asymptotic robustness factor (2.25):

$$\kappa_+(d;E) = \lim_{n_1,n_2,\ldots,n_L\to\infty} \kappa(d) \tag{6.47}$$

and the robustness order $\nu_0(d)$ defined by (2.26).

Corollary 6.8 *Under the conditions of Theorem 6.6 the plug-in decision rule $d = d_1(x;A)$ using ML-estimators is a robust decision rule of first order.*

Proof. By Theorem 6.5 and (6.46), (6.47), (6.25) we may evaluate the asymptotic robustness factor of the decision rule $d_1(\cdot)$:

$$\kappa_+(d_1;E) = \gamma(E)/r_0 + o(\epsilon_+^2).$$

From this fact by the definition (2.26) we find the plug-in decision rule robustness order $\nu_0(d_1)=1$.

■

Consider now the problem of robustness order increase for the decision rule $d_1(\cdot)$. It is seen from Theorems 6.4, 6.6 and definition (2.26) that the robustness order can be increased by decreasing the systematic error $\theta_i^* - \theta_i^0$ of the estimation of θ_i^0 $(i \in S)$. Let us investigate two ways of decreasing this error.

The first of them is functional transformation of the statistics $\{\hat{\theta}_i\}$.

Substituting $\{\hat{\theta}_i\}$ into (6.13) instead of true unknown values $\{\theta_i^0\}$, we define the following auxiliary statistics:

$$\hat{G}_{ij}^{(k)} = G_{ij}^{(k)}(\hat{\theta}_i;\hat{\theta}_j), \quad k=1,2; i,j \in S, \tag{6.48}$$

$\hat{G}_{ij}^{(1)}$ is an m-vector statistic and $\hat{G}_{ij}^{(2)}$ is an $(m\times m)$-matrix statistic. As $\tau_*\to 0$, one may consider $\hat{G}_{ij}^{(k)}$ to be a consistent estimator of $\overset{\circ}{G}{}_{ij}^{(k)}$.

Theorem 6.7 *Suppose that the conditions of Theorem 6.6 are fulfilled and*

$$\bar{\theta}_i = \hat{\theta}_i + (\hat{G}_{ii}^{(2)})^{-1} \sum_{j=1,j\neq i}^{L} \epsilon_{ij}\hat{G}_{ij}^{(1)} \quad (i\in S). \tag{6.49}$$

Then the decision rule

$$d = d_2(x;A) = \sum_{i=1}^{L} i\mathbf{I}_{V_i}(x), \quad \mathbf{I}_{V_i}(x) = \prod_{k=1,k\neq i}^{L} \mathbf{1}\left(\sum_{l=1}^{L} c_{lki}p(x;\bar{\theta}_l)\right) \tag{6.50}$$

is a robust decision rule of second order. Its unconditional risk admits the expansion

$$r(d_2) = r_0 + \sum_{i=1}^{L} \alpha_i / n_i + o(\tau_*^2). \tag{6.51}$$

Proof. By the condition of the theorem, $G_{ij}^{(k)}(\theta_i; \theta_j)$ is a continuously differentiable (matrix) function in θ_i, θ_j. Therefore from (6.48), (6.17), (6.18) it follows (as $n_i \to \infty$) that

$$\hat{G}_{ij}^{(1)} \overset{a.s.}{\to} \overset{\circ}{G}{}_{ij}^{(1)} + O(\epsilon_+)\mathbf{1}_m, \quad (\hat{G}_{ii}^{(2)})^{-1} \overset{a.s.}{\to} (\overset{\circ}{G}{}_{ii}^{(2)})^{-1} + O(\epsilon_+)\mathbf{1}_{m\times m}.$$

Then from (6.49), (6.18) we conclude:

$$\bar{\theta}_i \overset{a.s.}{\to} \theta_i^* + (\overset{\circ}{G}{}_{ii}^{(2)})^{-1} \sum_{j=1, j\neq i}^{L} \epsilon_{ij} \overset{\circ}{G}{}_{ij}^{(1)} + O(\epsilon_+^2)\mathbf{1}_m = \theta_i^0 + O(\epsilon_+^2)\mathbf{1}_m.$$

Thus, $\bar{\theta}_i - \theta_i^0$ has higher order of smallness with respect to ϵ_+ in comparison with $\hat{\theta}_i - \theta_i^0$; the coefficients at $\{\epsilon_{ij}\}$ become equal to zero. Therefore, instead of (6.44) we have

$$V_{ik} = \delta_{ik} (\overset{\circ}{G}{}_{ii}^{(2)})^{-1} \overset{\circ}{J}_{ii} (\overset{\circ}{G}{}_{ii}^{(2)})^{-1} + o(\tau_*^2)\mathbf{1}_{m\times m}.$$

This result is proved in the same way as Lemma 6.1. Using Theorem 6.5 now, we obtain the expansion (6.51) for risk of the adaptive decision rule $d_2(\cdot)$. From (6.51), (6.47), (6.25) it follows that $\kappa_+(d_2; E) = o(\epsilon_+^2)$. This means that the robustness order for the decision rule $d_2(\cdot)$ is $\nu_o(d_2) = 2$.

■

Note that if we neglect the remainder term in (6.51), then the robust decision rule $d_2(\cdot)$ based on training sample A with misclassifications has the same risk as a traditionally used adaptive decision rule $d_1(\cdot)$ in absence of contaminations in A (compare (6.51), (6.42) with (3.33)–(3.35)).

Another method of increasing the robustness order for adaptive decision rule $d_1(\cdot)$ is based on using ML-estimators $\{\check{\theta}_i\}$ for the mixture of the distributions $\{p(x; \theta_k) : k \in S\}$. These estimators are the solution of the extremum problem:

$$\sum_{i=1}^{L} \sum_{j=1}^{n_i} \ln \sum_{k=1}^{L} \epsilon_{ik} p(z_{ij}; \theta_k) \to \max_{\{\theta_i\}}. \tag{6.52}$$

Using $\{\check{\theta}_i\}$, we obtain the adaptive decision rule:

$$d = d_3(x; A) = \sum_{i=1}^{L} i \mathbf{I}_{V_i}(x), \quad \mathbf{I}_{V_i}(x) = \prod_{k=1, k\neq i}^{L} \mathbf{1}\Big(\sum_{l=1}^{L} c_{lki} p(x; \check{\theta}_l)\Big). \tag{6.53}$$

If $\{\check{\theta}_i\}$ coincide with the global maximum of the objective function (6.52), then as $\{n_i \to \infty\}$ under the Chibisov regularity conditions (Section 1.6), $\check{\theta}_i \overset{a.s.}{\to} \theta_i^0$ $(i \in S)$, therefore,

$$r(d_3) \to r_0, \quad \kappa(d_3; E) \to 0.$$

Thus, under the given conditions the adaptive decision rule (6.53) is absolutely robust. But the synthesis of the decision rule $d_3(\cdot)$ is computationally expensive because of global maximum search in (6.52) (the objective function is multivariate multiextremal). This problem is analytically unsolvable even in the Gaussian case. It should be noted that if the estimators corresponding to a local maximum in (6.52) are used in (6.53) instead of $\{\check{\theta}_i\}$, then the property of absolute robustness for decision rule $d_3(\cdot)$ is lost.

Let us apply the results of this section to the case of Gaussian model with type D.3.3 distortions:

$$p(x; \theta_i^0) = n_N(x \mid \mu_i^0, \Sigma_i^0), \quad x \in R^N$$

is the Gaussian probability density function; $\mu_i^0 = (\mu_{ij}^0)$ is the vector of mathematical expectation for class Ω_i; $\Sigma_i^0 = (\sigma_{ijk}^0)$ is the covariance matrix ($i \in S$); $L = 2; \pi_1 = \pi_2 = 0.5$ (two equiprobable classes); $w_{11} = w_{22} = 0, w_{12} = w_{21} = 1$ (in this case the risk r is in fact the unconditional classification error probability). Consider two situations distinguished by the level of prior information about the parameters $\{\mu_i^0, \Sigma_i^0\}$.

1. The matrix $\Sigma_1^0 = \Sigma_2^0 = \Sigma$ is known, $\{\theta_i^0 = \mu_i^0\}$ are unknown.

In this situation, according to Section 1.4, the Bayesian decision rule is linear:

$$d = d_0(x) = \mathbf{1}(b^T x - \gamma), \tag{6.54}$$

$$b = (\Sigma^0)^{-1}(\mu_2^0 - \mu_1^0), \quad \gamma = (\mu_2^0 + \mu_1^0)^T(\Sigma)^{-1}(\mu_2^0 - \mu_1^0)/2,$$

and the Bayesian error probability r_0 is $\Phi(-\Delta/2)$.

A traditionally used adaptive decision rule is the decision rule $d = d_1(x; A)$ derived by inserting the ML-estimator based on the subsample A_i into (6.54) instead of μ_i^0:

$$\hat{\mu}_i = \frac{1}{n_i}\sum_{j=1}^{n_i} z_{ij} \quad (i \in S = \{1, 2\}). \tag{6.55}$$

Using the Corollaries 6.5, 6.6, we may find an approximation for the robustness factor of the adaptive decision rule $d_1(\cdot)$:

$$\kappa(d_1) = \frac{\exp(-\Delta^2/8)}{16(2\pi)^{1/2}\Delta\Phi(-\Delta/2)} \times$$

$$\times \left(\Delta^4((\epsilon_2 - \epsilon_1)^2 + \frac{\epsilon_1}{n_1} + \frac{\epsilon_2}{n_2}) + (4(N-1) + \Delta^2)(\frac{1}{n_1} + \frac{1}{n_2}) \right). \tag{6.56}$$

According to Corollary 6.8, the asymptotic robustness factor for the adaptive decision rule $d_1(\cdot)$ is

$$\kappa_+(d_1; E) = \frac{\Delta^3 \exp(-\Delta^2/8)}{16(2\pi)^{1/2}\Phi(-\Delta/2)}(\epsilon_2 - \epsilon_1)^2 + o(\epsilon_+^2).$$

It is seen that if the classes Ω_1, Ω_2 are "equidistorted" ($\epsilon_1 = \epsilon_2$), then $d_1(\cdot)$ becomes robust decision rule of second order. In general case, if the distortion levels are different ($\epsilon_1 \neq \epsilon_2$), then for the adaptive decision rule $d_1(\cdot)$ the robustness order $\nu_0(d_1)$ is 1. Using the definition of the "breakdown point" from Section 2.3 and the main expansion term in $\kappa_+(d_1; E)$, we find the breakdown point value under D.3.3 type distortions:

$$\epsilon_+^* = 4\sqrt{(\Phi(\Delta/2) - 1/2)/\Delta\phi(\Delta/2)}/\Delta.$$

Figure 6.2 plots this dependence.

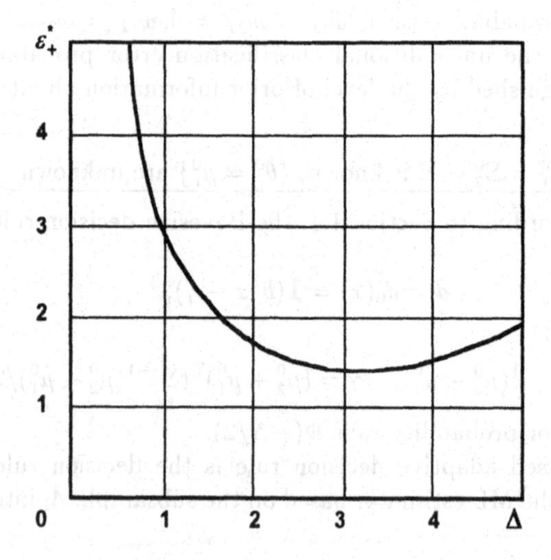

Figure 6.2: "Breakdown point" vs. Δ

According to Theorem 6.7, the robust decision rule of second order $d_2(\cdot)$ is constructed from (6.54) by replacing μ_i^0 by the statistic:

$$\bar{\mu}_i = (\bar{\mu}_{ij}) = \hat{\mu}_i + (-1)^i \epsilon_i (\hat{\mu}_2 - \hat{\mu}_1) \quad (i \in S). \tag{6.57}$$

The robustnes factor for this decision rule is

$$\kappa(d_2) = \frac{\exp(-\Delta^2/8)}{16(2\pi)^{1/2}\Delta\Phi(-\Delta/2)}(4(N-1) + \Delta^2)(\frac{1}{n_1} + \frac{1}{n_2}). \tag{6.58}$$

The decision rules $d_1(\cdot), d_2(\cdot)$ and $d_3(\cdot)$ were also investigated by computer modeling. In the computer implementation of decision rule $d_3(\cdot)$ defined by (6.53), (6.52), to compute the statistics $\{\check{\mu}_i\}$ 10 iterations of the Newton-Rafson procedure were used for finding the solution of (6.52) with initial values $\{\hat{\mu}_i\}$.

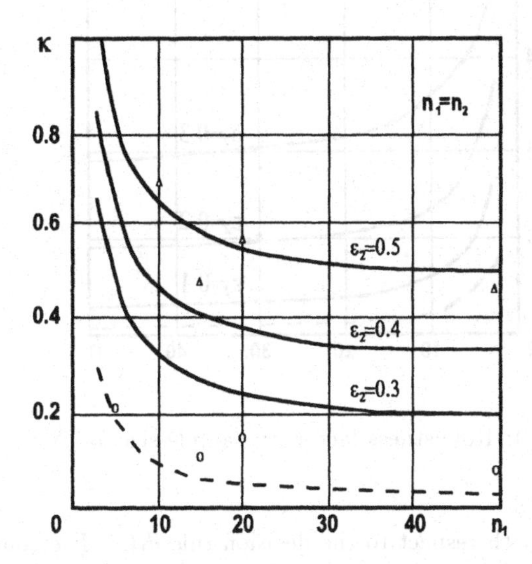

Figure 6.3: Robustness factor vs. sample size for $\Delta = 2.56$

At Figure 6.3 solid lines plot the dependence of the robustness factor $\kappa(d_1)$ on sample size $n_1 = n_2$ for observation space dimension $N = 2$, Mahalanobis distance $\Delta = 2.56$ ($r_0 = 0.1$) and different contamination levels $\epsilon_1 = 0; \epsilon_2 = 0.3; 0.4; 0.5$. Small triangles near the curve indicate the computer simulation results for the decision rule $d_1(\cdot)$ at $\epsilon_2 = 0.5$. The dashed line plots $\kappa(d_2)$ for the robust decision rule $d_2(\cdot)$ computed by formula (6.58), and the corresponding computer results at $\epsilon_2 = 0.5$ are indicated by small circles. Computer experiments revealed that the difference between $\kappa(d_3)$ and $\kappa(d_2)$ is insignificant.

In the same form, the plots for $\Delta = 4.65$ ($r_0 = 0.01$) are presented at Figure 6.4. It is seen from these figures that the accuracy of the approximations (6.56), (6.58) constructed by the asymptotic expansion method is sufficiently high. The comparison of (6.56) and (6.58) shows that if

$$(\epsilon_2 - \epsilon_1)^2 + \frac{\epsilon_1}{n_1} + \frac{\epsilon_2}{n_2} > \frac{4(N-1) + \Delta^2}{\Delta^4}\left(\frac{1}{n_1} + \frac{1}{n_2}\right)\delta, \tag{6.59}$$

then

$$\kappa(d_1) > (1 + \delta)\kappa(d_2).$$

This means that for any $\delta > 0$ the application of the robust decision rule $d_2(\cdot)$ increases the robustness more than $(1 + \delta)$ times: it decreases the relative bias κ

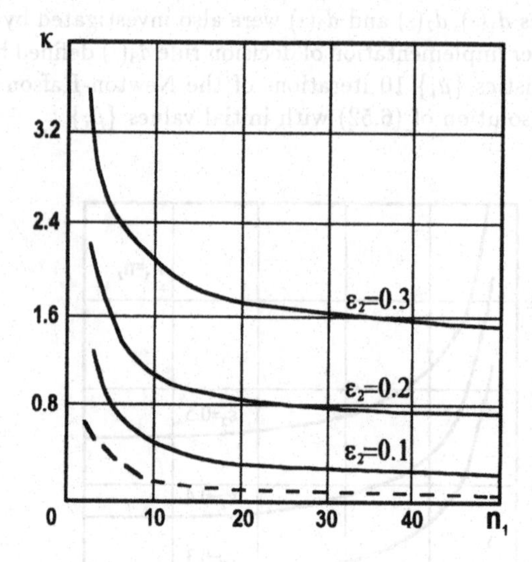

Figure 6.4: Robustness factor vs. sample size for $\Delta = 4.65$

of error probability with respect to the decision rule $d_1(\cdot)$. The condition (6.59) is convenient for practical use as a criterion of importance (δ-significance) of influence of contaminations in A on pattern recognition accuracy. To this end, in order to evaluate Δ in (6.59), the statistical estimator

$$\bar{\Delta} = \sqrt{(\bar{\mu}_2 - \bar{\mu}_1)^T (\Sigma^0)^{-1} (\bar{\mu}_2 - \bar{\mu}_1)}$$

should be used or its "expected" value $\bar{\Delta} = -2\Phi^{-1}(\bar{r})$, where $\bar{r}$ is an "*a priori* expected value" of error probability.

Let us consider two special cases important for applications.

(1) If $\epsilon_1 = \epsilon_2 = \epsilon_+$ (equidistorted classes), then (6.59) assumes a simple form:

$$\epsilon_+ > \epsilon_+(\delta, \Delta, N), \quad \epsilon_+(\delta, \Delta, N) = \delta(4(N-1) + \Delta^2)/\Delta^4. \qquad (6.60)$$

Here $\epsilon_+(\delta, \Delta, N)$ is the critical value of distortion level. Plots of the dependence $\epsilon_+ = \epsilon_+(\delta, \Delta, N)$ for $\delta = 1$ (in this case $\kappa(d_1) > 2\kappa(d_2)$, i.e., robustness gain is more than 100%) are presented at Figure 6.5.

(2) If $\epsilon_1 = 0, \epsilon_2 = \epsilon_+$ (only the sample from Ω_2 is misclassified), then the critical value of contamination level is

$$\epsilon_+(\delta, \Delta, N) = (\sqrt{1 + 4\delta n_2 \Delta^{-4}(4(N-1) + \Delta^2)(1 + n_1^{-1} n_2)} - 1)/(2n_2). \qquad (6.61)$$

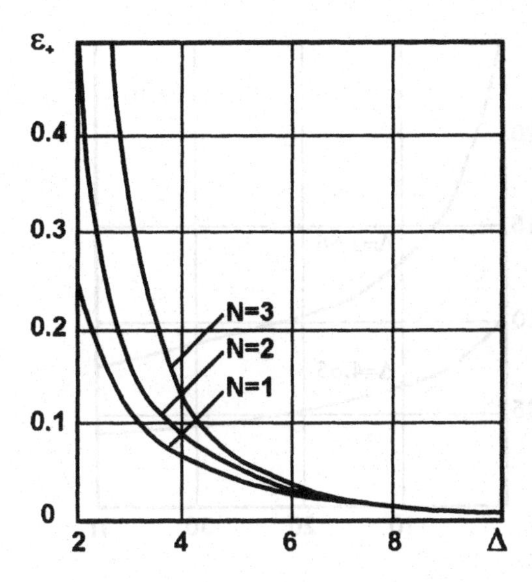

Figure 6.5: Critical contamination level vs. Δ

Plots of this dependence for $N = 2, \delta = 1, n_1 = n_2$ are shown at Figure 6.6. Under the Kolmogorov-Deev asymptotics ($n_1 = n_2 \to \infty, N \to \infty, n_1/N \to g > 0$) the formula (6.61) becomes simpler:

$$\epsilon_+(\delta, \Delta, N) = 2(2\delta)^{1/2}/(\Delta^2 g^{1/2}).$$

It is seen that with the increase of Δ and g the contaminations of the sample A become more significant.

For the considered situation let us analyze an additional version of the adaptive decision rule $d_4(\cdot)$, which is derived by the substitution of the so-called λ_i-*estimator* (Aivazyan and Meshalkin, 1989) $\bar{\mu}_i$ for μ_i^0 in (6.54); $\bar{\mu}_i$ is the solution of the following system:

$$\bar{\mu}_i = \sum_{j=1}^{n_i} z_{ij}(n_N(z_{ij} \mid \bar{\mu}_i, \Sigma_i^0))^{\lambda_i} / \sum_{j=1}^{n_i}(n_N(z_{ij} \mid \bar{\mu}_i, \Sigma_i^0))^{\lambda_i}.$$

The parameter $\lambda_i > -1/2$ is referred to as *exponential weighting parameter*. For $\lambda_i = 0$ we have the ML-estimator: $\bar{\mu}_i = \hat{\mu}_i \quad (i \in S)$. If $\lambda_i > 0$, then large weights are assigned to observations close to the "center" of the distribution and small weights are assigned to its "tails". An asymptotic analysis of the estimator $\bar{\mu}_i$ is given in (Shurygin, 1980) for homogeneous samples. Using these results for $E = \mathbf{I}_L$ ($\epsilon_+ = 0$) we find by Theorem 6.5:

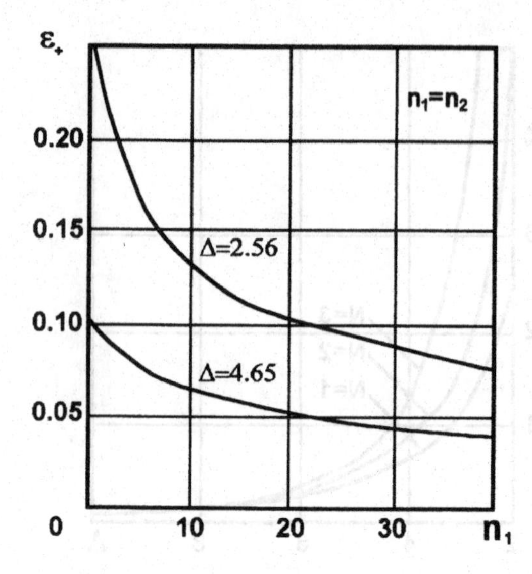

Figure 6.6: Critical contamination level vs. $n_1 = n_2$

$$\kappa(d_4) = \frac{\exp(-\Delta^2/8)}{16(2\pi)^{1/2}\Delta r_0}(4(N-1)+\Delta^2)\left(\frac{\psi(N,\lambda_1)}{n_1} + \frac{\psi(N,\lambda_2)}{n_2}\right), \quad (6.62)$$

where

$$\psi(N;\lambda_i) = (1+\lambda_i)^{N+2}/(1+2\lambda_i)^{N/2+1} \geq 1.$$

Using (6.56) in absence of contaminations ($\epsilon_1 = \epsilon_2 = 0$) and (6.62), we evaluate the relative error probability increment ratio for the decision rule $d_4(\cdot)$ in comparison with $d_1(\cdot)$ at $\lambda_1 = \lambda_2 = \lambda$:

$$\kappa(d_4 : d_1) = \frac{r(d_4) - r_0}{r(d_1) - r_0} = \psi(N,\lambda) \geq 1. \quad (6.63)$$

Some advantages of λ-estimators under sample contamination are known (Shurygin, 1980); but in absence of distortions these estimators, as it is seen from (6.63), lose to ML-estimators. In order to keep this loss value $\kappa(d_4 : d_1) - 1$ not greater than the given level $\tau > 0$, the parameter λ should be chosen subject the condition

$$\psi(N,\lambda) \leq 1 + \tau.$$

For $0 < \lambda << N$ this condition is equivalent to the following one:

$$\lambda \leq \lambda_+ = \sqrt{(2\ln(1+\tau))/(N+2)}.$$

Let us tabulate the values of λ_+ and $\kappa_+(d_4 : d_1) = \psi(N, \lambda_+)$ for $\tau = 1$ as a function of dimension N of the observation space:

N	1	3	5	8	20	50	100
λ_+	0.680	0.527	0.445	0.372	0.251	0.163	0.117
κ_+	1.308	1.373	1.417	1.466	1.572	1.678	1.749

2. $\{\Theta_i^0\} = \{\mu_i^0, \Sigma_i^0\}$ are unknown.

In this situation, according to the results of Section 1.4, the Bayesian decision rule is quadratic decision rule:

$$d = d_0(x) = \mathbf{1}(\sum_{i=1}^{2}(-1)^{i+1}((x-\mu_i^0)^T(\Sigma_i^0)^{-1}(x-\mu_i^0) + \ln \mid \Sigma_i^0 \mid)) + 1. \tag{6.64}$$

Consider the adaptive decision rule $d_1(\cdot)$, resulting from the substituting of ML-estimators $\{\hat{\mu}_i\}$ defined by (6.55) and $\{\hat{\Sigma}_i\}$:

$$\hat{\Sigma}_i = (\hat{\sigma}_{ijk}) = \frac{1}{n_i}\sum_{l=1}^{n_i}(z_{il}-\hat{\mu}_i)(z_{il}-\hat{\mu}_i)^T \quad (j,k = 1,\ldots,N)$$

into (6.64).

Let $\hat{\Sigma}_i^{-1} = (\bar{\sigma}_{ijk})$ denote the inverse matrix. If $j,k,p,q = 1,\ldots,N$ are some indices, then let (j,k) denote the two-dimensional index that varies in the following way:

$$(1,1),(1,2),\ldots,(1,N),(2,2),\ldots,(2,N),(3,3),\ldots,(N,N),$$

assuming $N(N+1)/2$ different values. Moreover, if (b_{jk}) is an $(N\times N)$-matrix, then $(b_{(j,k)}) = (b_{11}\ldots b_{1N}b_{22}\ldots b_{NN})$ is an $N(N+1)/2$-vector; if (b_{jkpq}) is an $N\times N\times N\times N$-tensor, then $(b_{(j,k),(p,q)})$ is an $(N(N+1)/2)\times(N(N+1)/2)$-matrix. Then according to (6.48), (6.13), (6.15) ($\bar{i} = 3-i$):

$$\hat{G}_{i\bar{i}}^{(1)} = \begin{pmatrix} \hat{\Sigma}_i^{-1}(\hat{\mu}_i - \hat{\mu}_{\bar{i}}) \\ \cdots\cdots\cdots\cdots\cdots\cdots\cdots\cdots\cdots\cdots \\ ((2-\delta_{jk})(\hat{\Sigma}_i^{-1} - \hat{\Sigma}_i^{-1}(\hat{\Sigma}_{\bar{i}} + (\hat{\mu}_1-\hat{\mu}_2)(\hat{\mu}_1-\hat{\mu}_2)^T)\hat{\Sigma}_i^{-1})/2)_{(j,k)} \end{pmatrix}$$

$$\hat{G}_{ii}^{(2)} = \hat{J}_{ii} = \begin{pmatrix} \hat{\Sigma}_i^{-1} & \vdots & \mathbf{0} \\ \cdots & \vdots & \cdots \\ \mathbf{0} & \vdots & \mathcal{A}_i \end{pmatrix}, \tag{6.65}$$

where

$$\mathcal{A}_i = \left(((2-\delta_{jk})(\bar{\sigma}_{ijq}\bar{\sigma}_{ikp} + (1-\delta_{pq})\bar{\sigma}_{ijp}\bar{\sigma}_{ikq}))_{(j,k),(p,q)}\right).$$

The robust decision rule $d_2(\cdot)$ is obtained by using the estimators $\{\bar{\mu}_i, \bar{\Sigma}_i\}$ defined by the expressions (6.49), (6.65) in the block-matrix form:

$$\left(\begin{pmatrix} \bar{\mu}_i \\ \cdots \\ ((\bar{\Sigma}_i)_{(j,k)}) \end{pmatrix}\right) = \left(\begin{pmatrix} \hat{\mu}_i \\ \cdots \\ ((\hat{\Sigma}_i)_{(j,k)}) \end{pmatrix}\right) + \epsilon_i \hat{J}_{ii} \hat{G}^{(1)}_{ii} \quad (i \in S). \tag{6.66}$$

In particular, if $\Sigma_i^0 = \mathrm{diag}\{\sigma_{ij}^0\}$ is a diagonal matrix (i.e., the components of X_i are independent), then (6.65), (6.66) imply

$$\bar{\sigma}_{ij} = \hat{\sigma}_{ij} + \epsilon_i(\hat{\sigma}_{ij} - \hat{\sigma}_{3-i,j} - (\bar{\mu}_{2j} - \bar{\mu}_{1j})^2), \quad j = 1, \ldots, N, i \in S.$$

Note that in order to avoid violation of variance nonnegativity condition $\bar{\sigma}_{ij} > 0$ in the case of large contamination levels $\{\epsilon_i\}$ and small sample sizes $\{n_i\}$, the estimator $\bar{\sigma}_{ij}$ was limited from below:

$$\bar{\sigma}_{ij+} = \max\{\bar{\sigma}_{ij}, q_{ij}\},$$

where the critical value $q_{ij} > 0$ was chosen using the interval estimator for σ_{ij}.

Let us tabulate pointwise the estimates $\hat{\kappa}(d_1)$, $\hat{\kappa}(d_2)$ of the robustness factor for the adaptive decision rules $d_1(\cdot)$, $d_2(\cdot)$ computed from a sample of 4000 realizations for the following example case:

$$N = 2, n_1 = n_2 = 100, \epsilon_1 = 0, r_0 = 0.01,$$

$$\mu_1^0 = \begin{pmatrix} 0 \\ 0 \end{pmatrix}, \mu_2^0 = \begin{pmatrix} 3.288 \\ 3.288 \end{pmatrix}, \Sigma_1^0 = \begin{pmatrix} 25 & 0 \\ 0 & 0.5 \end{pmatrix}, \Sigma_2^0 = \begin{pmatrix} 1 & 0 \\ 0 & 1 \end{pmatrix}.$$

Here is the table of computer calculations:

ϵ_2	0	0.1	0.2	0.3	0.4	0.5
$\hat{\kappa}(d_1)$	0.6	1.5	2.6	3.2	3.8	4.5
$\hat{\kappa}(d_2)$	0.6	0.8	0.8	1.0	1.6	2.3

One can see considerable gain for the robust decision rule $d_2(\cdot)$ as compared with the classical decision rule $d_1(\cdot)$.

6.3 Parametric ϵ–nonhomogeneity of Training Samples

The classical assumption about the homogeneity of the training sample from class Ω_i $(i \in S)$

$$A_i = \{z_{i1}, z_{i2}, \ldots, z_{in_i}\} \subset R^N$$

states that all these sample elements are identically distributed with the same regular probability density function $q(\cdot;\theta_i^0) \in Q$ and the same true parameter value $\theta_i^0 \in \Theta \subseteq R^m$. It turns out that this homogeneity assumption is usually violated in applications of adaptive pattern recognition, see, e.g., (Aivazyan *et al.*, 1974, 1983, 1989), (Roussas, 1965), Silvey, 1961).

Consider a situation where the violation consists in "ϵ-variability" of the parameter θ_i^0 during the recording of the training sample (distortions of type D.3.1 defined in Section 2.2).

Assume that a training sample A_i (in this section the class number $i \in S$ will be omitted from notations, if this does not lead to confusion) observed in the feature space R^N is a sequence of n mutually independent random vectors $z_1, \ldots z_n$; a random vector $z_t \in R^N$ has probability density function $q(\cdot;\theta_t) \in Q$, where $\{\theta_t\}$ are defined like in (5.83):

$$(\theta_t - \theta^0)^T B(\theta_t - \theta^0) = \epsilon_t^2, \quad 0 \le \epsilon_t \le \epsilon_+, \quad t = 1, \ldots, n. \tag{6.67}$$

Here B is a positive definite $(m \times m)$-matrix; $\epsilon_+ \ge 0$ is a given maximal admissible level of distortions of the hypothetical model (if $\epsilon_+ = 0$, then we have the hypothetical model of random sample A_i used in Chapter 3). In the parameter space R^m, the restrictions (6.67) determine an ellipsoid $\mathbf{U}_{\theta^0}(\epsilon_+)$ centered at point θ^0. Its size depends on ϵ_+ and its form depends on the matrix B. Note that the parametric ϵ-nonhomogeneity model (6.67) can be considered as a generalization of the Tukey-Huber distortion model described in Section 5.1. Indeed, according to (Huber, 1981) the sample A of size n is generated by the mixture of only two distributions: "hypothetical" and "contaminating". In the model (6.67) investigated here the sample A is generated by the mixture of n, generally speaking, different distributions, so that the number of "contaminating" distributions can increase when the sample size increases. From Sections 6.1, 6.2 it is seen that the risk robustness of the plug-in decision rule essentially depends on robustness of the statistical estimator $\hat{\theta}$ for the parameter vector θ^0. Therefore we shall investigate the robustness of the ML-estimator:

$$\hat{\theta} = \arg\min_{\theta\in\Theta} L_0(\theta), \quad L_0(\theta) = n^{-1}\sum_{t=1}^{n}(-\ln q(z_t;\theta)) \tag{6.68}$$

in presence of parametric ϵ-nonhomogeneity (6.67).

Assume the notations:

$$G_{01}(\theta', \theta) = \mathbf{E}_\theta\{-\ln q(x;\theta')\},$$

$$G_{02}(\theta, \theta_1, \ldots, \theta_n) = n^{-1}\sum_{t=1}^{n} G_{01}(\theta, \theta_t); \quad \bar{\theta} = \frac{1}{n}\sum_{t=1}^{n}\theta_t, \tag{6.69}$$

where $\mathbf{E}_\theta$ denotes expectation with respect to the distribution $q(x;\theta)$; Θ^* will denote the closure of Θ.

Theorem 6.8 *Suppose that the family Q satisfies the Chibisov regularity conditions (Section 1.6) of order (k,r), $k>1, r>4$, for the contrast function $g(x;\theta) = -\ln q(x;\theta)$ and the point*

$$\theta^* = \arg\min_{\theta\in\Theta^*} G_{02}(\theta,\theta_1,\ldots,\theta_n) \tag{6.70}$$

is unique. Then in presence of parametric ϵ-nonhomogeneity (6.67) the ML-estimator $\hat{\theta}$ converges almost surely as $n\to\infty$:

$$\hat{\theta}-\theta^* \overset{a.s.}{\to} \mathbf{0}_m, \quad \theta^* = \bar{\theta} + \mathcal{O}(\epsilon^2)\mathbf{1}_m. \tag{6.71}$$

Proof. According to (6.68), (6.69),

$$\mathbf{E}\{-\ln q(z_t;\theta)\} = G_{01}(\theta,\theta_t), \quad t=1,\ldots,n;$$

$$\mathbf{E}\{L_0(\theta)\} = G_{02}(\theta,\theta_1,\ldots,\theta_n),$$

where $\mathbf{E}\{\cdot\}$ denotes the mathematical expectation with respect to the distribution of an ϵ-nonhomogeneous training sample. By the Chibisov regularity conditions,

$$\mathbf{D}\{\ln q(z_t;\theta)\} \leq c < \infty,$$

therefore the Kolmogorov condition is fulfilled:

$$\sum_{t=1}^{\infty} t^{-2}\mathbf{D}\{\ln q(z_t;\theta)\} < \infty.$$

Consequently, the strong law of large numbers holds, which defines the asymptotics of the objective function for the estimator $\hat{\theta}$:

$$L_0(\theta) - G_{02}(\theta,\theta_1,\ldots,\theta_n) \overset{a.s.}{\to} 0, \quad \theta\in R^m.$$

Further, by using this limit relation and (6.70) the convergence $\hat{\theta}-\theta^* \overset{a.s.}{\to} \mathbf{0}$ may be proved in the same way as strong consistency of ML-estimators (Borovkov, 1984).

Let us prove now the expansion (6.71) for θ^*. A necessary condition for (6.70) is

$$G_{02}^{(1)}(\theta^*,\theta_1,\ldots,\theta_n) = 0.$$

Let us apply the linear Taylor formula to the left side of this equation in the neighborhood of point θ^0:

$$\theta^*-\theta = -(G_{02}^{(2)}(\theta^0;\theta_1,\ldots,\theta_n))^{-1}G_{02}^{(1)}(\theta^0;\theta_1,\ldots,\theta_n) + \mathcal{O}(\mid\theta^*-\theta^0\mid^2)\cdot\mathbf{1}_m.$$

Using the assumed notations we obtain:

$$G_{02}^{(1)}(\theta^0;\theta_1,\ldots,\theta_n) = \frac{1}{n}\sum_{t=1}^{n}\nabla_\theta G_{01}(\theta,\theta_t)\mid_{\theta=\theta^0} = \nabla_\theta G_{01}(\theta,\theta^0)\mid_{\theta=\theta^0} +$$

$$+\nabla_{\theta'}\nabla_{\theta}G_{01}(\theta,\theta')\mid_{\theta=\theta'=\theta^0}(\bar{\theta}-\theta^0)+\mathcal{O}(\epsilon^2)\cdot\mathbf{1}_m,$$

$$G_{02}^{(2)}(\theta^0;\theta_1,\ldots,\theta_n)=\nabla_{\theta}^2G_{01}(\theta,\theta^0)\mid_{\theta=\theta^0}+\mathcal{O}(\epsilon)\cdot\mathbf{1}_{m\times m}.$$

Differentiating with respect to θ,θ' under the regularity conditions results in:

$$\nabla_{\theta}G_{01}(\theta,\theta^0)\mid_{\theta=\theta^0}=\mathbf{O}_m,$$

$$\nabla_{\theta}^2G_{01}(\theta,\theta^0)\mid_{\theta=\theta^0}=J,$$

$$\nabla_{\theta'}\nabla_{\theta}G_{01}(\theta,\theta')=-\int_{R^N}\frac{1}{q(x;\theta)}\nabla_{\theta'}q(x;\theta')\nabla_{\theta}^Tq(x;\theta)dx,$$

$$\nabla_{\theta'}\nabla_{\theta}G_{01}(\theta,\theta')\mid_{\theta=\theta'=\theta^0}=-J,$$

where $J=\mathbf{E}_{\theta^0}\{-\nabla_{\theta^0}^2\ln q(x;\theta^0)\}$ is the positive definite Fisher information matrix. Using these expressions and selecting the terms of order $\mathcal{O}(\epsilon)$ we obtain (6.71):

$$\theta^*-\theta^0=\bar{\theta}-\theta^0+\mathcal{O}(\epsilon^2)\cdot\mathbf{1}_m.$$

■

Taking (6.71) into account, we shall characterize the accuracy of $\hat{\theta}$ by the deviation of θ^* from θ^0 in the generalized metric specified by the matrix B:

$$\rho=\rho(\epsilon_+,\theta^0)=\sup_{\{\theta_t\in U_{\theta^0}(\epsilon_+)\}}\parallel\theta^*-\theta^0\parallel_B=$$

$$=\sup_{\{\theta_t\in U_{\theta^0}(\epsilon_+)\}}\sqrt{(\theta^*-\theta^0)^TB(\theta^*-\theta^0)}. \tag{6.72}$$

As ρ increases, the accuracy and the robustness of the estimator (6.68) decrease in the case of parametric ϵ-nonhomogeneity of the training sample A.

Corollary 6.9 *Under the conditions of Theorem 6.8,*

$$\rho(\epsilon_+,\theta^0)=\epsilon_++\mathcal{O}(\epsilon_+^2).$$

Proof. From (6.71), (6.72) we have

$$\rho^2(\epsilon_+,\theta^0)=\sup_{\{\theta_t\in U_{\theta^0}(\epsilon_+)\}}\left(\frac{1}{n}\sum_{t=1}^n(\theta_t-\theta^0)\right)^TB\left(\frac{1}{n}\sum_{t=1}^n(\theta_t-\theta^0)\right)+$$

$$+\mathcal{O}(\epsilon_+^3)=\epsilon_+^2+\mathcal{O}(\epsilon_+^3),$$

hence the required relation holds.

■

Consider the problem of enhancing the robustness for the ML-estimator $\hat{\theta}$ under conditions of parametric ϵ-nonhomogeneity (6.67).

Let us construct a logarithmic likelihood function taking into account the nonhomogeneity of the training sample A:

$$l(\theta, \theta_1, \ldots, \theta_n) = \sum_{t=1}^{n} \ln q(z_t; \theta_t), \quad \theta_t \in U_\theta(\epsilon_+),$$

where the ϵ_+-neighborhood $U_\theta(\epsilon_+)$ of the point θ in the parameter space is determined by the restrictions of type (6.67). Let us coordinate the generalized metrics in parameter space with the family Q of probability density functions by choosing the matrix B in (6.67) in a special way:

$$B = J(\theta),$$

where $J(\theta) = \mathbf{E}_\theta \{-\nabla_\theta^2 \ln q(z; \theta)\}$ is nonsingular Fisher information matrix. By construction, the logarithmic likelihood function $l(\theta, \theta_1, \ldots, \theta_n)$ depends not only on the parameter θ to be estimated but also on n non-identifiable parameters $\theta_1, \ldots, \theta_n$. In connection with this, let us define the *worst-case likelihood function*:

$$l_-(\theta) = \min_{\{\theta_t \in U_\theta(\epsilon)\}} l(\theta, \theta_1, \ldots, \theta_n) =$$

$$= \sum_{t=1}^{n} \min_{\{\theta_t \in U_\theta(\epsilon_+)\}} \ln q(z_t; \theta_t).$$

The construction of the function $l_-(\theta)$ amounts to solving of n identical extremum problems ($t = 1, \ldots, n$) with the restrictions:

$$\ln q(z_t; \theta_t) \to \min_{\theta_t},$$

$$(\theta_t - \theta)^T J(\theta)(\theta_t - \theta) = \epsilon_t^2, \quad 0 \le \epsilon_t \le \epsilon_+. \tag{6.73}$$

To solve the t-th problem, let us perform the quadratic Taylor approximation of the objective function:

$$\ln q(z_t; \theta_t) = \ln q(z_t; \theta) + \nabla_\theta^T \ln q(z_t; \theta) \cdot (\theta_t - \theta) +$$

$$+ \frac{1}{2}(\theta_t - \theta)^T \cdot \nabla_\theta^2 \ln q(z_t; \theta) \cdot (\theta_t - \theta) + \mathcal{O}(\epsilon_+^3).$$

Taking into account at the moment only the restrictions of equality type in (6.73), we construct the Lagrange function:

$$\mathcal{L}(\theta_t, \lambda) = \ln q(z_t; \theta) + \nabla_\theta^T \ln q(z_t; \theta) \cdot (\theta_t - \theta) +$$

$$+ \frac{1}{2}(\theta_t - \theta)^T \cdot \nabla_\theta^2 \ln q(z_t; \theta) \cdot (\theta_t - \theta) + \frac{\lambda}{2}(\epsilon_t^2 - (\theta_t - \theta)^T J(\theta)(\theta_t - \theta)),$$

where λ is an indefinite Lagrange multiplier. The necessary condition of minimum $\nabla_{\theta_t}\mathcal{L}(\theta_t,\lambda)=\mathbf{0}_m$ leads to the expression

$$\theta_t-\theta=-(\nabla_\theta^2\ln q(z_t;\theta)-\lambda J(\theta))^{-1}\nabla_\theta \ln q(z_t;\theta)=$$

$$=\frac{1}{\lambda}(J(\theta))^{-1}(\mathbf{I}_m-\frac{1}{\lambda}\nabla_\theta^2\ln q(z_t;\theta)\cdot(J(\theta))^{-1})^{-1}\nabla_\theta\ln q(z_t;\theta).$$

Using the equality type restriction in (6.73), we obtain an asymptotic expansion for λ^{-1}:

$$\frac{1}{\lambda}=\pm\frac{\epsilon_+}{\sqrt{\nabla_\theta^T\ln q(z_t;\theta)(J(\theta))^{-1}\nabla_\theta\ln q(z_t;\theta)}}+\mathcal{O}(\epsilon_+^2).$$

Substituting $\theta_t-\theta$ into the expression for the objective function, minimizing this function with respect to $\epsilon_t\in[0,\epsilon_+]$ and keeping the terms of order $\mathcal{O}(\epsilon_+)$, we find

$$\min_{\theta_t}\ln q(z_t;\theta_t)=$$

$$=\ln q(z_t;\theta)-\epsilon_+\sqrt{\nabla_\theta^T\ln q(z_t;\theta)(J(\theta))^{-1}\nabla_\theta\ln q(z_t;\theta)}+\mathcal{O}(\epsilon_+^2).$$

As a result,

$$l_-(\theta)=\sum_{t=1}^n(\ln q(z_t;\theta)-$$

$$-\epsilon_+\sqrt{\nabla_\theta^T\ln q(z_t;\theta)(J(\theta))^{-1}\nabla_\theta\ln q(z_t;\theta)})+\mathcal{O}(\epsilon_+^2). \qquad (6.74)$$

Now we shall construct a robust statistical estimator $\tilde{\theta}$ of the parameter θ^0 in presence of nonhomogeneities (6.67) by maximizing the worst-case likelihood function:

$$l_-(\theta)\to\max_{\theta\in\Theta^*}.$$

According to (6.74), up to the terms of order $\mathcal{O}(\epsilon_+^2)$, we shall represent the statistical estimator $\tilde{\theta}$ in the form (Huber, 1981) traditional for M–estimators:

$$\tilde{\theta}=\arg\min_{\theta\in\tilde{\Theta}}L(\theta),\quad L(\theta)=\sum_{t=1}^n g(z_t;\theta)/n, \qquad (6.75)$$

$$g(z;\theta)=-\ln q(z;\theta)+\epsilon_+\sqrt{\nabla_\theta^T\ln q(z;\theta)(J(\theta))^{-1}\nabla_\theta\ln q(z;\theta)}.$$

The necessary condition of minimum in (6.75) generates a system of m equations with respect to $\tilde{\theta}$ (this system is given here under the assumption that $J(\theta)$ does not depend on θ):

$$\frac{1}{n}\sum_{t=1}^{n}\left(\mathbf{I}_m - \frac{\epsilon_+}{\sqrt{\nabla_\theta^T \ln q(z_t;\theta)(J)^{-1}\nabla_\theta \ln q(z_t;\theta)}}\times \right.$$
$$\left. \times J^{-1}\nabla_\theta^2 \ln q(z_t;\theta)\right) \nabla_\theta \ln q(z_t;\theta) = \mathbf{0}_m.$$

In applications, the family of probability density functions Q often has the property:

$$\nabla_\theta^2 \ln q(z;\theta) = -b(z,\theta)\cdot J,$$

$b(z,\theta)$ being a real function in variables $z \in R^N, \theta \in R^m$ such that $\mathbf{E}_\theta\{b(z,\theta)\} = 1$. In this case, the system of equations becomes simplified:

$$\frac{1}{n}\sum_{t=1}^{n}\left(1 + \frac{\epsilon_+ b(z_t,\theta)}{\sqrt{\nabla_\theta^T \ln q(z_t;\theta) J^{-1}\nabla_\theta \ln q(z_t;\theta)}}\right)\nabla_\theta \ln q(z_t;\theta) = \mathbf{0}_m. \qquad (6.76)$$

The method of successive approximations can be used for construction of an algorithm to solve (6.76).

Consider the case of N-dimensional Gaussian family Q with shift parameter $\theta \in R^N$, often used in applications. Here,

$$\nabla_\theta \ln q(z;\theta) = \Sigma^{-1}(z-\theta),$$

$$\nabla_\theta^2 \ln q(z;\theta) = -\Sigma^{-1}, \quad J = \Sigma^{-1}, \quad b(z,\theta) \equiv 1.$$

As a result, the system (6.76) assumes the form

$$\frac{1}{n}\sum_{t=1}^{n}\left(1 + \frac{\epsilon_+}{\sqrt{(z_t-\theta)^T\Sigma^{-1}(z_t-\theta)}}\right)(z_t-\theta) = \mathbf{0}_N.$$

Using the sample mean statistic $\bar{z} = n^{-1}\sum_{t=1}^n z_t$, we may transform this system as follows:

$$\bar{z} - \theta + \epsilon_+ \cdot \frac{1}{n}\sum_{t=1}^{n}\left(\frac{1}{\sqrt{(z_t-\theta)^T\Sigma^{-1}(z_t-\theta)}}\right)(z_t-\theta) = \mathbf{0}_N.$$

As an approximate solution of this system, we construct a so-called *one-step estimator* for θ^0:

$$\hat{\theta} = \bar{z} + \epsilon_+ \cdot \frac{1}{n}\sum_{t=1}^{n}\left(\frac{1}{\sqrt{(z_t-\bar{z})^T\Sigma^{-1}(z_t-\bar{z})}}\right)(z_t-\bar{z}). \qquad (6.77)$$

To clarify the sense of this new estimator (6.77), let us write down this estimator and its generating equation for the one-dimensional observation space ($N = m = 1$):

$$\bar{z} - \theta + \epsilon_+ \cdot \frac{1}{n}\sum_{t=1}^{n}\operatorname{sign}(z_t-\theta) = 0,$$

$$\hat{\theta} = \bar{z} + \epsilon_+ \cdot \frac{1}{n} \sum_{t=1}^{n} \text{sign}(z_t - \bar{z}).$$

The left side of the equation consists of two terms. If the second term is neglected (this case coincides with the classical model without any distortions, $\epsilon_+ = 0$), then we obtain the equation: $\bar{z} - \theta = 0$, generating the classical estimator $\hat{\theta}' = \bar{z}$. If the first term is neglected (the case of "large" distortion levels), then we arrive at the equation

$$\frac{1}{n} \sum_{t=1}^{n} \text{sign}(z_t - \theta) = 0,$$

which gives the sample median: $\hat{\theta}'' = \text{med}\{z_i\}$ as the estimator. The estimator $\hat{\theta}$ accounts for both these extreme cases.

To compare the robustness of estimators $\bar{z}, \hat{\theta}$, a series of computer experiments was performed. In the experiments Q was the family of bivariate ($N = 2$) normal distributions with shift parameter (vector of mathematical expectations) θ and a given covariance matrix

$$\Sigma = \begin{pmatrix} 1 & 0.5 \\ 0.5 & 1 \end{pmatrix}.$$

The sample size was $n = 20$. Nonhomogeneous samples were simulated, for which the vectors of mathematical expectations $\{\theta_t\}$ were chosen, according to (6.67), inside the ellipsoid centered at point $\theta^0 = (5, 5)$:

$$\theta_t \in U_{\theta^0}(\epsilon_+) = \{\theta : (\theta - \theta^0)^T \Sigma^{-1} (\theta - \theta^0) \leq \epsilon_+^2\}.$$

In the experiments, the distortion level ϵ_+ was set to the values: $\epsilon_+ = 0; 1; 2; 3; 4; 5; 6$. The family of ϵ_+-neighborhoods corresponding to these values of ϵ_+ is shown at Figure 6.7.

Robustness of the estimators $\bar{z}, \hat{\theta}$ was characterized by the robustness factors:

$$\kappa_1(\epsilon_+) = \frac{| V_{\epsilon_+}\{\bar{z}\} |}{| V_0\{\bar{z}\} |}, \quad \kappa_2(\epsilon_+) = \frac{| V_{\epsilon_+}\{\hat{\theta}\} |}{| V_0\{\hat{\theta}\} |},$$

where $V_{\epsilon_+}\{\hat{\theta}\}$ is the estimate of the variance matrix for $\hat{\theta}$ at distortion level ϵ_+ found by computer modeling using $M = 100$ independent random samples of size $n = 20$.

Figure 6.8 plots the dependence of robustness factors $\kappa_1(\epsilon_+)$, $\kappa_2(\epsilon_+)$ on distortion level ϵ_+, found by computer modeling. The sample distortions were significantly asymmetric in these experiments: 70% of the values $\{\theta_t\}$ correspond to the nondistorted parameter θ^0, and the rest of parameter values $\{\theta_t\}$ correspond to the "extreme" values, which are indicated by small circles at Figure 6.7.

It is seen from Figure 6.8 that for $\epsilon_+ > 3$ the estimator $\hat{\theta}$ constructed by (6.77) guarantees significant gain in robustness in comparison with the classical estimator $\bar{z}$. Considerable gain in robustness was also detected with respect to another criterion b_Σ, which is the sum of absolute values of biases of components of the estimator $\hat{\theta}$.

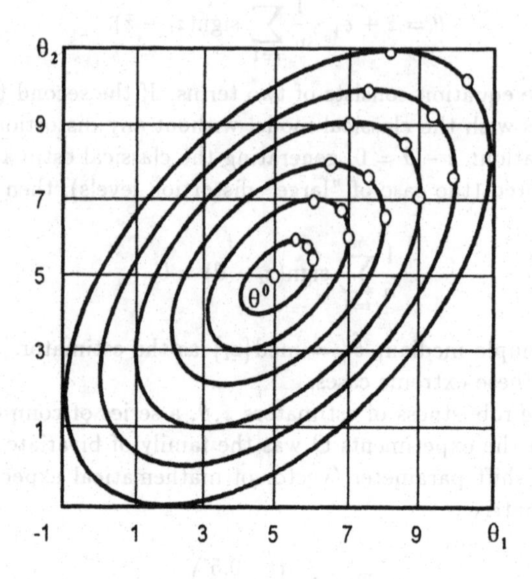

Figure 6.7: A family of ϵ_+-neighborhoods

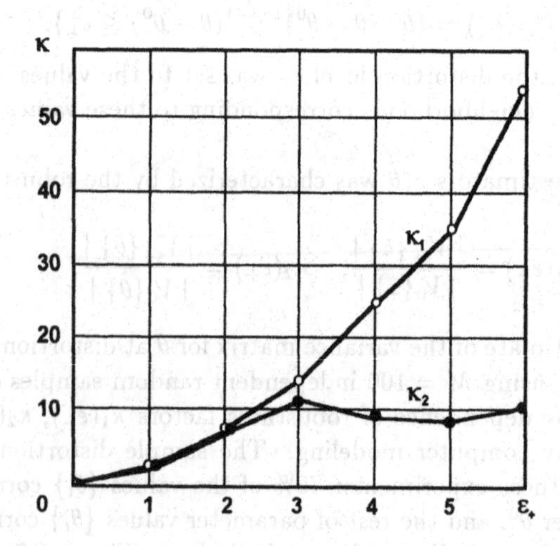

Figure 6.8: Robustness factors vs. distortion level

For example, at $\epsilon_+ = 5$, $b_\Sigma = 2.55$ for the estimator $\bar{z}$, and $b_\Sigma = 0.26$ for the new estimator $\hat{\theta}$, i.e., approximately 10 times smaller.

In conclusion we would like to mention high efficiency of the computer algorithm (6.77) for evaluation of the new robust estimator $\hat{\theta}$.

6.4 Classification of Gaussian Observations with Outliers in Training Sample

As it was noted in Section 1.4 and Chapter 3, the Gaussian model of data to be recognized is widespread in practice. In this situation, adaptive decision rules of "plug-in" type become linear or quadratic (see Section 3.3). Consider now the situation where the training samples $A_1, A_2, \ldots, A_L$ used for the construction of the adaptive decision rule $d = d(x; A)$ are *contaminated by outliers* (i.e., have distributions of type D.3.4). As it is noted in (Huber, 1981), (Ashikaga, 1981), (Balakrishnan, 1985, 1988), (Hampel *et al.*, 1986), (Tiku, 1986), (Aivazyan *et al.*, 1989), this case is frequent in applied pattern recognition problems. Under the Gaussian data model, the outliers are conveniently described by the multivariate Tukey model (Tukey, 1960): the probability density function of feature variables for the class Ω_i $(i \in S)$ is a mixture of two N-variate Gaussian distributions:

$$p_i(x) = (1 - \epsilon_i) n_N(x \mid \mu_i^0, \Sigma_i^0) + \epsilon_i n_N(x \mid \mu_i^+, \Sigma_i^+), \tag{6.78}$$

$0 \le \epsilon_i \le \epsilon_{+i}$. It means that during the registration of the training sample A_i a vector of feature variables, described by "hypothetical" Gaussian distribution $n_N(x \mid \mu_i^0, \Sigma_i^0)$ with mathematical expectation $\mu_i^0 \in R^N$ and covariance matrix Σ_i^0, is observed with probability $1 - \epsilon_i$, and a random outlier described by "contaminating" Gaussian distribution $n_N(x \mid \mu_i^+, \Sigma_i^+)$ with mathematical expectation $\mu_i^+ \in R^N$ and covariance matrix Σ_i^+ is observed with complementary probability ϵ_i.

We shall distinguish two special cases of the distortions (6.78):

a) contaminations are generated by "jumping" change of the mathematical expectation vector when

$$\Sigma_i^+ = \Sigma_i^0, \quad | \mu_i^+ - \mu_i^0 | \gg \max_{i \neq j} | \mu_j^0 - \mu_i^0 |;$$

b) contaminations are generated by "jumping" change of the covariance matrix when

$$\mu_i^+ = \mu_i^0, \quad \Sigma_i^+ = k^2 \Sigma_i^0,$$

where $k \gg 1$ is a scale factor.

The classical plug-in decision rule $d = d_1(x; A)$ using traditional statistical estimators

$$\hat{\mu}_i = n^{-1}\sum_{j=1}^{n_i} z_{ij}, \quad \hat{\Sigma}_i = (n_i - 1)^{-1}\sum_{j=1}^{n_i}(z_{ij} - \hat{\mu}_i)(z_{ij} - \hat{\mu}_i)^T, \quad i \in S, \tag{6.79}$$

optimal in absence of outliers, becomes essentially instable in their presence.

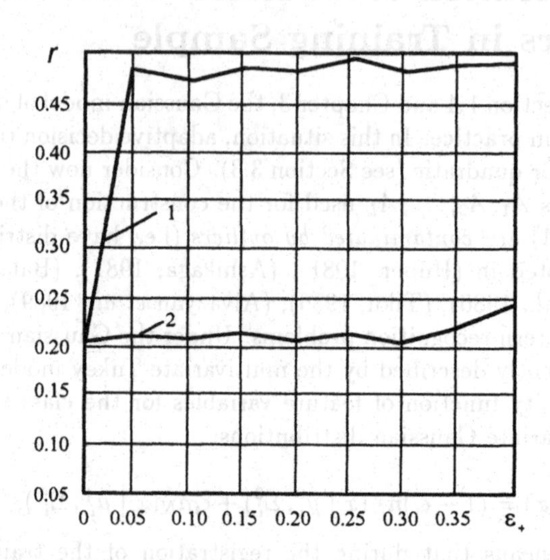

Figure 6.9: Error probability vs. distortion level

It is illustrated at Figure 6.9, where curve 1 indicates the dependence of error probability r on distortion level $\epsilon_+ = \epsilon_{+1} = \epsilon_{+2}$ in the following computer experiment: $L = 2, N = 2, \pi_1 = \pi_2 = 0.5, w_{11} = w_{22} = 0, w_{12} = w_{21} = 1,$ $n_1 = n_2 = 50, k = 100$ (case "b" of distortions),

$$\mu_1^0 = \begin{pmatrix} 0.59 \\ 0.59 \end{pmatrix}, \mu_2^0 = \begin{pmatrix} -0.59 \\ -0.59 \end{pmatrix}, \Sigma_1^0 = \Sigma_2^0 = \begin{pmatrix} 1 & 0 \\ 0 & 1 \end{pmatrix}.$$

Error probability was estimated basing on an examination sample of size $M = 30$ in a series of 20 experiments. In order to overcome the influence of outliers, a robust decision rule of "plug-in" type $d = d_*(x; A)$ was constructed with median type robust estimators (Huber, 1981), (Broffit *et al.*, 1980):

$$\check{\mu}_i = (\check{\mu}_{ij}), \quad \check{\mu}_{ij} = \text{med}_l\{z_{ilj}\}, \tag{6.80}$$

$$\check{\Sigma}_i = (\check{\sigma}_{ijk}), \quad \check{\sigma}_{ijj} = (1.483\,\text{med}_l\{|\, z_{ilj} - \check{\mu}_{ij}\,|\})^2,$$

$$\check{\sigma}_{ijk} = \sqrt{\check{\sigma}_{ijj}\check{\sigma}_{ikk}}\check{\rho}_{ijk}, \quad j \neq k,$$

$$\check{\rho}_{ijk} = M_{ijk}/N_{ijk},$$

$$M_{ijk} = \left(\operatorname{med}_l\{\bar{z}_{ilj} + \bar{z}_{ilk} - \operatorname{med}_s\{\bar{z}_{isj} + \bar{z}_{isk}\}\}\right)^2 -$$

$$-\left(\operatorname{med}_l\{\bar{z}_{ilj} - \bar{z}_{ilk} - \operatorname{med}_s\{\bar{z}_{isj} - \bar{z}_{isk}\}\}\right)^2,$$

$$N_{ijk} = \left(\operatorname{med}_l\{\bar{z}_{ilj} + \bar{z}_{ilk} - \operatorname{med}_s\{\bar{z}_{isj} + \bar{z}_{isk}\}\}\right)^2 +$$

$$+\left(\operatorname{med}_l\{\bar{z}_{ilj} - \bar{z}_{ilk} - \operatorname{med}_s\{\bar{z}_{isj} - \bar{z}_{isk}\}\}\right)^2,$$

where

$$\bar{z}_{ilj} = \frac{z_{ilj}}{\operatorname{med}_l\{\mid z_{ilj} - \operatorname{med}_t\{z_{itj}\}\mid\}} \quad (j,k = 1,\ldots,N;\ l,s,t = 1,\ldots,n_i,\ i \in S),$$

and $\operatorname{med}_l\{z_{ilj}\}$ denotes the sample median for the sample $z_{i1j},\ldots,z_{in_ij}$.

At Figure 6.9, curve 2 shows the dependence of error probability on the distortion level ϵ in the same series of computer experiments for the robust decision rule $d_*(\cdot)$. An essential gain is observed when the robust decision rule is used. For example, even at 5% contamination ($\epsilon_1 = \epsilon_2 = 0.05$) of the samples A_1, A_2 by outliers the error probability of the traditional adaptive decision rule using the estimators (6.79) increases two times and becomes equal to 0.48, while the error probability for the robust decision rule using the estimators (6.80) practically does not vary and equals to 0.21.

In conclusion, we present one of applied pattern recognition problems with Tukey-Huber outliers, which was successfully solved by robust algorithms. It is a problem of lung cancer recognition for hospital patients by biomedical data collected at the Belarussian Institute of Oncology and Medical Radiology from 1179 patients during years 1986–1993 (see (Mashevsky, 1994), (Abramovich, Kharin, and Mashevsky, 1993)). We shall use here $N = 3$ most informative features (from the total 44 features) derived by immuno-ferment biochemical blood tests: x_1 is the concentration of cancer-embryonic antigen (CEA), x_2 is the concentration of neuron-specific enalaza (NSE), and x_3 is a feature measured by the serum blood ESR-spectrometry method (ALPHA). The sample contains some observations with missing feature values. After deletion of observations with missing values of the features x_1, x_2, x_3, we had a sample of 890 3-dimensional observations from $L = 2$ classes:

$\Omega_1 =$ {patient has lung cancer},

$\Omega_2 =$ {patient has chronical lung disease}.

This sample was split into two subsamples: the training sample A of size 870 (567 observations (A_1) from class Ω_1 and 303 observations (A_2) from Ω_2) and the examination subsample A' of size 20 (11 cases from Ω_1 and 9 cases from Ω_2). Some scatter diagrams of training subsamples are given at Figures 2.3, 2.4 for two-dimensional feature subspaces: (x_3, x_1) and (x_1, x_2) respectively (see description of the scatter diagrams in Section 2.2). As it was previously noted in Section 2.2, the training subsample A contains some outliers; specialists in oncology and medical radiology explain them by individuality of patients and anomality of observations measured for terminal stages of cancer process.

The software package ROSTAN (RObust STatistical ANalysis), published by the Belarussian State University (see (Kharin *et al.*, 1994)), was used for construction of decision rules $d = d(x; A)$ from the training sample A and also for classification of the examination sample A'. Using the hypothetical Gaussian model (1.48) we calculated the ML-estimators of the conditional mean vectors:

$$\hat{\mu}_1 = \begin{pmatrix} 16.41 \\ 35.55 \\ 2.82 \end{pmatrix}, \ \hat{\mu}_2 = \begin{pmatrix} 2.77 \\ 15.52 \\ 1.79 \end{pmatrix}$$

and the inverse covariance matrices:

$$\hat{\Sigma}_1^{-1} = \begin{pmatrix} 0.0001 & -0.0000 & -0.0005 \\ -0.0000 & 0.0009 & 0.0011 \\ -0.0005 & 0.0011 & 1.0657 \end{pmatrix},$$

$$\hat{\Sigma}_2^{-1} = \begin{pmatrix} 0.28 & -0.05 & -0.04 \\ -0.05 & 0.02 & -0.06 \\ -0.04 & -0.06 & 3.01 \end{pmatrix}$$

for classes Ω_1 and Ω_2. The hypothesis about equality of conditional covariance matrices ($\Sigma_1 = \Sigma_2$) was rejected by Wilks test, so the classical quadratic decision rule $d = d_1(x; A)$ with ML-estimates $\{\hat{\mu}_i, \hat{\Sigma}_i\}$ was constructed. Pointwise estimates of error probability by reclassification of the training sample A and classification of the examination sample A' for classical decision rule $d_1(\cdot)$ are presented in Table 6.2.

Table 6.2: Performance of $d_1(\cdot)$ and $d_*(\cdot)$

Decision rule	Point estimates of error probability	
	by reclassification of training sample	by classification of examination sample
Classical	0.40	0.80
Robust	0.17	0.20

To reduce the influence of outliers we calculated the Huber robust estimates (see (Huber, 1981)) of the conditional mean vectors:

$$\tilde{\mu}_1 = \begin{pmatrix} 4.79 \\ 26.47 \\ 2.61 \end{pmatrix}, \ \tilde{\mu}_2 = \begin{pmatrix} 1.95 \\ 12.22 \\ 1.61 \end{pmatrix}$$

and the inverse covariance matrices:

$$\tilde{\Sigma}_1^{-1} = \begin{pmatrix} 0.27 & -0.01 & -0.25 \\ -0.01 & 0.02 & -0.01 \\ -0.25 & -0.01 & 4.85 \end{pmatrix},$$

$$\tilde{\Sigma}_2^{-1} = \begin{pmatrix} 1.70 & -0.21 & -1.19 \\ -0.21 & 0.11 & -0.53 \\ -1.19 & -0.53 & 14.67 \end{pmatrix},$$

which significantly differ from the classical ones. For robust recognition of Ω_1, Ω_2 we constructed the plug-in decision rule $d = d_*(x; A)$ by the substitution of $\{\tilde{\mu}_i, \tilde{\Sigma}_i\}$ for unknown true values of parameters $\{\mu_i, \Sigma_i\}$. Performance characteristics of this decision rule are also presented in Table 6.2. One can see considerable superiority of the decision rule $d_*(\cdot)$. Note that robust decision rules allowed not only to detect cancer but also to recognize its form and stage.

Chapter 7

Cluster Analysis under Distorted Model Assumptions

This chapter is devoted to new problems of robust statistical pattern recognition with unclassified training samples (robust cluster analysis) under distortions of hypothetical models of data. We investigate here four types of distortions, common in applications: 1) small-sample effects; 2) presence of "run structure" in the observed samples; 3) Markov dependence of class indices; 4) presence of outliers in the observed samples. Estimates for robustness of traditional decision rules are found, and new cluster analysis procedures are constructed for the indicated types of distortions.

7.1 Small-sample Effects and Robustness of Unsupervised Decision Rules

Up to now in construction and robustness analysis of adaptive decision rules we assumed that the training sample A is classified and consists of L subsamples $A_1, \ldots, A_L$, corresponding to the classes $\Omega_1, \ldots, \Omega_L$. Let us consider now a higher level of prior uncertainty, when the training sample A of size n is unclassified: $A^T = (x_1^T \vdots \ldots \vdots x_n^T)$. The process of adaptation of a pattern recognition algorithm by unclassified training samples is called *self-learning (learning without teacher, unsupervised learning)*, and the corresponding decision rules are called *unsupervised decision rules.* In this section, we shall investigate problems of robustness analysis for unsupervised decision rules with respect to the effect of size finiteness of the unclassified training sample (type D.1.1 distortions).

Suppose that in the feature space R^N random observations of objects from two classes Ω_1, Ω_2 are registered with probabilities $\pi_1, \pi_2 = 1 - \pi_1$ (note that the conclusions made in this section may also be generalized to $L > 2$ classes). An observation from Ω_i is a random vector of feature variables $X_i \in R^N$ with probability density function $q(x, \theta_i^0)$ $(i \in S = \{1, 2\})$. The true values of parameters $\theta_1^0, \theta_2^0 \in \Theta \subset R^N$ and, probably, π_1, π_2 are unknown. An unclassified training sample A of size n from Ω_1, Ω_2 is observed; the observations $x_1, \ldots, x_n \in R^N$ are mutually independent. Let us use a $(0-1)$-loss matrix (1.18).

Now we shall define the plug-in decision rule using the sample A. Let $\hat{\theta}_1, \hat{\theta}_2$ be any consistent estimators of θ_1^0, θ_2^0 by the sample A. Then the plug-in decision rule is constructed from the Bayesian decision rule similarly to (3.4), (3.5):

$$d = d(x; \hat{\theta}) = \mathbf{1}(G(x; \hat{\theta})) + 1, \quad x \in R^N, \tag{7.1}$$

where $\hat{\theta}^T = (\hat{\theta}_1^T \vdots \hat{\theta}_2^T) \in R^{2m}$ is an estimator for the composed vector of parameters $\theta^T = (\theta_1^T \vdots \theta_2^T) \in R^{2m}$;

$$G(x; \hat{\theta}) = (1 - \pi_1) q(x; \hat{\theta}_2) - \pi_1 q(x; \hat{\theta}_1) \tag{7.2}$$

is the estimator of the Bayesian discriminant function. Note that the plug-in decision rule (7.1) can be used to classify A:

$$\hat{d}_t = d(x_t; \hat{\theta}), t = \overline{1, n}. \tag{7.3}$$

When classifying a random observation $X \in R^N$ independent from A and belonging to the class Ω_{ν^0} ($\mathbf{P}\{\nu^0 = i\} = \pi_i, i \in S$) we shall evaluate the effectiveness of the plug-in decision rule (7.1) and of the decisions (7.3) by the error probability r :

$$r = \mathbf{P}\{d(X; \hat{\theta}) \neq \nu^0\}. \tag{7.4}$$

Consider the following problems, whose importance was noted in (Patrick, 1972), (Milenkij, 1975), (Repin *et al.*, 1977), (Van Ryzin, 1977), (Bock, 1989):

1) evaluation of error probability r as a function of sample size n and finding conditions for asymptotic robustness of the plug-in decision rule (7.1);

2) comparison of the plug-in decision rule robustness in the cases of classified and unclassified training sample, i.e., for supervised and unsupervised classifiers;

3) finding an admissible sample size n that guarantees a fixed level of robustness.

In order to solve these problems, we shall construct an asymptotic expansion of the error probability r. The training sample A is a random sample of size n from the mixture of two distributions:

$$p(x; \theta^0) = (1 - \pi_1) q(x; \theta_2^0) + \pi_1 q(x; \theta_1^0). \tag{7.5}$$

Because A is unclassified and we want to avoid an ambiguity arising from different indexing of classes (see, e.g., (Milenkij, 1975)), we shall assume that $\theta_2^0 \succ \theta_1^0$ (here "$\succ$" is the symbol of lexicographic comparison). In the assumed model $p(\cdot; \theta^0)$ is an unknown element of the family of mixtures:

$$\mathcal{P} = \{p(x; \theta) : \theta^T = (\theta_1^T \vdots \theta_2^T); \theta_1, \theta_2 \in \Theta, \theta_2 \succ \theta_1\}.$$

As an estimator $\hat{\theta}$, we shall use (taking into account the results developed in Section 3.2) the ML-estimator:

$$\hat{\theta} = \arg\max_{\theta} \sum_{t=1}^{n} \ln p(x_t; \theta). \tag{7.6}$$

Note that in (Patrik, 1972), (Milenkij, 1975) some numerical solution methods for the multiextremum problem (7.6) are formulated and investigated.

First, let us investigate statistical properties of the estimators $\hat{\theta}_1, \hat{\theta}_2$, determined by (7.6), and compare them with the ML-estimators of θ_1^0, θ_2^0 in the situation where the classification of A is known *a priori*. We shall assume that the family of probability densities $\{q(\cdot;\theta_1) : \theta_1 \in \Theta\}$ satisfies the following regularity conditions:

C_1) θ_1 is an identifiable parameter, i.e.,

$$\mathbf{E}_{\theta_*}\{\ln q(X;\theta_*)\} > \mathbf{E}_{\theta_*}\{\ln q(X;\theta)\} \quad (\theta_* \neq \theta);$$

C_2) for any compact set $K \subset \Theta$ and any points $\theta_1^0, \theta_2^0 \in K$ there exist neighborhoods $U_{\theta_1^0}, U_{\theta_2^0} \subset K$ such that for some $a, c > 1, b > 2$, for any neighborhood $U \subset U_{\theta_2^0}$ and for any $\theta_1 \in U_{\theta_1^0}, \theta_2 \in U_{\theta_2^0}$ the functions

$$|\ln q(x;\theta_k)|^a, \quad (\sup_{\theta' \in U} |\ln q(x;\theta')|)^a,$$

$$\left|\frac{\partial^2 \ln q(x;\theta_k)}{d\theta_{ki} d\theta_{kj}}\right|^b, \left|\frac{\partial \ln q(x;\theta_k)}{\partial\theta_{ki}} \cdot \frac{\partial \ln q(x;\theta_*)}{\partial\theta_{*j}}\right|^b,$$

$$\left|\frac{\partial^3 \ln q(x;\theta_k)}{\partial\theta_{ki}\partial\theta_{kj}\partial\theta_{kt}}\right|^c, \left|\frac{\partial \ln q(x;\theta_s)}{\partial\theta_{st}} \cdot \frac{\partial^2 \ln q(x;\theta_k)}{\partial\theta_{ki}\partial\theta_{kj}}\right|^c$$

are uniformly integrable with respect to probability density function $q(x;\theta_*)$, $\theta_* \in K$; $k, s \in \{1,2\}; i, j, t = \overline{1,m}$; uniform integrability of $f(x;\theta_k)$ means that as $z \to \infty$,

$$\int_{|f(x;\theta_k)|>z} |f(x;\theta_k)| \, q(x;\theta_*)dx \to 0;$$

C_3) $\mathbf{E}_{\theta_k^0}\{\nabla_{\theta_k^0} \ln q(X_k;\theta_k^0)\} = 0, \quad \theta_k^0 \in \Theta;$

C_4) The Fisher information matrices

$$H_k = \mathbf{E}_{\theta_k^0}\{-\nabla^2_{\theta_k^0} \ln q(X_k;\theta_k^0)\}, \quad J = J(\theta^0) = \mathbf{E}_{\theta^0}\{-\nabla^2_{\theta^0} \ln p(X;\theta^0)\}$$

are positive definite, and moreover, the minimal eigenvalues of these matrices are separated from zero.

As in Lemma 3.1, the following statement can be proved.

Lemma 7.1 *If the regularity conditions $C_1 - C_4$ hold, then a random deviation $\Delta\theta = \hat{\theta} - \theta^0$ of the ML-estimator (7.6) has third order moments and as $\tau = 1/\sqrt{n} \to 0$ the following asymptotic expansions hold:*

– *for the bias:*

$$B(\theta^0) = \mathbf{E}_{\theta^0}\{\Delta\theta\} = \mathbf{1}_{2m} \cdot \mathcal{O}(\tau^3);$$

– *for the covariance matrix:*

$$\mathbf{E}_{\theta^0}\{\Delta\theta(\Delta\theta)^T\} = \tau^2 V + \mathbf{1}_{2m\times 2m} \cdot \mathcal{O}(\tau^3), \quad V = J^{-1};$$

– *for the third order moments ($k, l, s \in S; i, j, t = \overline{1, m}$):*

$$\mathbf{E}_{\theta^0}\{\mid (\hat{\theta}_{ki} - \theta^0_{ki})(\hat{\theta}_{lj} - \theta^0_{lj})(\hat{\theta}_{st} - \theta^0_{st}) \mid\} = \mathcal{O}(\tau^3);$$

where $\mathbf{1}_{2m}, \mathbf{1}_{2m\times 2m}$ are the $(2m)$-vector-column and the $(2m \times 2m)$-matrix whose all elements are equal to 1; $\tau^2 V$ is an asymptotic expression of the covariance matrix for the estimator $\hat{\theta}$.

Let us assume the notations: $G(x; \theta^0)$ is the Bayesian discriminant function determined by (7.1);

$$r_0 = \mathbf{P}\{d(X; \theta^0) \neq \nu^0\}$$

is the Bayesian error probability (for the Bayesian decision rule $d = d(x; \theta^0) = \mathbf{1}(G(x; \theta^0)) + 1$);

$$\Gamma = \{x : G(x; \theta^0) = 0\} \subset R^N$$

is the Bayesian discriminant hypersurface;

$$Q(x) = (\nabla_{\theta^0} G(x; \theta^0))^T J^{-1} \nabla_{\theta^0} G(x; \theta^0) \geq 0, \quad x \in R^N. \tag{7.7}$$

Theorem 7.1 *If the conditions $C_1 - C_4$ are satisfied, the probability density function $q(x; \theta^0_k)$ has derivatives with respect to x ($x \in R^N, k \in S$) and*

$$\int_\Gamma Q(x) \mid \nabla_x G(x; \theta^0) \mid^{-1} ds_{N-1} < \infty,$$

then the error probability of the decision rule (7.1) admits the asymptotic expansion

$$r = r_0 + \frac{\alpha}{n} + \mathcal{O}(n^{-3/2}), \tag{7.8}$$

where

$$\alpha = \frac{1}{2} \int_\Gamma Q(x) \mid \nabla_x G(x; \theta^0) \mid^{-1} ds_{N-1} \geq 0. \tag{7.9}$$

Proof is conducted by lemma 7.1 in the same way as the proof of Theorems 3.1, 3.3.

■

Corollary 7.1 *Under the conditions of Theorem 7.1 the decision rule (7.1) is consistent: $r \to r_0$ at $n \to \infty$.*

For comparison let us present the asymptotic expansion of error probability r_-^*, which follows from (3.33), for the case when the sample A is classified and the number of observations in the sample A from class Ω_i is equal to $n_i = n \cdot \pi_i$:

$$r_-^* = r_0 + \frac{\rho}{n} + \mathcal{O}(n^{-3/2}), \tag{7.10}$$

where

$$\rho = \frac{\rho_1^*}{\pi_1} + \frac{\rho_2^*}{\pi_2},$$

$$\rho_k^* = \frac{1}{2} \int\limits_{\Gamma} Q_k(x) \mid \nabla_x G(x;\theta^0) \mid^{-1} ds_{N-1} \geq 0, \tag{7.11}$$

$$Q_k(x) = \pi_k^2 (\nabla_{\theta_k^0} q(x;\theta_k^0))^T H_k^{-1} \nabla_{\theta_k^0} q(x;\theta_k^0) \geq 0.$$

It is seen from comparison of (7.8) and (7.10) that the convergence orders of the error probability to r_0 in the cases both of classified sample A and of unclassified sample are the same: $\mathcal{O}(n^{-1})$. But the convergence rates are different and are determined by the coefficients ρ and α respectively. Let us find a relation between these coefficients.

We shall write the Fisher information matrix $J = J(\theta^0)$ for the composed vector of parameters $\theta^{0T} = (\theta_1^{0T} \vdots \theta_2^{0T}) \in R^{2m}$ in block form:

$$J = \begin{pmatrix} J_{11} & \vdots & J_{12} \\ \dots & \vdots & \dots \\ J_{21} & \vdots & J_{22} \end{pmatrix}$$

and define auxiliary $2m \times 2m$-matrices:

$$J_* = \begin{pmatrix} \pi_1 H_1 & \vdots & \mathbf{0}_{m\times m} \\ \dots\dots & \vdots & \dots\dots \\ \mathbf{0}_{m\times m} & \vdots & \pi_2 H_2 \end{pmatrix}$$

$$E = \begin{pmatrix} E_{11} & \vdots & -E_{12} \\ \dots\dots & \vdots & \dots\dots \\ -E_{21} & \vdots & E_{22} \end{pmatrix} = (\varepsilon_{ij}), \quad i,j = \overline{1,2m}, \tag{7.12}$$

where $\mathbf{0}_{m\times m}$ is the zero $(m \times m)$-matrix, and

$$E_{kl} = \int\limits_{R^N} \frac{\pi_1 q(x;\theta_1^0)\pi_2 q(x;\theta_2^0)}{\pi_1 q(x;\theta_1^0) + \pi_2 q(x;\theta_2^0)} F_{kl}(x)dx, \tag{7.13}$$

$$F_{kl}(x) = \nabla_{\theta_k^0} \ln q(x;\theta_k^0)(\nabla_{\theta_l^0} \ln q(x;\theta_l^0))^T \quad (k,l \in S).$$

Theorem 7.2 *The Fisher information matrix J for a $(2m)$-dimensional vector of parameters θ^0 may be represented as follows:*

$$J = J_* - E. \tag{7.14}$$

Proof. First, note that

$$\pi_1 q(x;\theta_1^0)/(\pi_1 q(x;\theta_1^0) + \pi_2 q(x;\theta_2^0)) \le 1,$$

therefore, by the regularity condition C_2, the integrals (7.13) exist. Let us now verify (7.14) blockwise. For any $k, l \in S$, according to the condition C_3, we have:

$$J_{kl} = \mathbf{E}_{\theta^0}\{\nabla_{\theta_k^0} \ln p(X;\theta^0)(\nabla_{\theta_l^0} \ln p(X;\theta^0))^T\} =$$

$$= \int\limits_{R^N} \frac{\pi_k q(x;\theta_k^0)\pi_l q(x;\theta_l^0)}{\pi_1 q(x;\theta_1^0) + \pi_2 q(x;\theta_2^0)} F_{kl}(x)dx.$$

From this fact for $k \ne l$, using (7.12) and (7.13) we conclude that (7.14) holds for nondiagonal blocks. For diagonal blocks for $k = l$ we shall use the notation $t = 3-k$ and the equality $H_k = \mathbf{E}_{\theta_k^0}\{F_{kk}(X_k)\}$:

$$J_{kk} = \int\limits_{R^N} \frac{\pi_k q(x;\theta_k^0)(\pi_k q(x;\theta_k^0) + \pi_t q(x;\theta_t^0)) - \pi_k q(x;\theta_k^0)\pi_t q(x;\theta_t^0)}{\pi_k q(x;\theta_k^0) + \pi_t q(x;\theta_t^0)} \times$$

$$\times F_{kk}(x)dx = \pi_k H_k - E_{kk},$$

and the latter expression corresponds to (7.14). ■

Let us analyze the properties of the matrix E.

Lemma 7.2 *The matrix E is symmetric and is nonnegative definite.*

Proof. The symmetry immediately follows from (7.12), (7.13). For any row-vector $z^T = (z_1^T \vdots z_2^T) \in R^{2m}$ we have:

$$z^T E z = \sum_{k,l=1}^{2} z_k^T E_{kl} z_l =$$

$$= \int\limits_{R^N} \frac{\pi_1 q(x;\theta_1^0)\pi_2 q(x;\theta_2^0)}{\pi_1 q(x;\theta_1^0) + \pi_2 q(x;\theta_2^0)} \left(\sum_{k=1}^{2} (-1)^{k+1} z_k^T \nabla_{\theta_k^0} \ln q(x;\theta_k^0) \right)^2 dx \geq 0,$$

and this means that E is nonnegative definite.

■

Lemma 7.3 *If in (7.13) $F_{kl}(x)$ is a matrix, with all elements equal to 1, then for any $i, j = \overline{1, 2m}$,*

$$\mid \varepsilon_{ij} \mid \leq r_0.$$

Proof. Denote

$$\Gamma = \frac{\pi_1 q(x;\theta_1^0)\pi_2 q(x;\theta_2^0)}{\pi_1 q(x;\theta_1^0) + \pi_2 q(x;\theta_2^0)}.$$

According to (7.12), (7.13),

$$\mid \varepsilon_{ij} \mid \leq \int\limits_{R^N} \Gamma dx = \int\limits_{V_1} \Gamma dx + \int\limits_{V_2} \Gamma dx, \tag{7.15}$$

where

$$V_1 = \{x : \pi_1 q(x;\theta_1^0) > \pi_2 q(x;\theta_2^0)\},$$
$$V_2 = \{x : \pi_1 q(x;\theta_1^0) < \pi_2 q(x;\theta_2^0)\},$$

are the regions of the Bayesian decision $d(x;\theta^0)$ making in favor of the classes Ω_1 and Ω_2 respectively. Further, if we denote $t = 3 - k$, then:

$$\mid \varepsilon_{ij} \mid \leq \sum_{k=1}^{2} \pi_k \int\limits_{V_t} \left(1 + \frac{\pi_k q(x;\theta_k^0)}{\pi_t q(x;\theta_t^0)} \right)^{-1} q(x;\theta_k^0) dx \leq$$

$$\leq \sum_{k=1}^{2} \pi_k \mathbf{P}\{d(X_k;\theta^0) \neq k\} = r_0.$$

■

Corollary 7.2 *The matrix E_{kl} can be represented as:*

$$E_{kl} = \sum_{s=1}^{2} \pi_s \int\limits_{V_t} \left(1 + \frac{\pi_s q(x;\theta_s^0)}{\pi_t q(x;\theta_t^0)} \right)^{-1} F_{kl}(x) q(x;\theta_s^0) dx, \tag{7.16}$$

$$E_{kl} = \sum_{j=0}^{\infty} (-1)^j \sum_{s=1}^{2} \int\limits_{V_t} \frac{(\pi_s q(x;\theta_s^0))^{j+1}}{(\pi_t q(x;\theta_t^0))^j} F_{kl}(x) dx, \quad t = 3 - s.$$

Corollary 7.3 *If the classes do not overlap:*

$$q(x;\theta_1^0)q(x;\theta_2^0) \equiv 0,$$

then $J = J_*$.

It is seen from Lemma 7.3 and Corollaries 7.2, 7.3 that if the "overlapping of classes" (Milenkij, 1975) decreases, i.e., the interclass distance increases (and, consequently, the Bayesian error probability decreases), then

$$\varepsilon_+ = \max_{i,j=\overline{1,2m}} |\varepsilon_{ij}| \to 0. \tag{7.17}$$

The asymptotics (7.17) is of practical importance, because under "large overlapping of classes" (when (7.17) is violated) the value r_0 is large and it is not recommended to apply the decision rule (7.1).

Theorem 7.3 *If the conditions of the Theorem 7.1 are satisfied, then under the asymptotics (7.17), the following expansion takes place:*

$$\alpha = \rho + \lambda + \mathcal{O}(\varepsilon_+^2), \tag{7.18}$$

where

$$\lambda = \frac{1}{2}\int_{\Gamma}\sum_{k,l=1}^{2}(H_k^{-1}\nabla_{\theta_k^0}q(x;\theta_k^0))^T E_{kl} H_l^{-1}\nabla_{\theta_l^0}q(x;\theta_l^0)\times$$
$$|\nabla_x G(x;\theta^0)|^{-1}\, ds_{N-1} \geq 0. \tag{7.19}$$

Proof. By the regularity condition C_4, the inverse matrix exists:

$$J_*^{-1} = \text{diag}\{\frac{1}{\pi_1}H_1^{-1}, \frac{1}{\pi_2}H_2^{-1}\},$$

therefore, according to (7.14),

$$J^{-1} = (\mathbf{I}_{2m} - J_*^{-1}E)^{-1}J_*^{-1} = J_*^{-1} + J_*^{-1}EJ_*^{-1} + \mathcal{O}(\varepsilon_+^2)J_*^{-1}.$$

Further, by (7.2),

$$\nabla_{\theta^0}G(x;\theta^0) = \begin{pmatrix} -\pi_1\nabla_{\theta_1^0}q(x;\theta_1^0) \\ \cdots\cdots\cdots\cdots \\ \pi_2\nabla_{\theta_2^0}q(x;\theta_2^0) \end{pmatrix}.$$

Substituting these expressions into (7.7) and using (7.11), we find:

$$Q(x) = \sum_{k=1}^{2}\frac{1}{\pi_k}Q_k(x) +$$

$$+\sum_{k,l=1}^{2}(\nabla_{\theta_k^0}q(x;\theta_k^0))^T H_k^{-1}E_{kl}H_l^{-1}\nabla_{\theta_l^0}q(x;\theta_l^0)+\mathcal{O}(\varepsilon_+^2).$$

Using this relation in (7.9), according to (7.11), we obtain (7.18), (7.19). The nonnegativity of the term λ in (7.18) follows from the fact determined by Lemma 7.2: the matrix E is nonnegative definite.

■

Corollary 7.4 *The following expansion holds:*

$$r = r_-^* + \frac{\lambda}{n} + \mathcal{O}(n^{-3/2} + \frac{\varepsilon_+^2}{n}).$$

Corollary 7.5 *The decision rule (7.1) that uses an unclassified sample loses to the decision rule that uses a classified sample in convergence rate of the error probability to the minimal Bayesian error probability r_0. The value of this loss is $\lambda = \alpha - \rho = \mathcal{O}(\varepsilon_+) \geq 0$, and it is less when the "overlapping of classes" is less.*

Using (7.8), (7.10), (1.18), we shall obtain the following asymptotic expression for the relative increment of the classification error probability for the unsupervised decision rule with respect to the supervised decision rule:

$$\gamma = \frac{r - r_0}{r_-^* - r_0} = 1 + \frac{\lambda}{\rho} \geq 1. \tag{7.20}$$

The coefficient γ indicates how much times the supervised decision rule is more robust than the unsupervised decision rule. Let us use the constructed expansions to compare the robustness of supervised and unsupervised classifiers using the values of minimal δ-admissible sample size ($\delta > 0$). For the supervised classifier, according to (3.47), (7.10), we shall find the minimal δ-admissible sample size n_δ^- from the condition:

$$\kappa^* = \frac{r_-^* - r_0}{r_0} < \delta, \quad n_\delta^- \approx \left\lfloor \frac{1}{\delta r_0}\sum_{k=1}^{2}\frac{\rho_k}{\pi_k}\right\rfloor + 1. \tag{7.21}$$

For the unsupervised classifier, according to (7.8), (7.18), the δ-admissible sample size n_δ can be determined from the condition:

$$\kappa = \frac{r - r_0}{r_0} < \delta, \quad n_\delta \approx n_\delta^- + \left\lfloor \frac{\lambda}{r_0\delta}\right\rfloor + 1. \tag{7.22}$$

Comparing (7.21) and (7.22), we can conclude that to achieve the same level $(1+\delta)r_0$ of error probability, the unsupervised classifier requires a sample of size larger by

$$n_\delta - n_\delta^- \approx \left\lfloor \frac{\lambda}{r_0\delta}\right\rfloor + 1. \tag{7.23}$$

than for the supervised classifier.

This increase of sample size is larger, if δ is smaller and λ/r_0, which depends on the level of "classes overlapping", is larger.

Let us consider now the case often used in applications: both $\{\theta_i^0\}$ and prior probabilities $\{\pi_i\}$ are unknown. In this case, the composed vector of parameters is a $(2m+1)$-dimensional block-vector:

$$\theta^0 = \begin{pmatrix} \theta_1^0 \\ \dots \\ \theta_2^0 \\ \dots \\ \pi_1 \end{pmatrix} \in R^{2m+1},$$

$$p(x;\theta^0) = (1-\pi_1)q(x;\theta_2^0) + \pi_1 q(x;\theta_1^0),$$

$$G(x;\theta^0) = (1-\pi_1)q(x;\theta_2^0) - \pi_1 q(x;\theta_1^0).$$

Theorem 7.4 *The information matrix J for the $(2m+1)$–dimensional vector of parameters θ^0 can be expressed in the form (7.14), where*

$$J = \begin{pmatrix} J_{11} & \vdots & J_{12} & \vdots & J_{13} \\ .. & \vdots & .. & \vdots & .. \\ J_{21} & \vdots & J_{22} & \vdots & J_{23} \\ .. & \vdots & .. & \vdots & .. \\ J_{31} & \vdots & J_{32} & \vdots & J_{33} \end{pmatrix},$$

$$J_* = \begin{pmatrix} \pi_1 H_1 & \vdots & \mathbf{0}_{m\times m} & \vdots & \mathbf{0}_{m\times 1} \\ \dots\dots & \vdots & \dots\dots & \vdots & \dots\dots\dots \\ \mathbf{0}_{m\times m} & \vdots & \pi_2 H_2 & \vdots & \mathbf{0}_{m\times 1} \\ \dots\dots & \vdots & \dots\dots & \vdots & \dots\dots\dots \\ \mathbf{0}_{1\times m} & \vdots & \mathbf{0}_{1\times m} & \vdots & (\pi_1\pi_2)^{-1} \end{pmatrix},$$

$$E = \begin{pmatrix} E_{11} & \vdots & -E_{12} & \vdots & E_{13} \\ \dots\dots & \vdots & \dots\dots & \vdots & \dots\dots \\ -E_{21} & \vdots & E_{22} & \vdots & -E_{23} \\ \dots\dots & \vdots & \dots\dots & \vdots & \dots\dots \\ E_{31} & \vdots & -E_{32} & \vdots & E_{33} \end{pmatrix},$$

E_{kl} are determined by (7.13) and by the following equalities:

$$F_{s3}(x) = \frac{1}{\pi_2\pi_1}\nabla_{\theta_s^0} \ln q(X;\theta_s^0) \quad (s \in S), \quad F_{33}(x) = (\pi_1^2\pi_2^2)^{-1}.$$

Proof is conducted as the proof of Theorem 7.2. ■

As in the case with known $\{\pi_i\}$, the asymptotic expansion of error probability is determined by Theorem 7.1 (with the corresponding modification of the function $Q(x)$) and has the form:

$$r' = r_0 + \frac{\beta}{n} + \mathcal{O}(n^{-3/2}). \tag{7.24}$$

Theorem 7.5 *If the conditions of Theorem 7.1 are satisfied, then, under the asymptotics (7.17), the following expansion holds:*

$$\beta = \alpha + \lambda' + \mathcal{O}(\varepsilon_+^2), \tag{7.25}$$

where

$$\lambda' = \frac{1 + \pi_1\pi_2 E_{33}}{2}\left(\frac{\pi_1}{\pi_2}\right)\int_\Gamma q^2(x;\theta_1^0) \mid \nabla_x G(x;\theta^0)\mid^{-1} ds_{N-1} + \tag{7.26}$$

$$+\pi_1^2 \sum_{s=1}^{2} \pi_s^{-1} \int_\Gamma q^2(x;\theta_1^0)(\nabla_{\theta_s^0} \ln q(x;\theta_s^0))^T H_s^{-1} E_{s3} \mid \nabla_x G(x;\theta^0)\mid^{-1} ds_{N-1}.$$

Proof is conducted as the proof of Theorem 7.3. ■

It is seen from Theorem 7.5 that in the case of unknown $\{\pi_i\}$, the deceleration of error probability convergence to r_0 is

$$\beta - \alpha = \lambda' = \mathcal{O}(1) \geq 0.$$

Using (7.8), (7.10), (7.24), (7.25), we shall find, similarly to (7.20), an asymptotic expression of relative error probability increment for the unsupervised decision rule (when $\theta_1^0, \theta_2^0, \pi_1$ are unknown) with respect to the supervised decision rule:

$$\gamma' = \frac{r' - r_0}{r_-^* - r_0} = 1 + \frac{\lambda + \lambda'}{\rho} = \gamma + \frac{\lambda'}{\rho} \geq 1. \tag{7.27}$$

From the condition $\kappa' = (r' - r_0)/r_0 < \delta$, similarly to (7.21) - (7.23), we obtain the expressions for the δ-admissible sample size:

$$n_\delta' \approx n_{\bar{\delta}} + \left\lfloor \frac{\lambda + \lambda'}{r_0\delta} \right\rfloor \approx n_\delta + \left\lfloor \frac{\lambda'}{r_0\delta} \right\rfloor + 1. \tag{7.28}$$

It follows from (7.28) that for unknown $\{\pi_i\}$, in order to achieve the same level $(1+\delta)r_0$ of error probability, the unsupervised classifier needs the following increase of sample size:

$$n_\delta' - n_{\bar{\delta}} \approx \left\lfloor \frac{\lambda + \lambda'}{r_0\delta} \right\rfloor.$$

Let us present now some results of computer experiments. Let the classes Ω_1, Ω_2 be equiprobable ($\pi_1 = \pi_2 = 0.5$) and described by Gaussian probability distributions:

$$q(x; \theta_k^0) = n_N(x \mid \theta_k^0, \Sigma), \quad k \in S,$$

where Σ is a known covariance matrix. The results of Section 3.3 and (7.10) together imply that

$$r_0 = \Phi(-\Delta/2), \quad \Gamma = \{x : (\theta_1^0 - \theta_2^0)^T \Sigma^{-1}(x - (\theta_1^0 + \theta_2^0)/2) = 0\},$$

$$\rho = \frac{N - 1 + \Delta^2/4}{\sqrt{2\pi}\Delta} e^{-\Delta^2/8},$$

where Δ is the interclass Mahalanobis distance.

Now let us use Theorems 7.1–7.5. Here it is convenient to exploit the following property in (7.16):

$$\frac{(\pi_s q(x; \theta_s^0))^{j+1}}{(\pi_t q(x; \theta_t^0))^j} = e^{j(j+1)\Delta^2/2} n_N(x \mid \theta_t^0 + (j+1)(\theta_s^0 - \theta_t^0), \Sigma),$$

where $t = 3 - s$.

Performing computations using (7.16), we shall obtain:

$$E_{11} = E_{22} = \sqrt{\pi/2}\frac{e^{-\Delta^2/8}}{\Delta}\left((1 - \frac{4}{\Delta^2}(1 + \frac{\pi^2}{8}))\Sigma^{-1}(\theta_2^0 - \theta_1^0)(\theta_2^0 - \theta_1^0)^T\Sigma^{-1} + \right.$$

$$\left. +4\Sigma^{-1} + \mathcal{O}(\Delta^{-2})\mathbf{1}_{m\times m}\right),$$

$$E_{12} = E_{21} = -\sqrt{\pi/2}\frac{e^{-\Delta^2/8}}{\Delta}\left((1 + \frac{4}{\Delta^2}(1 - \frac{\pi^2}{8}))\times \right.$$

$$\left. \times\Sigma^{-1}(\theta_2^0 - \theta_1^0)(\theta_2^0 - \theta_1^0)^T\Sigma^{-1} - 4\Sigma^{-1} + \mathcal{O}(\Delta^{-2})\mathbf{1}_{m\times m}\right),$$

$$E_{33} = (4(2\pi)^{1/2}/\Delta)\frac{e^{-\Delta^2/8}}{\Delta}\left(1 - \frac{\pi^2}{2\Delta^2} + \mathcal{O}(\Delta^{-4})\right),$$

$$E_{s3} = (-1)^{s+1}E_{22}\Sigma^{-1}(\theta_2^0 - \theta_1^0)/8, \quad s \in S.$$

Taking into account $H_k = \Sigma^{-1}$ and using (7.19), (7.26), we find the main expansion terms:

$$\lambda = \frac{\Delta^2}{16}e^{-\Delta^2/4}(1 + \mathcal{O}(\Delta^{-2})),$$

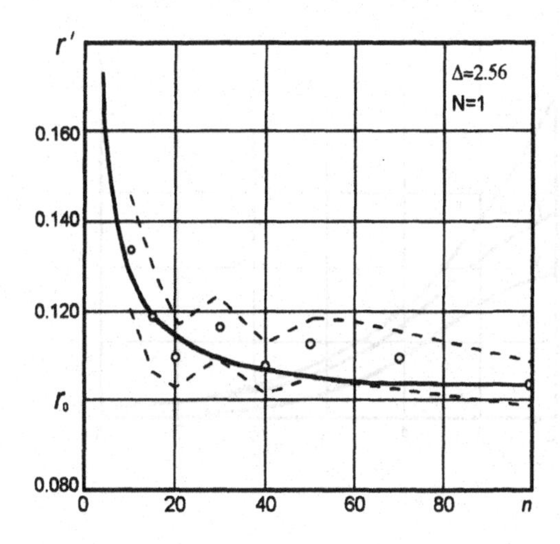

Figure 7.1: Error probability vs. sample size

$$\lambda' = e^{-\Delta^2/8}/((2\pi)^{1/2}\Delta) + e^{-\Delta^2/4}\Delta^{-2} + \mathcal{O}(e^{-\Delta^2/4}\Delta^{-3}).$$

Figure 7.1. plots the dependence of error probability $r' = r_0 + (\rho + \lambda + \lambda')/n$ on n by solid line for the decision rule (7.1), for $N = 1$, $\Delta = 2.56$ ($r_0 = 0.1$). The plot is computed by the asymptotic expansions constructed above. The points indicate the estimations of r' by computer modeling (50 independent experiments); the dashed lines indicate 90%–confidence intervals for r'. It is seen that the accuracy of the approximation formula for r' based on the constructed expansions is sufficiently high.

According to (7.20), (7.27), let us give asymptotic expressions for relative error probability increment for the unsupervised classifier in comparison with the supervised classifier (for the cases of known $\{\pi_i\}$ and unknown $\{\pi_i\}$):

$$\gamma \approx 1 + \sqrt{\pi/2}\frac{e^{-\Delta^2/8}\Delta}{2}, \quad \gamma' \approx \gamma + \frac{1}{N - 1 + \Delta^2/4}. \tag{7.29}$$

Figure 7.2 plots the dependencies of γ and γ' on Δ determined by (7.29). It is seen that if the interclass distance Δ increases then the effectivenesses of the supervised and unsupervised classifiers become close to each other.

By means of (7.21), (7.22), (7.28) let us determine the δ-admissible sample sizes for the supervised decision rule and the unsupervised decision rule (for two variants: when $\{\pi_i\}$ are known and unknown):

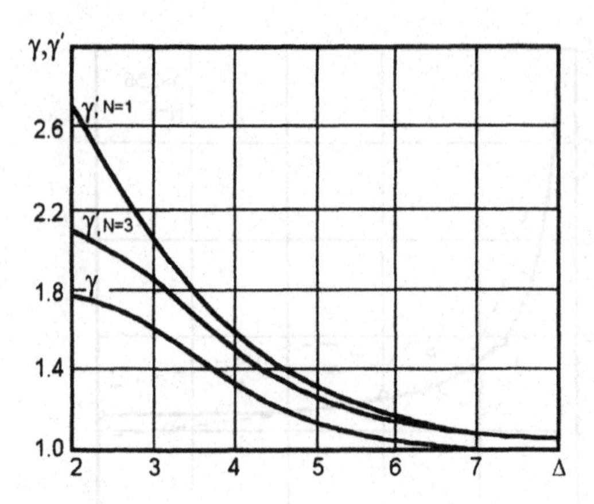

Figure 7.2: Relative increments of error probability vs. interclass distance

$$n_\delta^- \approx \left\lfloor \frac{1}{\delta}\left(\frac{\Delta^2}{8} + \frac{N}{2}\right)\right\rfloor + 1,$$

$$n_\delta \approx n_\delta^- + \left\lfloor \frac{\Delta^3}{16\delta}\sqrt{\pi/2}e^{-\Delta^2/8}\right\rfloor + 1, \tag{7.30}$$

$$n_\delta^{'} \approx n_\delta + \left\lfloor \frac{1}{\delta}\left(\frac{1}{2} + \frac{2}{\Delta^2} + \sqrt{\pi/2}\frac{1}{\Delta}e^{-\Delta^2/8}\right)\right\rfloor + 1.$$

Figure 7.3 plots the dependencies (7.30) for $N = 2, \delta = 0.2$. Figure 7.3 shows, for example, that for the supervised classifier error probability to be less than $(1+\delta)r_0 = 0.08$ (when $\Delta = 3$, $r_0 = 0.067$), the sample size n_δ^- should be at least 11, whereas for the unsupervised classifier it should be at least 15 (if $\{\pi_i\}$ are known) and at least 20 (if $\{\pi_i\}$ are unknown). If Δ increases, the plots $n_\delta, n_\delta^{'}$ come close to n_δ^- : for example, for $\Delta = 6 : n_\delta^- = 28$, $n_\delta = 29$, $n_\delta^{'} = 32$.

7.2 Cluster Analysis for Random-length Runs of Observations

Many problems of unsupervised statistical pattern recognition are stated as problems of cluster analysis (see e.g., (Fukunaga, 1972), (Patric, 1972), (McLachnan, 1992)).

In the existing theory of cluster analysis of multivariate statistical data (Aivazyan *et al.*, 1989), (Bock, 1989) the traditional model assumption is the assumption about

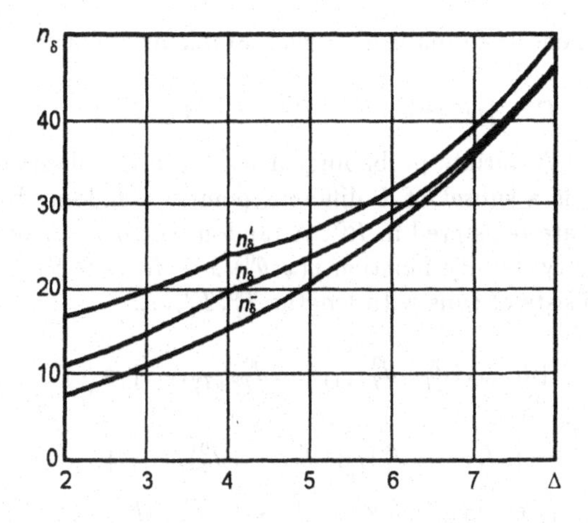

Figure 7.3: δ–admissible sample sizes vs. interclass distance

independence of observations, so that their arbitrary permutation is admissible. But in classification of meteorological (Anderson, 1958) and geophysical data (Gorjan *et al.*, 1978), (Devore, 1973), in medical and technical diagnostics (Artemjev, 1979), (Kazakov *et al.*, 1980) the observations are often essentially time-ordered (or ordered by other parameters) and form a time series (Anderson, 1971).

To illustrate this, consider a problem of technical diagnostics that consists in identification of a nonstationary dynamic system with $L \geq 2$ different modes of functioning by means of indirect observations (feature vectors) $x_1, x_2, \ldots, x_n \in R^N$ registered at discrete time moments $t = \overline{1,n}$ (Artemjev, 1979), (Kazakov *et al.*, 1980). An i-th mode (class) is described by its intrinsic probability distribution, and the system possesses the property of *piecewise stationarity,* or *inertiality:* if at moment $t+1$ the i-th mode started, then it keeps existing at the moments $t+1, \ldots, t+T_1^0$. Further, at moment $t+T_1^0+1$ this mode can be changed to the j-th mode ($j \neq i$), which keeps to exist during T_2^0 time units, and so on. The *inertiality interval lengths (run lengths)* $T_1^0, T_2^0, \ldots$ are unknown and assumed to be independent random variables with given probability distribution. This application problem of cluster analysis consists in estimating the true sequence of indices $d_1^0, \ldots, d_n^0 \in S = \{1, 2, \ldots, L\}$ of functioning modes.

For the first time this type of problems of cluster analysis was considered by (Kharin, 1985) in the particular case when the lengths $T_1^0, T_2^0, \ldots$ are divisible by the same unknown number $T^0 (T_- \leq T^0 \leq T_+)$. This case will be considered in the next subsection.

Let us present a general formalism for such problems of cluster analysis. Let

$$\mathcal{P} = \{p(x;\theta), \quad x \in R^N : \theta \in \Theta \subseteq R^m\}$$

be a regular family of N-variate probability densities in the observation space R^N; $\{\theta_1^0, \theta_2^0, \ldots, \theta_L^0\} \subset \Theta$ is a subset of L different points. Random observations from L classes $\Omega_1, \ldots, \Omega_L$ are registered in R^N. A random vector of N observed features from Ω_i has probability density function $p(x;\theta_i^0), x \in R^N (i \in S)$. The sequence of true class indices consists of runs with lengths $T_1^0, T_2^0, \ldots$:

$$\{d_1^0, \ldots, d_{T_1^0}^0, d_{T_1^0+1}^0, \ldots, d_{T_1^0+T_2^0}^0, \ldots\} = \tag{7.31}$$

$$= \{(J_1^0, \cdots, J_1^0)_{T_1^0}, (J_2^0, \cdots, J_2^0)_{T_2^0}, \ldots\},$$

where $J_k^0 \in S$ is a common class index for the k-th run ($k = 1, 2, \ldots$):

$$d_{T_1^0+\ldots+T_{k-1}^0+1}^0 = \ldots = d_{T_1^0+\ldots+T_k^0}^0 = J_k^0;$$

and T_k^0 is the length of the k-th run. Here $\{T_k^0, J_k^0\}$ are mutually independent random variables with probability distribution

$$\mathbf{P}\{T_k^0 = i\} = q_i \quad (i = 1, 2, \ldots); \quad \mathbf{P}\{J_k^0 = l\} = \pi_l \quad (l \in S), \tag{7.32}$$
$$k = 1, 2, \ldots .$$

A random sequence of n observations $X = (x_1, x_2, \ldots, x_n) \in R^N$ is observed; true classification of its elements is determined by (7.31): an observation $x_t \in R^N$ at moment t is an observation from the class $\Omega_{d_t^0}$; with fixed $d_1^0, d_2^0, \ldots$, the observations $x_1, x_2, \ldots$ are conditionally independent. The true values of the number of runs K^0, class indices $\{J_k^0\}$, lengths of runs $\{T_k^0\}$ and, maybe, parameters $\{\theta_i^0\}$ are unknown.

Let us consider the following important problems:

1) synthesis of decision rules for classification of a sample $X = (x_1, x_2, \ldots, x_n)$, or, equivalently, for estimation of K^0, $\{T_k^0\}$, $\{J_k^0\}$;

2) decision rule performance analysis.

First, let us consider the case when the parameters of the classes $\{\theta_i^0\}$ are known. Assume the notations: $T_{(k)}^0 = T_1^0 + \ldots + T_k^0$ denotes the moment of possible class index change ($k = \overline{1, K^0}$, $T_{(K^0)}^0 = n$); $T^0 = (T_{(1)}^0, \cdots, T_{(K^0)}^0)$; $J^0 = (J_1^0, \ldots, J_{K^0}^0)$ denotes the vector of classification; K, $J = (J_1, \ldots, J_K)$, and $T = (T_{(1)}, \ldots, T_{(K)})$ denote admissible values of parameters K^0, J^0, T^0 respectively. Let us indicate the discrete ranges for the parameters K, J, T ($T_{(0)} = 0$):

$$2 \le K \le n; \quad J_1, \ldots, J_K \in S; \quad T_{(K)} = n, \tag{7.33}$$
$$T_{(k)} \in A_k(K) = \{k, k+1, \ldots, n - K + k\};$$

$$T_{(k)} \geq T_{(k-1)} + 1 (k = \overline{1, K-1});$$

$X_k = (x_{T_{(k-1)}+1}, \ldots, x_{T_{(k)}})$ denotes the k-th run of observations ($k = \overline{1, K}$).

Let us also denote the statistics:

$$G(X_k; T_{(k-1)}, T_{(k)}) = \max_{i \in S} \left(\ln \pi_i + \sum_{\tau=1}^{T_{(k)}-T_{(k-1)}} \ln p(x_{T_{(k-1)}+\tau}; \theta_i^0) \right);$$

$$f_i(T_{(i)}, T_{(i+1)}) = \ln q_{T_{(i+1)}-T_{(i)}} + G(X_i; T_{(i)}, T_{(i+1)}) +$$

$$+ \delta_{i1} \left(\ln q_{T_{(1)}} + G(X_1; 0, T_{(1)}) \right) +$$

$$+ \delta_{i,K-2} \left(\ln q_{n-T_{(K-1)}} + G(X_K; T_{(K-1)}, n) \right), \quad i = \overline{1, K-2};$$

$$l_1(X; K, T) =$$

$$\begin{cases} \ln(q_{T_{(1)}} q_{n-T_{(1)}}) + G(X_1; 0, T_{(1)}) + G(X_2; T_{(1)}, n), & K = 2, \\ \sum_{i=1}^{K-2} f_i(T_{(i)}, T_{(i+1)}), & K \in \{3, \ldots, n\}; \end{cases}$$

$$l_2(X; K) = \max_T l_1(X; K, T), \tag{7.34}$$

where δ_{ij} is the Kronecker symbol.

Theorem 7.6 *If a random sample X to be classified has the mathematical model (7.31), (7.32), then the minimal error probability for estimation of parameters K^0, J^0, T^0 (this set of parameters uniquely corresponds to the vector of true class indices $D^0 = (d_1^0, \ldots, d_n^0)$) is attained by the decision rule:*

$$\hat{K} = \arg \max_{2 \leq K \leq n} l_2(X; K), \tag{7.35}$$

$$\hat{T} = (\hat{T}_{(1)}, \ldots, \hat{T}_{(\hat{K})}) = \arg \max_T l_1(X; \hat{K}, T), \tag{7.36}$$

$$\hat{J}_k = \arg \max_{i \in S} \left(\ln \pi_i + \sum_{t=\hat{T}_{(k-1)}+1}^{\hat{T}_{(k)}} \ln p(x_t; \theta_i^0) \right) \quad (k = \overline{1, \hat{K}}). \tag{7.37}$$

Proof. The minimality condition for error probability is of the form:

$$\mathbf{P}\{\hat{K} \neq K^0, \hat{T} \neq T^0, \hat{J} \neq J^0\} =$$

$$= 1 - \int_{R^{Nn}} \mathbf{P}\{K^0 = \hat{K}, T^0 = \hat{T}, J^0 = \hat{J}\} p(X \mid \hat{K}, \hat{T}, \hat{J}) dX \to \min_{\hat{K}, \hat{T}, \hat{J}},$$

where

$$p(X \mid K, T, J) = \prod_{k=1}^{K} \prod_{t=T_{(k-1)}+1}^{T_{(k)}} p(x_t; \theta^0_{J_k})$$

is the conditional probability density function of the sample X for fixed values of K^0, T^0, J^0, and the probability $\mathbf{P}\{K^0 = K, T^0 = T, J^0 = J\}$ is found according to (7.32).

This minimization problem is equivalent to maximization of the integrand at each point $X \in R^{nN}$. Taking the logarithms, we obtain the maximization problem

$$\ln \mathbf{P}\left\{K^0 = \hat{K}(X),\ T^0 = \hat{T}(X),\ J^0 = \hat{J}(X)\right\} +$$

$$+ \ln p\left(X|\hat{K}(X),\ \hat{T}(X), \hat{J}(X)\right) \longrightarrow \max_{\hat{K}(\cdot), \hat{T}(\cdot), \hat{J}(\cdot)} .$$

Using the notations (7.34) we get optimal decision functions (7.35)-(7.37).

■

The extremum problems (7.35) and (7.37) can be easily solved by examining $n-1$ and L values of the objective function respectively. The main difficulties are connected with problems (7.34), (7.36), (7.33). For $K = 2$ (the two-run case) these problems are solved by examining $n-1$ values of the objective function $l_1(\cdot)$. For $K \geq 3$ the problems (7.34), (7.36) assume the form:

$$l_1(X; K, T) = \sum_{i=1}^{K-2} f_i(T_{(i)}, T_{(i+1)}) \rightarrow \max_{\{T_{(i)} \in A_i(K), T_{(i)} \geq T_{(i-1)}+1\}} . \tag{7.38}$$

The special structure of the objective function and of the restrictions allow to apply the dynamic programming approach (Gabasov *et al.*, 1975) for problem (7.38). This approach consists in the following. First we define $K-2$ Bellman functions:

$$B_2(y) = \max_{1 \leq i \leq y-1} f_1(i, y), y \in A_2(K);$$

$$B_{s+1}(y) = \max_{s \leq i \leq y-1} (B_s(i) + f_s(i, y)), \quad y \in A_{s+1}(K), \quad s = \overline{2, K-2}.$$

Then, using the Bellman functions, we obtain a solution of the problem (7.38):

$$l_2(X; K) = \max_{y \in A_{K-1}(K)} B_{K-1}(y); \quad \hat{T}_{(K-1)} = \arg \max_{y \in A_{K-1}(K)} B_{K-1}(y),$$

$$\hat{T}_{(K-l-1)} = \arg \max_{K-l-1 \leq i \leq \hat{T}_{(K-l)}-1} (B_{K-l-1}(i) + f_{K-l-1}(i, \hat{T}_{(K-l)}))$$

$$(l = \overline{1, K-3}),$$

$$\hat{T}_{(1)} = \arg \max_{1 \le i \le \hat{T}_{(2)}-1} f_1(i, \hat{T}_{(2)}).$$

Let us investigate now the case where the parameters of the classes $\{\theta_i^0\}$ are unknown. Let

$$\bar{\theta}_{(k)} = \arg \max_{\theta} \sum_{t=T_{(k-1)}+1}^{T_{(k)}} \ln p(x_t; \theta)$$

denote the ML-estimator of the parameter θ^0 from the k-th run of observations $X_{(k)}(k = \overline{1, K})$. The function $G(\cdot)$ determined by (7.34) is bounded from the above by

$$G(\cdot) \le G_+(X_k; T_{(k-1)}, T_{(k)}) = \max_{i \in S}(\ln \pi_i) + \sum_{t=T_{(k-1)}+1}^{T_{(k)}} \ln p(x_t; \bar{\theta}_{(k)}).$$

Let us determine the statistic

$$l_4(X; K) = \max_T l_3(X; K, T),$$

where $l_3(\cdot)$ is obtained from $l_1(\cdot)$ by substitution of $G_+(\cdot)$ for $G(\cdot)$. Then similarly to (7.35), (7.36), we shall obtain the decision rules for estimation of K^0, T^0:

$$\hat{K} = \arg \max_{2 \le K \le n} l_4(X; K), \quad \hat{T} = \arg \max_T l_3(X; \hat{K}, T). \tag{7.39}$$

For estimation of J^0 we shall additionally assume that run lengths are sufficiently large in comparison with the number m of unknown parameters θ:

$$T^0_{(k)} - T^0_{(k-1)} \gg m \quad (k = \overline{1, K^0}).$$

Let us assume that estimates $\{\hat{T}_{(k)} - \hat{T}_{(k-1)}\}$ of run lengths satisfy the similar condition. If the estimators (7.39) are of sufficient accuracy, then by asymptotic properties of ML-estimators the set $\{\bar{\theta}_{(1)}, \ldots, \bar{\theta}_{(\hat{K})}\} \subset R^m$ can be considered as a random nonhomogeneous and unclassified sample of size $\hat{K}$; $\bar{\theta}_{(k)}$ has m-variate Gaussian distribution with mathematical expectation $\theta^0_{J^0_k}$ and covariance matrix

$$(\hat{T}_{(k)} - \hat{T}_{(k-1)})^{-1}(B_{J^0_k})^{-1},$$

where B_i is the Fisher information matrix for parameter θ_i^0 $(i \in S, k = \overline{1, \hat{K}})$. The arising problem of $\{\theta_i^0\}$ estimation from $\{\bar{\theta}_{(k)}\}$ is a classical problem in cluster analysis (Anderberg, 1973), (Milenkij, 1975), and it is solved by known algorithms (for example, algorithms which use the compactness hypothesis). Let $\{\hat{\theta}_i\}$ be the estimates for $\{\theta_i^0\}$ obtained from $\{\bar{\theta}_{(k)}\}$ in this way. Then we estimate the classification vector J^0 using the decision rule (7.37) with estimates $\{\theta_i^0\}$ replaced by $\{\hat{\theta}_i\}$.

We shall characterize the accuracy of classification of random sample X by the error probability averaged over the n observations:

$$Q = \mathbf{E}\left\{\frac{1}{n}\sum_{t=1}^{n}\left(1-\delta_{\hat{d}_t,d_t^0}\right)\right\}, \tag{7.40}$$

where $\hat{D} = (\hat{d}_1, \ldots, \hat{d}_n)$ is the vector of decisions composed in the same way as (7.31) was with the help of the statistics $\hat{K}, \hat{J}, \hat{T}$.

Let us evaluate a potential classification accuracy, which is attained by the decision rule determined by Theorem 7.6 for *a priori* known parameters $K^0, T^0, \{\theta_i^0\}$. Consider the case of two equiprobable classes: $L = 2, \pi_1 = \pi_2 = 0.5$.

Assume the notations: $\mathbf{E}_\theta\{\cdot\}$, $\mathbf{D}_\theta\{\cdot\}$ denote mean and variance with respect to density $p(\cdot;\theta)$; the Kullback information divergence (Kullback, 1967) is

$$H_i = \mathbf{E}_{\theta_i^0}\{\ln \frac{p(x_t;\theta_i^0)}{p(x_t;\theta_{3-i}^0)}\} > 0;$$

$$\sigma_i^2 = \mathbf{D}_{\theta_i^0}\{\ln \frac{p(x_t;\theta_i^0)}{p(x_t;\theta_{3-i}^0)}\} > 0, \quad \delta_i = H_i/\sigma_i;$$

$\Phi(z)$, $\varphi(z) = \Phi'(z)$ are the distribution function and probability density function of the standard normal distribution law.

Theorem 7.7 *Consider the model of random length runs (7.31), (7.32) with known* K^0, T^0, $\{\theta_i^0\}$ *for* $L = 2$ *equiprobable classes* $(\pi_1 = \pi_2 = 1/2)$, *and*

$$n = \sum_{k=1}^{K^0} T_k^0,$$

where K^0 *is the number of runs,* $T_1^0, \ldots, T_{K^0}^0$ *are independent random variables identically distributed in the set* $\{T_-, T_- + 1, \ldots, T_+\}$. *Suppose that* $\sigma_i > 0$ $(i = 1, 2)$, *and the minimal run length tends to infinity:* $T_- \to \infty$. *Then the error probability* Q *is:*

$$Q = \frac{1}{2}\sum_{i=1}^{2}\mathbf{E}\left\{\frac{\sum_{k=1}^{K^0} T_k^0\, q(T_k^0, i)}{\sum_{k=1}^{K^0} T_k^0}\right\}, \tag{7.41}$$

where $q(\cdot)$ *satisfies the asymptotics :*

$$q(\tau, i) - \Phi(-\delta_i\sqrt{\tau}) \longrightarrow 0 \ (i = 1, 2; \ \tau \to \infty). \tag{7.42}$$

Proof. Using the run representation (7.31) of the sequences D^0, $\hat{D}$, we obtain from (7.40) :

$$Q = \mathbf{E}\left\{\frac{1}{n}\sum_{k=1}^{K^0} T_k^0 \mathbf{P}\{\hat{J}_k \neq J_k^0 \mid T_k^0, J_k^0\}\right\}.$$

For fixed values T^0, J^0 we define random variables :

$$\zeta_k = (-1)^{J_k^0} \sum_{t=T^0_{(k-1)}+1}^{T^0_{(k)}} \ln \frac{p(x_t;\theta_2^0)}{p(x_t;\theta_1^0)} \quad (k=\overline{1,K^0}). \tag{7.43}$$

Under the assumptions of the theorem, the decision rule (7.37) leads to the equivalence:

$$\left\{\hat{J}_k \neq J_k^0\right\} \leftrightarrow \{\zeta_k < 0\},$$

therefore we come to the expression (7.41) with

$$q(T_k^0, i) = \mathbf{P}\{\,\zeta_k < 0 \,|\, T_k^0, J_k^0 = i\,\}. \tag{7.44}$$

Let us calculate conditional moments for (7.43) using the introduced notations for H_i, σ_i, δ_i :

$$\mathbf{E}\{\,\zeta_k \,|\, T_k^0, J_k^0 = i\,\} = T_k^0 \cdot H_i \quad ,$$

$$\mathbf{D}\{\,\zeta_k \,|\, T_k^0, J_k^0 = i\,\} = T_k^0 \cdot \sigma_i^2 \quad .$$

According to (7.43), ζ_k is the sum of $T_k^0 \geq T_-$ independent identically distributed random variables. Therefore, (7.44) and the Levy–Lindeberg central limit theorem together imply the asymptotic expression (7.42).

■

Corollary 7.6 *If $T_-, T_+ \to \infty$, $\delta_1, \delta_2 \to 0$ in such a way that*

$$\delta_i\sqrt{T_-} \to a_{i-},\ \delta_i\sqrt{T_+} \to a_{i+}\ (i=1,2) \quad ,$$

then the asymptotic value of error probability Q has the following lower and upper bounds :

$$\frac{1}{2}\sum_{i=1}^{2} \Phi(-a_{i+}) \leq Q \leq \frac{1}{2}\sum_{i=1}^{2} \Phi(-a_{i-}) \quad . \tag{7.45}$$

Proof. Under the indicated asymptotics, (7.42) imply:

$$q(\tau, i) \to \Phi(-a_i),\ a_{i-} < a_i < a_{i+} \quad .$$

By the monotonicity of $\Phi(\cdot)$ we have :

$$\Phi(-a_{i+}) \leq \lim_{\tau\to\infty} q(\tau, i) \leq \Phi(-a_{i-}) \quad .$$

Using of these inequalities in (7.41) we obtain (7.45).

■

Note that in some cases (for example, in the Gaussian case, investigated at the end of this section) the probability $q(\tau, i)$ can be calculated by (7.44) in explicit form. Then it is possible to find the exact value of Q by (7.41), and asymptotics $T_- \to \infty$ becomes unnecessary.

Let us specify the asymptotic expressions for error probability (7.41), (7.42) using additional information about the probability distribution of run length T_k^0.

Theorem 7.8 *Let the conditions of Theorem 7.7 be satisfied, and*

$$\bar{T} = \mathbf{E}\{T_1^0\}, \; \alpha = \sqrt{\mathbf{D}\{T_1^0\}}/\bar{T} \quad .$$

Then under the asymptotics :

$$\bar{T} \to \infty, \; \alpha \to 0, \; \frac{\mathbf{E}\{|T_1^0 - \bar{T}|^3\}}{(\bar{T})^3} = \mathcal{O}(\alpha^3), \; \delta_i\sqrt{\bar{T}} \to a_i \; (i = 1, 2) \tag{7.46}$$

the following asymptotic expansion of error probability Q is possible:

$$Q = \frac{1}{2}\sum_{i=1}^{2}\left(\Phi(-a_i) + \frac{\alpha^2}{2}\left(\frac{a_i(a_i^2 - 3)}{4}\varphi(a_i) + \right.\right.$$
$$\left.\left. + \frac{1}{K^0}(a_i\varphi(a_i) - \Phi(-a_i))\right)\right) + \mathcal{O}(\alpha^3) \quad . \tag{7.47}$$

Proof. Let us introduce auxiliary random variables ($k = \overline{1, K^0}$) :

$$\xi_k = (T_k^0 - \bar{T})/\bar{T}, \quad \bar{\xi} = \sum_{k=1}^{K^0} \xi_k / K^0 \quad .$$

In asymptotics (7.46) we have :

$$\mathbf{E}\{\xi_k\} = 0, \, \mathbf{D}\{\xi_k\} = \alpha^2 \to 0, \; \mathbf{E}\{|\xi_k|^3\} = \mathcal{O}(\alpha^3) \to 0,$$
$$\mathbf{E}\{\bar{\xi}\} = 0, \, \mathbf{D}\{\bar{\xi}\} = \alpha^2/K^0 \to 0. \tag{7.48}$$

Then by (7.42),

$$q(T_k^0, i) - \Phi\left(-a_i\sqrt{1 + \xi_k}\right) \to 0 \; (T_k^0 \in \{T_-, \ldots, T_+\}, \, i = 1, 2),$$

and (7.41) gives the asymptotic expression :

$$Q - \frac{1}{2}\sum_{i=1}^{2} \mathbf{E}\{f_i(\xi_1, \ldots, \xi_{K^0})\} \to 0 \quad , \tag{7.49}$$

where

$$f_i(\xi_1,\ldots,\xi_{K^0}) = \frac{\frac{1}{K^0}\sum_{k=1}^{K^0}(1+\xi_k)\Phi(-a_i\sqrt{1+\xi_k})}{1+\bar{\xi}} .$$

Because of (7.48), we have mean square convergence :

$$\xi_k \overset{m.s.}{\to} 0, \quad \bar{\xi} \overset{m.s.}{\to} 0.$$

Then applying the Taylor formula to $f_i(\cdot)$ in the neighborhood of point $\xi = 0$, we have

$$f_i(\xi_1,\ldots,\xi_{K^0}) = \Phi(-a_i) - \frac{a_i}{2}\varphi(a_i)\bar{\xi} + \frac{1}{2}(a_i\varphi(a_i) - \Phi(-a_i))(\bar{\xi})^2 +$$

$$+ \frac{a_i(a_i^2-3)}{8}\varphi(a_i)\frac{1}{K^0}\sum_{k=1}^{K^0}\xi_k^2 + \mathcal{O}_{\mathbf{E}}(|\xi_1|^3),$$

where $\eta = \mathcal{O}_{\mathbf{E}}(|\xi_1|^3)$ means a random variable for which $\mathbf{E}\{\eta\} = \mathcal{O}(\mathbf{E}\{|\xi_1|^3\})$. Taking the expectation for both sides of this equation and using (7.48) we obtain:

$$\mathbf{E}\{f_i(\xi_1,\ldots,\xi_{K^0})\} = \Phi(-a_i) + \frac{1}{2}\left(\frac{a_i(a_i^2-3)}{4}\varphi(a_i) + \right.$$

$$\left. + \frac{1}{K^0}(a_i\varphi(a_i) - \Phi(-a_i))\,\alpha^2\right) + \mathcal{O}(\alpha^3) .$$

Substituting this expansion into (7.49) we have (7.47).

■

Note that the asymptotics (7.46) means the increasing of run lengths and the decreasing of interclass distances.

It is useful for practice to apply the expansion (7.47) for some typical probability distributions of random run length T_k^0 $(k = \overline{1, K^0})$.

Corollary 7.7 *If the length of random run T_1^0 has uniform discrete distribution on the set $\{T_-, T_-+1, \ldots, T_+\}$, and*

$$T_- \to \infty,\ T_+ - T_- \to \infty,\ \frac{T_+ - T_-}{T_+ + T_-} \to 0,$$

then the error probability admits the asymptotic expansion

$$Q = \frac{1}{2}\sum_{i=1}^{2}\left(\Phi(-a_i) + \frac{1}{24}\left(\frac{a_i(a_i^2-3)}{4}\varphi(a_i) + \right.\right.$$

$$\left.\left. + \frac{1}{K^0}(a_i\varphi(a_i) - \Phi(-a_i))\right)\left(\frac{T_+ - T_-}{\bar{T}}\right)^2\right) + o\left(\left(\frac{T_+ - T_-}{\bar{T}}\right)^2\right) .$$

Proof. The uniform distribution of T_1^0 implies

$$\bar{T} = \frac{T_+ + T_-}{2}, \quad \mathbf{D}\{T_1^0\} = \frac{(T_+ - T_- + 1)^2 - 1}{12},$$

$$\alpha^2 = \frac{1}{12}\left(\frac{T_+ - T_-}{\bar{T}}\right)^2 \left(1 + \frac{2}{T_+ - T_-}\right) .$$

Substituting these values into (7.47) we come to the required result.

■

Corollary 7.8 *If the random variable $T_k^0 - 1$ has the Poisson probability distribution with $\lambda \to \infty$, and $\delta_i\sqrt{\lambda} \to a_i > 0$ $(i = 1,2)$, then the error probability admits the asymptotic expansion*

$$Q = \frac{1}{2}\sum_{i=1}^{2}\left(\Phi(-a_i) + \frac{1}{2\lambda}\left(\frac{a_i(a_i^2 - 3)}{4}\varphi(a_i) + \right.\right.$$

$$\left.\left. + \frac{1}{K^0}\left(a_i\varphi(a_i) - \Phi(-a_i)\right)\right)\right) + \mathcal{O}\left(\lambda^{-3/2}\right) . \qquad (7.50)$$

Proof. The Poisson probability distribution of $T_1^0 - 1$ implies

$$\bar{T} = \lambda + 1, \mathbf{D}\{T_1^0\} = \lambda, \ \alpha = \frac{1}{\sqrt{\lambda}} \cdot \frac{1}{1 + \frac{1}{\sqrt{\lambda}}} .$$

Substituting these expressions into (7.47) we come to (7.50).

■

The practical value of Theorems 7.7, 7.8 and of their corollaries is in the fact that they give us formulas for approximate evaluation of potential classification accuracy (these formulas are produced from asymptotic expansions by neglecting the remainder terms).

For example, consider the situation with N-dimensional Gaussian observations (see Section 1.4):

$$p(x;\theta_i^0) = (2\pi)^{-N/2} \mid \Sigma \mid^{-1/2} \exp\{-\frac{1}{2}(x - \theta_i^0)^T\Sigma^{-1}(x - \theta_i^0)\},$$

$$x \in R^N, (i = 1,2),$$

where θ_i^0 is the vector of mathematical expectations of features for class Ω_i; Σ is a nonsingular covariance matrix.

According to the notations we have for the Gaussian case :

$$\delta_i = \frac{\Delta}{2}, \quad a_i = \frac{\Delta}{2}\sqrt{\bar{T}} = a,$$

where $\Delta = \sqrt{(\theta_2^0 - \theta_1^0)^T \Sigma^{-1} (\theta_2^0 - \theta_1^0)}$ is the interclass Mahalanobis distance. According to (7.43) ζ_k is a linear function of observations :

$$\zeta_k = (-1)^{J_k^0} (\theta_2^0 - \theta_1^0)^T \Sigma^{-1} \left(\sum_{t=T^0_{(k-1)}+1}^{T^0_{(k)}} x_t - T_k^0 \cdot \frac{\theta_1^0 + \theta_2^0}{2} \right),$$

therefore instead of the asymptotic relation (7.42) we obtain the exact formula :

$$q(\tau, i) = \Phi\left(-\frac{\Delta}{2}\sqrt{\tau}\right), \ \tau \in \{T_-, \dots, T_+\}, \ i \in \{1, 2\}.$$

In the case of Poisson–distributed runs and $K^0 \to \infty$, by (7.50) we obtain the approximation :

$$Q \approx \Phi(-\sqrt{\lambda}\frac{\Delta}{2}) + \frac{\lambda^{1/2}\Delta^3}{64}\left(1 - \frac{12}{\lambda\Delta^2}\right)\varphi(\sqrt{\lambda}\frac{\Delta}{2}). \tag{7.51}$$

Note for comparison that if we classify the observations $x_1, \dots, x_n$ independently of each other (without using the run structure (7.31)), the error probability Q assumes the value (see Section 1.4)

$$Q_0 = \Phi(-\frac{\Delta}{2}). \tag{7.52}$$

The solid curves at Figure 7.4 plot the dependence (7.51) of error probability Q on interclass distance Δ for the decision rule that uses run structure (7.31) for different values of the parameter λ of the Poisson distribution (the expected length of a run is $\lambda + 1$) : $\lambda = 5; 15; 60$. The dashed line plots the dependence (7.52) of error probability Q_0 on Δ for the traditional decision rule, which ignores the run structure (7.31).

In conclusion, let us present some computer results of performance analysis for the decision rule (7.35)–(7.37) formulated in Theorem 7.6. It was performed by Monte-Carlo computer modeling for Gaussian two–dimensional observations ($N = 2$) and Poisson–distributed random runs with $\lambda = 10$. For different values of Δ the Table 7.1 contains theoretical values Q_0, Q (calculated by (7.51), (7.52)), statistical estimates $\hat{K}$, $\hat{T}$, $\hat{J}$, and also the error rates $\hat{Q}$ in 8 computer experiments for $n = 60$, $K^0 = 6$, $T^0 = (10, 16, 29, 37, 51)$, $J^0 = (1\ 2\ 1\ 2\ 1\ 1)$.

Figure 7.5 presents two plots of the dependence of the statistic $l_2 = l_2(X, K)$ on K (see (7.34), (7.35)): the dashed line connects the experimental points fixed for $\Delta = 1$ ($Q_0 = 0.31$), and the solid line is for $\Delta = 1.7$ ($Q_0 = 0.20$).

Computer comparison of performances for two classification methods is presented in the Table 7.2: 1) traditional pointwise classification, which ignores the run structure; 2) classification by the decision rule (7.35)–(7.37) that exploits the run structure of the sample.

It is seen from Figure 7.4 and Tables 7.1, 7.2 that exploiting the run structure (7.31) increases the classification accuracy: the larger the expected run length, the higher the classification accuracy.

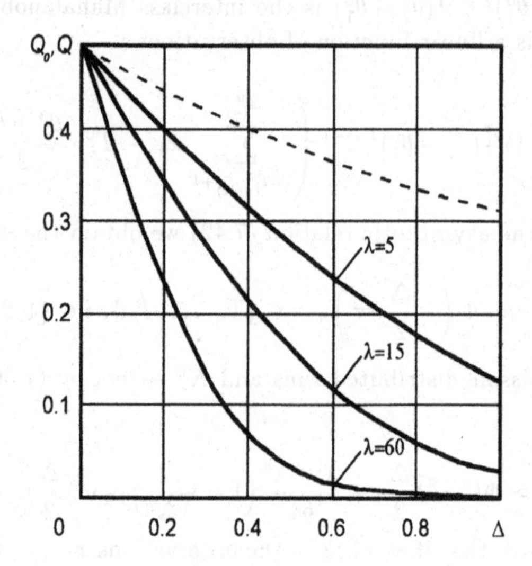

Figure 7.4: Error probabilities Q_0 and Q vs. interclass distance

7.3 Cluster Analysis of T^0-runs

7.3.1 Mathematical Model

In this section we shall investigate the problems of cluster analysis for the samples described by the *T^0–run model.* This model is a special case of the random run length model presented in Section 7.2: the run lengths in the sequence of the class indices $\{d_t^0 : t = 1, 2, \ldots\}$ are equal, i.e.,

$$T_1^0 = T_2^0 = \ldots = T_k^0 = \ldots = T^0.$$

Table 7.1: Results of computer experiments

N	Δ	$100Q_0$	$100Q$	$\hat{K}$	$\hat{T}$	$\hat{J}$	$100\hat{Q}$
1	0.6	38	18	8	(17, 30, 31, 32, 33, 34, 59)	(11222212)	18
2	0.6	38	17	5	(22, 29, 40, 50)	(11211)	15
3	0.8	34	10	6	(12, 16, 28, 38, 49)	(121211)	7
4	1.0	31	6	6	(10, 16, 25, 38, 49)	(121211)	8
5	1.7	20	0.5	6	(11, 16, 29, 38, 49)	(121211)	3
6	1.8	18	0.4	6	(4, 16, 28, 37, 49)	(121211)	11
7	2.0	16	0.1	6	(12, 16, 29, 37, 49)	(121211)	3
8	2.6	10	0.01	6	(12, 16, 29, 37, 49)	(121211)	3

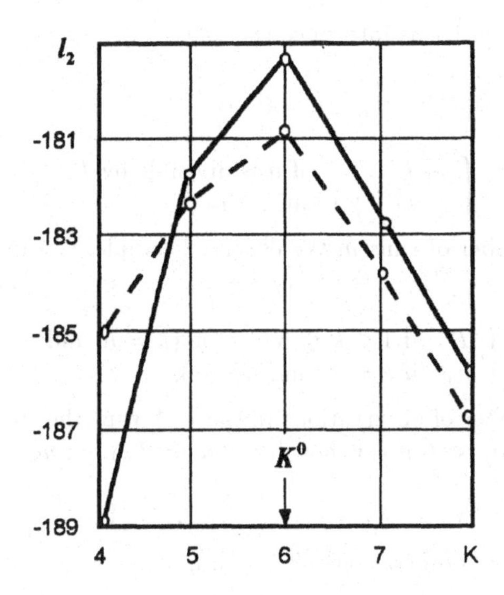

Figure 7.5: Plots of statistic l_2

The true value of T^0 is unknown and belongs to the set of admissible run lengths :

$$T^0 \in A = \{T_-, T_- + 1, \dots, T_+\}, \tag{7.53}$$

where T_- (T_+) is the minimal (maximal) admissible value of run length. Note that if $T_+ = n$, then the observed sample X is homogeneous : all observations belong to the same class $\Omega_{d_1^0}$. Applied problems described by this T^0–run model are usually encountered in technical diagnostics with "inertiality period" T^0 (see the introduction to Section 7.2).

Note that the uncertainty of T^0 is a new and significant feature of the investigated cluster analysis problem. If the true value T^0 is known *a priori*, then the investigated problem can be transformed to the classical cluster analysis problem.

Table 7.2: Point estimates of error probability

Type of DR	Δ							
	0.57	0.60	0.80	1.00	1.70	1.80	2.00	2.60
Run structure is not used	0.33	0.33	0.21	0.29	0.14	0.17	0.15	0.07
Run structure is used	0.18	0.15	0.07	0.08	0.03	0.11	0.03	0.03

7.3.2 Classification by Maximum Likelihood Rule

For any run length $T \in A$ let us introduce the following notations :

$$n_T = n - \left\lfloor \frac{n}{T} \right\rfloor T \in \{0, 1, \ldots, T-1\}; \tag{7.54}$$

$$K = K(T) = \begin{cases} n/T & \text{if } n \text{ is divisible by } T \ (n_T = 0), \\ \lfloor n/T \rfloor + 1 & \text{otherwise,} \end{cases}$$

where $K(T)$ is the number of runs in the observed sample X split into the runs of length T;

$$\tau_k = \tau_k(T) = \begin{cases} T & \text{if } 1 \le k \le K-1 \text{ or } \{k = K \text{ and } n_T = 0\}, \\ n_T & \text{if } k = K \text{ and } n_T > 0, \end{cases}$$

where $\tau_k(T)$ is the number of observations in the k-th run (the K-th run may have incomplete length $0 < n_T < T$ if n is not divisible by T, i. e., $n_T > 0$);

$$X_k = (x_{(k-1)T+1}, \ldots, x_{(k-1)T+\tau_k})$$

is the k-th random run of observations ($k = \overline{1, K}$);

$$f(x;\theta) = \ln p(x;\theta);$$

$$F_k(X_k;\theta_i,T) = \sum_{\tau=1}^{\tau_k} f(x_{(k-1)T+\tau};\theta_i),$$

where $p(x;\theta)$ is a probability density function from a regular family $\mathcal{P}$ (see Section 7.2); $K^0 = K(T^0)$ is the true number of runs in X, when the sample X is split into runs of true length T^0;

$$J^0 = (J_1^0, \ldots, J_{K^0}^0) \in S^{K^0},$$

where $J_k^0 \in S$ is the true class index for the k-th run in the true classification vector $D^0 = (d_1^0, \ldots, d_n^0) \in S^n$.

Note that because of (7.31) D^0 and J^0 are uniqiuely defined from each other for the described T^0–run model :

$$d_t^0 = J^0_{\lfloor (t-0.5)/T^0 \rfloor + 1}, \ t = \overline{1, n}. \tag{7.55}$$

By (7.55), our problem of classification of X, which amounts to the estimation of D^0, is equivalent to the problem of estimation of T^0, J^0.

Let us construct statistical estimators for T^0, J^0 by the maximum likelihood criterion. At the beginning we shall assume that the true values of $\{\theta_i^0\}$ for the conditional p.d.f. $\{p(x;\theta_i^0)\}$ are known. Let us define the statistics :

$$\begin{gathered} \xi_k = \xi_k(T) = \max_{i \in S} F_k(X_k;\theta_i^0,T), \ k = \overline{1,K}, \ K = K(T), \\ l_1(T) = \sum_{k=1}^{K} \xi_k(T), T \in A. \end{gathered} \tag{7.56}$$

Theorem 7.9 *If the observed random sample X is described by the T^0–run model with known $\{\theta_i^0\}$, then the ML-estimators of the true run length T^0 and the true run classification vector J^0 are the statistics :*

$$\tilde{T} = \arg\max_{T \in A} l_1(T),\ \tilde{J} = (\tilde{J}_k),\ \tilde{J}_k = \arg\max_{i \in S} F_k(X_k; \theta_i^0, \tilde{T}), k = \overline{1, K(\tilde{T})}. \quad (7.57)$$

Proof. Define the logarithmic likelihood function in the notations (7.54) :

$$l = l(T, J, \{\theta_i^0\}) = \sum_{k=1}^{K} F_k(X_k; \theta_{J_k}^0, T),\ K = K(T),\ T \in A,\ J \in S^K,$$

where $J = (J_k)$ is an admissible vector of class indices. We find ML–estimators for T, J as the solution of the maximization problem :

$$l(\tilde{T}, \tilde{J}, \{\theta_i^0\}) = \max_{T \in A} \max_{J \in S^K} l(T, J, \{\theta_i^0\}).$$

By the separability of the logarithmic likelihood function with respect to the variables $J_1, \ldots, J_K$ and by (7.56) we have :

$$\max_{J \in S^K} l(T, J, \{\theta_i^0\}) = \sum_{k=1}^{K} \max_{J_k \in S} F_k(X; \theta_{J_k}^0, T) = l_1(T),$$

where this maximum is attained at the optimal value $J^* = (J_k^*)$:

$$J_k^* = \arg\max_{i \in S} F_k(X_k; \theta_i^0, T), k = \overline{1, K(T)}.$$

Maximizing $l_1(T)$ w.r.t. T we come to the ML–estimator $\tilde{T}$ in (7.57). Substituting $T^* = \tilde{T}$ into the last equation we get the ML–estimator $\tilde{J}$ in (7.57).

■

Note that the decision rules (7.57) may be easily implemented in computer software. Maximization problems in (7.57) are easily solved by examining the values of the objective functions. The computational complexity of this algorithm is of order $\mathcal{O}(nL(T_+ - T_- + 1))$.

Let us investigate some properties of the decision rules defined by Theorem 7.9. First, if the true run length is estimated without error ($\tilde{T} = T^0$), then the decision $\tilde{J}_k$ is the maximum likelihood decision about the belonging of the run $X_k = (x_{(k-1)T^0+1}, \ldots, x_{(k-1)T^0+\tau_k})$ to one of the L classes $\{\Omega_i\}$. This decision is known to be optimal (see Chapter 1) : it minimizes the error probability for the case of equiprobable classes.

Second, we would like to emphasize a new particular property of the estimator $\tilde{T}$ defined by (7.57). To formulate this property, let us assume that τ^0 ($\tau^0 < T^0$) is a positive integer which is a divisor of the true run length T^0, i.e., there exists a positive integer $p > 1$ such that $T^0 = p\tau^0$. Then the k-th run X_k of size τ_k consists of

$$\gamma_k = \lfloor(\tau_k - 0.5)/\tau^0\rfloor + 1$$

shorter runs (of size τ^0 and smaller). From (7.54), (7.56) we obtain the inequalities ($k = \overline{1, K^0}$) :

$$\xi_k(T^0) = \max_{i \in S} \sum_{j=1}^{\gamma_k} F_{(k-1)p+j}(X_{(k-1)p+j}; \theta_i^0, \tau^0) \leq \sum_{j=1}^{\gamma_k} \xi_{(k-1)p+j}(\tau^0).$$

Thus, because of the last expression in (7.56), we obtain an inequality for an arbitrary sample X :

$$l_1(T^0) \leq l_1(\tau^0). \tag{7.58}$$

Therefore, according to (7.57), (7.58), if any divisor τ^0 of T^0 is in A, then the ML–estimator $\tilde{T}$ loses its consistency. Let us call this undesirable property of the ML–estimator $\tilde{T}$ the *multiplicity effect.* The multiplicity effect reduces the accuracy of the statistical estimation of run length T^0, and, as a result, reduces the accuracy of estimation J^0 (and D^0). Detrimental influence of the multiplicity effect is clear for the situations where the minimal admissible run length T_- is 1. Then the maximum likelihood decision rule (7.57) assumes a simple form:

$$\tilde{T} \equiv 1, \, \tilde{d}_t = \tilde{J}_t = \arg\max_{i \in S} p(x_t; \theta_i^0), t = \overline{1, n}.$$

This decision rule defines the well known algorithm of "pointwise classification" of n observations $x_1, \ldots, x_n$. This algorithm completely ignores the fact of presence of runs in the sample. As a result, the classification accuracy decreases significantly (the illustrating numerical results are given in Section 7.2).

7.3.3 Classification by Homogeneity Criterion

Let us see how to overcome the multiplicity effect detected for the maximum likelihood decision rule (7.57).

For each admissible value $T \in A$ of run length let us define the hypothesis of homogeneity of the observed sample X

$$H_{0T} : J_1 = J_2 = \ldots = J_K \, (K = K(T)),$$

and the nonhomogeneity alternative $H_{1T} = \overline{H_{0T}}$, which means that at least one of J_i differs from the remaining ones. If the hypothesis H_{0T^0} is true, then by (7.55) all elements of the true classification vector D^0 are equal :

$$d_1^0 = d_2^0 = \ldots = d_n^0 = J_1^0,$$

and X is a random sample of size n from the same class $\Omega_{J_1^0}$.

As it follows from the definition, the hypotheses H_{0T}, H_{1T} are composite hypotheses. Let us construct a homogeneity test for these hypotheses by the approach using the maximum likelihood ratio statistic (see, e.g, (Cox and Hinkley, 1974)):

$$\text{to accept} \begin{cases} H_{0T}, \text{ if } \lambda(T) < \gamma, \\ H_{1T}, \text{ if } \lambda(T) \geq \gamma, \end{cases} \tag{7.59}$$

where

$$\lambda(T) = \max_{J \in S^K} l(T, J, \{\theta_i^0\}) - \max_{J \in H_{0T}} l(T, J, \{\theta_i^0\})$$

is the statistic of logarithm of the maximum likelihood ratio; $\gamma \geq 0$ is the critical value.

Using the notations from the Subsection 7.3.2 and (7.56), we have from (7.59) :

$$\lambda(T) = l_1(T) - \max_{j \in S} \sum_{t=1}^{n} f(x_t; \theta_j^0). \tag{7.60}$$

Let us find the critical value γ in (7.59) such that the preassigned significance level $\alpha \in (0, 1)$ will be reached :

$$\mathbf{P}_{H_{0T}}\{\lambda(T) \geq \gamma\} = \alpha. \tag{7.61}$$

Lemma 7.4 *The statistic $\lambda(T)$ defined by (7.59), (7.60) has a special presentation*

$$\lambda(T) = \min_{j \in S} \lambda_j,$$

$$\lambda_j = \sum_{k=1}^{K} \max_{i \in S} \zeta_{kij}, \; K = K(T),$$

$$\zeta_{kij} = \sum_{\tau=1}^{\tau_k} \ln \frac{p(x_{(k-1)T+\tau}; \theta_i^0)}{p(x_{(k-1)T+\tau}; \theta_j^0)}.$$

Proof. We have from (7.60) by (7.56) and equivalent transformations :

$$\lambda(T) = \min_{j \in S} \left(l_1(T) - \sum_{t=1}^{n} \ln p(x_t; \theta_j^0) \right) =$$

$$= \min_{j \in S} \left(\sum_{k=1}^{K} \max_{i \in S} \sum_{\tau=1}^{\tau_k} \ln p(x_{(k-1)T+\tau}; \theta_i^0) - \sum_{k=1}^{K} \sum_{\tau=1}^{\tau_k} \ln p(x_{(k-1)T+\tau}; \theta_j^0) \right).$$

This expression is equivalent to the required representation.

■

As it is seen from (7.61), in order to find the threshold γ in the homogeneity test (7.59), one needs to construct the probability distribution of the statistic $\lambda(T)$ under the true hypothesis H_{0T}. According to Lemma 7.4, to solve this problem, one needs to find the probability distributions of the statistics $\{\zeta_{kij}\}$, $\{\lambda_j\}$, and $\lambda(T)$. For an arbitrary L this problem turns out to be highly difficult, since it leads to the well known problem of calculation of probability distributions for order statistics from dependent random sample (see, e.g., (David, 1975)). Therefore, we shall investigate the case with two classes Ω_1, Ω_2 ($L = 2$, $S = \{1,2\}$), usual in practice.

Let us formulate some auxiliary results. Denote : $\bar{j} = 3 - j$ is the class index complementary to the index $j \in S$; $\Phi(z)$, $\varphi(z) = \Phi'(z)$ are the standard Gaussian distribution function and probability density function respectively.

Lemma 7.5 *If $L = 2$, then the statistic λ_j defined in Lemma 7.4 admits the representation :*

$$\lambda_j = \sum_{k=1}^{K} \zeta_{k\bar{j}j} \mathbf{1}(\zeta_{k\bar{j}j}),\ j \in S.$$

Proof. By definition, $\zeta_{kjj} \equiv 0$, therefore we have a sequence of equalities :

$$\lambda_j = \sum_{k=1}^{K} \max_{i \in \{j,\bar{j}\}} \zeta_{kij} = \sum_{k=1}^{K} \max\{0, \zeta_{k\bar{j}j}\} = \sum_{k=1}^{K} \zeta_{k\bar{j}j} \mathbf{1}(\zeta_{k\bar{j}j}).$$

■

Lemma 7.6 *If ζ is a random variable with Gaussian distribution $\mathcal{N}_1(a, B)$, then the random variable*

$$\zeta_+ = \zeta \cdot \mathbf{1}(\zeta) = \max\{0, \zeta\}$$

has the following moments ($\delta = |a|/\sqrt{B} \geq 0$) :

$$\mathbf{E}\{\zeta_+\} = a\Phi(\delta \mathrm{sign}(a)) + \sqrt{B}\varphi(\delta),$$

$$\mathbf{E}\{\zeta_+^2\} = (a^2 + B)\Phi(\delta \mathrm{sign}(a)) + a\sqrt{B}\varphi(\delta).$$

Proof is by the immediate calculation of the moments from their definitions and by properties of the Gaussian probability distribution.

■

Corollary 7.9 *If $\delta \to \infty$, then the following asymptotic expansions for the moments of ζ_+ take place :*

$$\mathbf{E}\{\zeta_+\} = \sqrt{B}\left(\delta \mathbf{1}(a) + \delta^{-2}\varphi(\delta)\right) + \mathcal{O}\left(\delta^{-3}\varphi(\delta)\right),$$

$$\mathbf{D}\{\zeta_+\} = B \begin{cases} 1 - 2\delta^{-1}\varphi(\delta) + \mathcal{O}(\delta^{-3}\varphi(\delta)) & \text{if } a > 0, \\ 2\delta^{-3}\varphi(\delta) + \mathcal{O}(\delta^{-5}\varphi(\delta)) & \text{if } a < 0. \end{cases}$$

Corollary 7.10 *If $\delta \to 0$, then the following asymptotic expansions for the moments of ζ_+ take place :*

$$\mathbf{E}\{\zeta_+\} = \sqrt{B}\left(\frac{1}{\sqrt{2\pi}} + \frac{\delta}{2}\mathrm{sign}(a)\right) + \mathcal{O}(\delta^2),$$

$$\mathbf{D}\{\zeta_+\} = \frac{B}{2}\left(1 - \frac{1}{\pi}\right) + \frac{B\delta\mathrm{sign}(a)}{\sqrt{2\pi}} + \mathcal{O}(\delta^2).$$

Proofs of Corollaries 7.9, 7.10 are based on asymptotic expansions of the standard normal distribution and density functions $\Phi(\delta)$, $\varphi(\delta)$ (see, e.g, (Bolshev and Smirnov, 1983)).

■

Let us assume the notations : $g_{1-\alpha} = \Phi^{-1}(1-\alpha)$ is the $(1-\alpha)$–quantile of the standard normal distribution; $\mathbf{E}_i\{\cdot\}$ denotes the expectation for the distribution with density $p(x;\theta_i^0)$;

$$J(i,l) = \mathbf{E}_i\left\{\ln\frac{p(x_t;\theta_i^0)}{p(x_t;\theta_l^0)}\right\} \geq 0$$

is the well known Kullback–Leibler information (Kullback, 1967) for the densities $p(\cdot;\theta_i^0)$ and $p(\cdot;\theta_l^0)$ $(i \neq l \in S)$;

$$\sigma^2(i,l) = \mathbf{E}_i\left\{\left(\ln\frac{p(x_t;\theta_i^0)}{p(x_t;\theta_l^0)}\right)^2\right\} - J^2(i,l) \geq 0;$$

$$\delta(i,l) = \frac{J(i,l)}{\sigma(i,l)};$$

$$\nu_i(T) = \frac{J(i,\bar{i})}{\delta^3(i,\bar{i})}\varphi\left(\sqrt{T}\delta(i,\bar{i})\right), \quad \mu_i(T) = \frac{2\sigma^2(i,\bar{i})\nu_i(T)}{J(i,\bar{i})};$$

$$s = \arg\max_{j\in S}\frac{1}{n}\sum_{t=1}^{n} f(x_t;\theta_j^0).$$

Theorem 7.10 *If $L = 2$, $0 < J(i,\bar{i}) < \infty$, $0 < \sigma(i,\bar{i}) < \infty$ $(i \in S)$, and the threshold is*

$$\gamma = \gamma(s,\alpha,T) = nT^{-3/2}\nu_s(T) + g_{1-\alpha}\sqrt{nT^{-3/2}\mu_s(T)}, \tag{7.62}$$

then as $T \to \infty$, $n_T \to \infty$, $n/T \to \infty$, the asymptotic value of the attained size of the test (7.59) coincides with the preassigned significance level $\alpha \in (0,1)$.

Proof. Let us find the asymptotic probability distribution of the statistic λ_j (defined in Lemma 7.4) under the true hypothesis H_{0T}. By Lemma 7.4 and the introduced notations we find expressions for the moments of first and second orders of the statistic $\zeta_{k\bar{j}j}$ $(k = \overline{1,K}$, $K = K(T)$, $j \in S)$:

$$a_{kj} = \mathbf{E}_{J_k^0}\{\zeta_{k\bar{J}j}\} = (-1)^{j-J_k^0}\tau_k J(J_k^0, \overline{J_k^0}),$$
$$B_{kj} = \tau_k^2 \sigma^2(J_k^0, \overline{J_k^0}),\ 0 < B_{kj} < \infty. \tag{7.63}$$

Here it is taken into account that under the true hypothesis H_{0T} the statistic $\zeta_{k\bar{J}j}$ is the sum of independent identically distributed random variables. According to the conditions of this theorem, $T \to \infty$ and $n_T \to \infty$, thus (7.54) implies $\tau_k \to \infty$. Therefore, by the Levy-Lindeberg central limit theorem (see, e.g., (Koroljuk, 1984)) we come to the asymptotic normality of $\zeta_{k\bar{J}j}$:

$$\mathcal{L}\left\{\frac{\zeta_{k\bar{J}j} - a_{kj}}{\sqrt{B_{kj}}}\right\} \to \mathcal{N}_1(0,1).$$

The random runs $X_1, X_2, \ldots, X_K$ are independent (because of their construction). This implies the independence of the statistics $\zeta_{1\bar{J}j}, \zeta_{2\bar{J}j}, \ldots, \zeta_{K\bar{J}j}$. Because of (7.54) under the asymptotics declared in this Theorem, we have the increase of the number of runs : $K = K(T) \to \infty$. Then from Lemma 7.5 and the Levy–Lindeberg central limit theorem we infer the fact of asymptotic normality of the statistics $\{\lambda_j\}$. Using (7.63) and the introduced notations we have :

$$\delta = |a_{kj}|/\sqrt{B_{kj}} = \sqrt{\tau_k}\delta(J_k^0, \overline{J_k^0}) \to \infty.$$

We can calculate the moments of λ_j by Corollary 7.9. Let us select the main terms of asymptotic expansions for the moments and use the notations :

$$\mathbf{E}\{\lambda_j\} = nT^{-3/2}\nu_{J_1^0}(T) + nJ(J_1^0, \overline{J_1^0})\delta(j, \overline{J_1^0}), \tag{7.64}$$

$$\mathbf{D}\{\lambda_j\} = \begin{cases} n\sigma^2(J_1^0, \overline{J_1^0})\left(1 - 2T^{-1/2}\delta^{-1}(J_1^0, \overline{J_1^0})\varphi\left(\sqrt{T}\delta(J_1^0, \overline{J_1^0})\right)\right), & \text{if } j = \overline{J_1^0}, \\ nT^{-3/2}\mu_{J_1^0}(T), & \text{if } j = J_1^0. \end{cases}$$

According to Lemma 7.5, (7.64) and the Kolmogorov theorem (see, e.g., (Koroljuk, 1984)), the strong law of large numbers holds :

$$\frac{1}{n}\lambda_{\overline{J_1^0}} - \frac{1}{n}\lambda_{J_1^0} \xrightarrow{a.s.} J(J_1^0, \overline{J_1^0}) > 0,$$

therefore

$$\arg\min_{j \in S} \lambda_j \xrightarrow{a.s.} J_1^0.$$

Then by Lemma 7.4, the asymptotic distributions of $\lambda(T)$ and $\lambda_{J_1^0}$ are equivalent. Taking this fact into account, we get the limit distribution of the statistic $\lambda(T)$ under H_{0T} :

$$\mathcal{L}\left\{\frac{\lambda(T) - nT^{-3/2}\nu_{J_1^0}(T)}{\sqrt{nT^{-3/2}\mu_{J_1^0}(T)}}\right\} \to \mathcal{N}_1(0,1).$$

Once again by the Kolmogorov theorem we have :

$$s \xrightarrow{a.s.} J_1^0,$$

therefore

$$\mathcal{L}\left\{\frac{\lambda(T) - nT^{-3/2}\nu_s(T)}{\sqrt{nT^{-3/2}\mu_s(T)}}\right\} \to \mathcal{N}_1(0,1).$$

Using this asymptotical distribution and the expression (7.62) for the threshold, let us calculate the attained size of the test (7.59) :

$$\mathbf{P}_{H_{0T}}\{\lambda(T) \geq \gamma\} = 1 - \mathbf{P}_{H_{0T}}\{\lambda(T) < \gamma(s,\alpha,T)\} \to 1 - \Phi(g_{1-\alpha}) = \alpha.$$

Thus the asymptotic value of the attained size of the test (7.59) satisfies the expression (7.61).

■

The proved theorem allows to construct estimators for run length T^0 and classification vector D^0 which are free of the multiplicity effect revealed in Section 7.3.2.

As it follows from Theorem 7.10, the statistic $\lambda(T)-\gamma(s,\alpha,T)$ can be considered as a statistical measure of distinguishability of the hypotheses H_{0T}, H_{1T} for any fixed significance level α. If $T = T^0$ (true value of run length), then the difference between the hypotheses H_{0T}, H_{1T} is maximal; therefore, we define a new statistical estimator of run length T^0 as the statistic of the homogeneity test:

$$\hat{T} = \arg\max_{T\in A} (\lambda(T) - \gamma(s,\alpha,T)). \tag{7.65}$$

Substituting this estimator $\hat{T}$ into (7.57) for $\tilde{T}$, we define the estimator $\hat{J} = (\hat{J}_k)$ for class indices J^0 :

$$\hat{J}_k = \arg\max_{i\in S} F_k(X_k;\theta_i^0;\hat{T}),\ k = \overline{1, K(\hat{T})}. \tag{7.66}$$

Now let us compare the estimators $\{\hat{T},\hat{J}\}$ based on the homogeneity test, and the estimators $\{\tilde{T},\tilde{J}\}$ constructed in the Subsection 7.3.2 by the maximum likelihood test. Their difference is caused only by difference of the estimators $\hat{T}$, $\tilde{T}$ for the true run length T^0. Since the second term in the objective function $\lambda(T)$ defined by (7.60) does not depend on T, we may represent the estimators $\hat{T}$, $\tilde{T}$ in a convenient form using (7.57), (7.60) and (7.65) :

$$\tilde{T} = \arg\max_{T\in A} l_1(T),\ \hat{T} = \arg\max_{T\in A} (l_1(T) - \gamma(s,\alpha,T)).$$

It is seen that the objective function of the estimator $\hat{T}$ differs from the objective function of the estimator $\tilde{T}$ by the term $-\gamma(s,\alpha,T) < 0$. As it follows from (7.62), $\gamma(s,\alpha,T)$ is a decreasing function of $T \in A$ (at fixed α, X). Therefore, along with (7.58), we have the inequality :

$$\gamma(s, \alpha, T^0) < \gamma(s, \alpha, \tau^0),$$

which allows to overcome the multiplicity effect. This advantage of the estimator $\tilde{T}$ will be analyzed in Section 7.3.5 in more detail. Note that the computational complexity of implementation of the estimators $\hat{T}, \hat{J}$ and $\tilde{T}, \tilde{J}$ is almost the same.

7.3.4 The Case of Unknown Parameters $\{\theta_i^0\}$

Let us generalize the decision rules $\hat{T}$, $\hat{J}$, based on homogeneity statistics, to the situation where the parameters $\{\theta_i^0\}$ are *a priori* unknown.

In this situation the homogeneity hypothesis assumes the form

$$H_{0T} : \theta_1 = \theta_2 = \ldots = \theta_L,\ J_1 = J_2 = \ldots = J_K = 1.$$

The nonhomogeneity alternative H_{1T} is that at least one of these $(L-1)m + K$ scalars differs from the remaining ones of the same type.

Let us introduce the notations : $G_M(z)$ is the standard χ^2–distribution function with M degrees of freedom; $G_M^{-1}(1-\alpha)$ is the $(1-\alpha)$-quantile for this probability distribution; Θ^* is the closure of the set $\Theta \subseteq R^m$;

$$\begin{aligned} l_2(T) &= \max_{\theta_1,\ldots,\theta_L \in \Theta^*} \sum_{k=1}^{K(T)} \max_{i \in S} F_k(X_k; \theta_i, T); \\ l_3 &= \max_{\theta \in \Theta^*} \sum_{t=1}^{n} f(x_t; \theta); \\ \lambda_1(T) &= l_2(T) - l_3. \end{aligned} \tag{7.67}$$

Theorem 7.11 *If $\mathcal{P} = \{p(x;\theta), x \in R^N : \theta \in \Theta\}$ is a regular family of probability densities (satisfying the Chibisov regularity conditions) and*

$$\gamma_1 = \gamma_1(\alpha, T) = \frac{1}{2} G^{-1}_{(L-1)m+K(T)}(1-\alpha),$$

then as $n \to \infty$ the test

$$\text{to accept} \begin{cases} H_{0T} \text{ if} \lambda_1(T) < \gamma_1, \\ H_{1T} \text{ if} \lambda_1(T) \geq \gamma_1 \end{cases} \tag{7.68}$$

has asymptotic size $\alpha \in (0,1)$.

Proof. According to the definition, the hypotheses H_{0T}, H_{1T} are composite. Let us use the maximum likelihood ratio test (see, e.g., (Cox and Hinkley, 1974)) for their testing :

$$\text{to accept} \begin{cases} H_{0T}\,, \text{ if } \Lambda(T) < 2\gamma, \\ H_{1T}\,, \text{ if } \Lambda(T) \geq 2\gamma, \end{cases}$$

where

$$\Lambda(T) = 2 \max_{J \in S^K} \max_{\theta_1,\ldots,\theta_L \in \Theta^*} l(T, J, \{\theta_i\}) - 2 \max_{\theta \in \Theta^*} l(T, J, \{\theta, \ldots, \theta\}).$$

According to (7.67) and the definitions used in Theorem 7.9,

$$\Lambda(T) = 2(l_2(T) - l_3) = 2\lambda_1(T),$$

therefore we get the test in the form (7.68).

Let us calculate the asymptotic size of the test (7.68) :

$$\alpha_1 = \lim_{n\to\infty} \mathbf{P}_{H_{0T}}\{\lambda_1(T) \geq \gamma\} = 1 - \lim_{n\to\infty} \mathbf{P}_{H_{0T}}\{\Lambda(T) < 2\gamma\}.$$

It is known (Cox and Hinkley, 1974), that under the regularity conditions if the hypothesis H_{0T} holds, the asymptotic probability distribution of the statistic $\Lambda(T)$ is the standard χ^2–distribution with M degrees of freedom, where M is equal to the number of independent scalar equalities in the definition of H_{0T} :

$$\lim_{n\to\infty} \left(\mathbf{P}_{H_{0T}}\{\Lambda(T) < 2\gamma\} - G_{(L-1)m+K(T)}(2\gamma)\right) = 0.$$

Therefore, if $\gamma = \gamma_1$, then

$$\alpha_1 = 1 - G_{(L-1)m+K(T)}\left(G^{-1}_{(L-1)m+K(T)}(1-\alpha)\right) = \alpha.$$

It means that the asymptotic size of the test (7.68) coincides with the given significance level α.

■

By Theorem 7.11 let us define the estimator of run length similarly to (7.65) :

$$\hat{T} = \arg\max_{T\in A} \left(\lambda_1(T) - \gamma_1(\alpha, T)\right). \tag{7.69}$$

To estimate $\{\theta_i^0\}$, J^0 we shall use the conditional ML–estimators $\{\hat{\theta}_i\}$, $\hat{J} = (\hat{J}_k)$ (conditional on $T = \hat{T}$) defined by the expressions :

$$\begin{aligned} \Psi(\hat{T}; \hat{\theta}_1, \ldots, \hat{\theta}_L) &= \max_{\theta_1,\ldots,\theta_L\in\Theta^*} \Psi(\hat{T}; \theta_1, \ldots, \theta_L), \\ \Psi(T; \theta_1, \ldots, \theta_L) &= \sum_{k=1}^{K(T)} \max_{J_k\in S} F_k(X_k; \theta_{J_k}, T); \end{aligned} \tag{7.70}$$

$$\hat{J}_k = \arg\max_{i\in S} F_k(X_k; \hat{\theta}_i, \hat{T}),\ k = \overline{1, K(\hat{T})}. \tag{7.71}$$

The maximization problems (7.69), (7.71) are easily solved by examining the values of the objective functions : $|A| = T_+ - T_- + 1$ values in the problem (7.69) and $L \cdot K(\hat{T})$ values in the problem (7.71). The problem (7.70) is the well known Steiner–Weber problem of optimization with $K = K(T)$ discrete variables $J \in S^K$ and mL continuous variables $\theta_1, \ldots, \theta_L \in \Theta \subseteq R^m$ (see, e.g., (Mouder, 1981)). We propose a suboptimal clusterization algorithm to solve the maximization problem (7.70).

Let us introduce the notations : $\omega_l \subset R^{mK}$ is the set of composite vectors $\theta' = \left(\theta^T_{(1)}, \ldots, \theta^T_{(K)}\right)^T$ such that $\theta_{(k)} \in \Theta^*$, and among K points $\theta_{(1)}, \ldots, \theta_{(K)}$ there are at most l different ones ($1 \leq l \leq K$);

$$\psi\left(\theta_{(1)},\ldots,\theta_{(K)}\right)=\sum_{k=1}^{K}F_k\left(X_k;\theta_{(k)},T\right),$$

$$\psi_l^*=\max_{\theta'\in\omega_l}\psi\left(\theta_{(1)},\ldots,\theta_{(K)}\right).$$

By definition, $\Theta^{*K}=\omega_K\supset\omega_{K-1}\supset\ldots\supset\omega_{L+1}\supset\omega_L$, therefore (7.67), (7.70) imply the following chain of inequalities :

$$l_2(T)=\psi_L^*\leq\psi_{L+1}^*\leq\cdots\leq\psi_K^*.$$

It means that the solution of the problems (7.67), (7.70), (7.71) can be found by stepwise maximization of $\psi(\cdot)$ for monotone decreasing sequence of the sets $\{\omega_l\}$ (in other words, by stepwise "merging of clusters"). Let us define the ν-th step of this procedure ($\nu=1,2,\ldots$).

Suppose that the result of the previous $(\nu-1)$-th step is the following statistics : $L^{(\nu-1)}$ is the number of the clusters; $\hat{\theta}_1^{(\nu-1)},\ldots,\hat{\theta}_{L^{(\nu-1)}}^{(\nu-1)}\in\Theta^*$ are estimators of the parameters for each cluster; $\hat{J}^{(\nu-1)}=\left(\hat{J}_k^{(\nu-1)}\right)$, $k=\overline{1,K}$ is the decision vector. Let us select a pair of indices (l_1,l_2) (where $l_1,l_2\in\{1,2,\ldots,L^{(\nu-1)}\}$; $l_1<l_2$) and define the logarithmic likelihood function for "merging" the clusters number l_1 and number l_2 :

$$\psi_{l_1,l_2}=\sum_{\substack{k=1,\\ \hat{J}_k^{(\nu-1)}\notin\{l_1,l_2\}}}^{K}F_k\left(X_k;\hat{\theta}_{\hat{J}_k^{(\nu-1)}}^{(\nu-1)},T\right)+\max_{\theta\in\Theta^*}\sum_{\substack{k=1,\\ \hat{J}_k^{(\nu-1)}\in\{l_1,l_2\}}}^{K}F_k(X_k;\theta,T).$$

By examining $L^{(\nu-1)}\left(L^{(\nu-1)}-1\right)/2$ values of this objective function we find the optimal index pair (l_1^*,l_2^*) for "cluster merging" :

$$\psi_{l_1^*,l_2^*}=\max_{l_1,l_2}\psi_{l_1,l_2}.$$

Now let us calculate for $j=\overline{1,K}$:

$$\theta_j^{(\nu)}=\begin{cases}\hat{\theta}_j^{(\nu-1)}\text{ if }\hat{J}_j^{(\nu-1)}\notin\{l_1^*,l_2^*\},\\ \arg\max\limits_{\theta\in\Theta^*}\sum\limits_{k=1}^{K}\mathbf{I}_{\{l_1^*,l_2^*\}}\left(\hat{J}_k^{(\nu-1)}\right)F_k(X_k;\theta,T)\text{ otherwise};\end{cases}$$

$$J_j^{(\nu)}=\begin{cases}\hat{J}_j^{(\nu-1)}\text{ if }\hat{J}_j^{(\nu-1)}\notin\{l_1^*,l_2^*\},\\ \arg\max\limits_{l\neq l_2^*}\sum\limits_{k=1}^{K}\mathbf{I}_{\{l_1^*,l_2^*\}}\left(\hat{J}_k^{(\nu-1)}\right)F_k(X_k;\theta_l^{(\nu)},T)\text{ otherwise}.\end{cases}$$

By construction, $\left(\theta_1^{(\nu)},\ldots,\theta_K^{(\nu)}\right)\in\omega_{L^{(\nu)}}$, $L^{(\nu)}\leq L^{(\nu-1)}-1$. Let us select $L^{(\nu)}$ different points among $\{\theta_j^\nu\}$ and renumber them : $\hat{\theta}_1^{(\nu)},\ldots,\hat{\theta}_{L^{(\nu)}}^{(\nu)}$. By a similar transformation of the vector $J^{(\nu)}$ we obtain $\hat{J}^{(\nu)}\in\{1,2,\ldots,L^{(\nu)}\}^K$.

This process of optimal "cluster merging" must be continued until the step at which the inequality $L^\nu \leq L$ becomes valid for the first time. The final values of this step : $\hat{\theta}_1^{(\nu)}, \ldots, \hat{\theta}_{L^{(\nu)}}^{(\nu)}, \hat{J}^{(\nu)}, \psi_{l_1^*, l_2^*} = l_2(T)$, are taken to be the solution of the problems (7.67), (7.70), (7.71). The starting values of this stepwise procedure can be defined by the expressions :

$$L^{(0)} = K,\ \hat{\theta}_j^{(0)} = \arg\max_{\theta \in \Theta^*} F_j(X_j; \theta, T),\ \hat{J}_j^{(0)} = j\ (j = \overline{1, K}).$$

Note in conclusion, that if the statistics $\{\hat{\theta}_i^{(\nu)}\}$ are continuous random variables, then $L^{(\nu)} = L^{(\nu-1)} - 1$ with probability one, and the number of steps of the procedure is equal to $K - L$.

7.3.5 Performance Analysis

This subsection is devoted to investigation of performance of the clusterization procedures developed in the Subsections 7.3.2 - 7.3.4.

The accuracy of the estimators $\hat{T}$, $\hat{J}$ (determining the classification vector $\hat{D}$) will be characterized by probabilities of errors:

$$Q_1 = \frac{1}{n} \sum_{t=1}^{n} \mathbf{P}\{\hat{d}_t \neq d_t^0\}$$

is the probability of classification error (averaged over all n observations);

$$Q_2 = \mathbf{P}\{\hat{T} \neq T^0\}$$

is the probability of the estimation error for the true run length. Here $\hat{D} = (\hat{d}_1, \hat{d}_2, \ldots, \hat{d}_n)$ is the classification vector, which is constructed using $\hat{J}$ similarly to (7.55) :

$$\hat{d}_t = \hat{J}_{\lfloor (t-0.5)/\hat{T} \rfloor + 1}\ (t = \overline{1, n}).$$

First, let us evaluate the potential accuracy of classification (7.66) attainable for known true values $\{\theta_i^0\}$, T^0. In this situation, as it is seen from the comparison of decision rules (7.57), (7.66), the decisions $\tilde{J}$ and $\hat{J}$ are equivalent. Let us use the notations (7.54) and formulate an auxiliary result.

Lemma 7.7 *For any random variables $\zeta_1, \ldots, \zeta_L$ the following inequalities are valid:*

$$\max_{j \in S} \mathbf{P}\{\zeta_j > 0\} \leq \mathbf{P}\{\max_{j \in S} \zeta_j > 0\} \leq \sum_{j=1}^{L} \mathbf{P}\{\zeta_j > 0\},$$

$$\sum_{j=1}^{L} \mathbf{P}\{\zeta_j > 0\} - \sum_{j=1}^{L-1} \sum_{l=j+1}^{L} \mathbf{P}\{\zeta_j > 0, \zeta_l > 0\} \leq$$

$$\leq \mathbf{P}\{\max_{j \in S} \zeta_j > 0\} \leq \sum_{j=1}^{L} \mathbf{P}\{\zeta_j > 0\}.$$

Proof. Let us define the following random events :

$$A_j = \{\zeta_j > 0\} \; (j \in S).$$

Then

$$\mathbf{P}\{\max_{j \in S} \zeta_j > 0\} = \mathbf{P}\left(\bigcup_{j \in S} A_j\right).$$

For any $j \in S$

$$A_j \subset \bigcup_{k \in S} A_k,$$

therefore the first chain of the inequalities (being proved) follows from the properties of probability measure. The second chain of these inequalities results from applying the Bonferroni inequality (see, e.g., (Koroljuk, 1984)). ■

Theorem 7.12 *If the true values of parameters* $\{\theta_i^0\}$ *are* a priori *known,* $\hat{T} = T^0$, $0 < J(i,j) < \infty$, $0 < \sigma(i,j) < \infty \, (j \neq i)$, *and there exists a constant* $a > 0$ *such that*

$$\mathbf{E}_{\theta_l^0} \left\{ \left(p(x_t; \theta_i^0) / p(x_t; \theta_j^0) \right)^a \right\} < \infty \, (i, j, l \in S),$$

then for $\delta(i,j) = o\left((T^0)^{-1/3}\right)$ *and as* T^0, $n_{T^0} \to \infty$ *the following two-sided estimate of* Q_1 *takes place :*

$$\frac{1}{K^0} \sum_{k=1}^{K^0} \max_{j \neq J_k^0} \Phi\left(-\sqrt{\tau_k}\delta(J_k^0, j)\right) \leq Q_1 \leq \frac{1}{K^0} \sum_{k=1}^{K^0} \sum_{j \neq J_k^0} \Phi\left(-\sqrt{\tau_k}\delta(J_k^0, j)\right). \quad (7.72)$$

Proof. According to (7.54), (7.55), (7.57) and Lemma 7.4 we have :

$$Q_1 = \tfrac{1}{n} \sum_{k=1}^{K} \tau_k Q_{1k},$$
$$Q_{1k} = \mathbf{P}\{\hat{J}_k \neq J_k^0\} = \mathbf{P}\left\{\max_{j \neq J_k^0} \zeta_{kjJ_k^0} > 0\right\}. \quad (7.73)$$

From the first group of inequalities in Lemma 7.7 we find :

$$\max_{j \neq J_k^0} Q_{1kj} \leq Q_{1k} \leq \sum_{j \neq J_k^0} Q_{1kj},$$

where $Q_{1kj} = \mathbf{P}\left\{\zeta_{kJ_k^0 j} < 0\right\}$. By Lemma 7.4 and the conditions of the theorem, the sequence of random variables $\zeta_{kJ_k^0 j}$ satisfies the Lindeberg–Feller central limit theorem :

$$\mathcal{L}\left\{\frac{\zeta_{kJ_k^0j} - \tau_k J(J_k^0, j)}{\sqrt{\tau_k}\sigma(J_k^0, j)}\right\} \to \mathcal{N}_1(0, 1).$$

Then using the asymptotics of probabilities of large deviations (see, e.g., (Koroljuk, 1984)), we obtain

$$Q_{1kj} = \Phi\left(-\sqrt{\tau_k}\delta(J_k^0, j)\right)(1 + o(1)).$$

Substituting this asymptotic expression into (7.73) by (7.54) we come to (7.72). ∎

One can see from (7.72) that the potential accuracy of classification increases with the increase of the Kullback interclass distances $\{\delta(i,j) : i \neq j, i,j \in S\}$ and of the run length T^0. The two–sided estimate (7.72) shows that even if the Kullback interclass distances for a given alphabet of classes $\{\Omega_i\}$ are "small", for sufficiently large run length T^0 a sufficiently small probability of classification error can be achieved.

Note in addition that the accuracy of the two–sided estimate (7.72) can be increased using the second group of inequalities in Lemma 7.7.

Now let us evaluate the potential accuracy of decision rules (7.57) and (7.65) for the estimation of run length T^0.

Lemma 7.8 *Suppose that a random vector $\zeta = (\zeta_i) \in R^4$ has 4–variate Gaussian probability distribution $\mathcal{N}_4(a, B)$ with vector mean $a = (a_i)$ and covariance matrix $B = (B_{ij})$; $a_1 > a_2$, $a_3 > a_4$,*

$$a_{ij} = |a_i - a_j| / \sqrt{B_{ii} + B_{jj} - 2B_{ij}}\ (i,j = \overline{1,4}),$$

$$a_{12} \to 0,\ a_{34} \to \infty.$$

Then the moments of the order statistics admit the following asymptotic expansions :

$$\mathbf{E}\{\max(\zeta_1, \zeta_2)\} = \frac{a_1 + a_2}{2} + \sqrt{\frac{B_{11} + B_{22} - 2B_{12}}{2\pi}} + \mathcal{O}(a_{12}),$$

$$\mathbf{E}\{\max(\zeta_3, \zeta_4)\} = a_3 + \frac{\sqrt{B_{33} + B_{44} - 2B_{34}}}{a_{34}^2}\varphi(a_{34}) + \mathcal{O}(a_{34}^{-3}\varphi(a_{34})),$$

$$\mathbf{D}\{\max(\zeta_1, \zeta_2)\} = \frac{\pi - 1}{2\pi}(B_{11} + B_{22}) - \frac{1}{\pi}B_{12} + \mathcal{O}(a_{12}),$$

$$\mathbf{D}\{\max(\zeta_3, \zeta_4)\} = B_{33} + \frac{2(B_{34} - B_{33})}{a_{34}}\varphi(a_{34}) + \mathcal{O}(a_{34}^{-3}\varphi(a_{34})),$$

$$\mathbf{Cov}\{\max(\zeta_1, \zeta_2), \max(\zeta_3, \zeta_4)\} = \frac{B_{13} + B_{23}}{2} + \mathcal{O}(a_{34}^{-3/2}\varphi^{1/2}(a_{34})).$$

Proof is based on the evident identities $(i = 1, 2)$:

$$\max(\zeta_{2i-1}, \zeta_{2i}) \equiv \zeta_{2i-1} + (\zeta_{2i} - \zeta_{2i-1})\mathbf{1}(\zeta_{2i} - \zeta_{2i-1}),$$

and also on Lemma 7.6 and its corollaries.

■

Let us investigate the case of $L = 2$ classes and introduce the notations : $\bar{A} = A \setminus \{T^0\}$ is the set of admissible run lengths (7.53) without the true run length T^0;

$$\sigma^2 = \frac{1}{n}\sum_{t=1}^{n} \sigma^2(d_t^0, \overline{d_t^0}),\ J = \frac{1}{K^0}\sum_{k=1}^{K^0} J(J_k^0, \overline{J_k^0}),$$

$$\mu(T) = \frac{1}{K^0}\sum_{k=1}^{K^0} \mu_{J_k^0}(T) > 0,\ \nu(T) = \frac{1}{K^0}\sum_{k=1}^{K^0} \nu_{J_k^0}(T) > 0,$$

$$\beta = \frac{1}{K(T)}\sum_{k=1}^{K(T)}\left(\frac{1}{T}\sum_{\tau=1}^{\tau_k}\sigma^2\left(d^0_{\tau+(k-1)T}, \overline{d^0_{\tau+(k-1)T}}\right)\right)^{1/2} > 0.$$

One can consider these values as the values $\sigma^2(d_t^0, \overline{d_t^0})$, $J(J_k^0, \overline{J_k^0})$, $\mu_{J_k^0}$, $\nu_{J_k^0}$ averaged over time moments or run indices.

Theorem 7.13 *If the number L of classes is 2, the true parameter values $\{\theta_i^0\}$ are* a priori *known, $0 < \sigma(i, \bar{i}) < \infty$ $(i = 1, 2)$, and $T_- \to \infty$, $n/T_+ \to \infty$, then the following two-sided estimate of error probability $Q_2 = \mathbf{P}\{\tilde{T} \neq T^0\}$ for the decision rule (7.56), (7.57) is valid:*

$$\max_{T \in \bar{A}} Q_2(T) \le Q_2 \le \sum_{T \in \bar{A}} Q_2(T), \tag{7.74}$$

where $Q_2(T)$ admits the approximation

$$Q_2(T) \approx \begin{cases} \Phi\left(\sqrt{n}T^{-3/4}\nu(T)/\sqrt{\mu(T)}\right), & if T^0 = pT, \\ \Phi\left(-\sqrt{\frac{\pi n}{2(\pi - 1)}}\left(J - \sqrt{\frac{2}{\pi T}}\beta + \frac{2\nu(T^0)}{(T^0)^{3/2}}\right)\frac{1}{\sigma}\right), & otherwise, \end{cases} \tag{7.75}$$

and $p > 1$ is a natural number (the condition $T^0 = pT$ means that the number $T < T^0$ is a divisor of the true run length).

Proof. Let us define the values :

$$T \in \bar{A},\ \zeta_T = l_1(T) - l_1(T^0),\ Q_2(T) = \mathbf{P}\{\zeta_T > 0\}.$$

Then the two-sided estimate (7.74) follows from (7.57) and the first group of inequalities in Lemma 7.7.

Let us prove now the validity of approximation (7.75). For $T^0 = pT$, the expression (7.75) is deduced using the asymptotic normality of the random variable

ζ_T. This property can be proved in the same way as it was made in the proof of Theorem 7.10 using Lemma 7.6 and the following equivalent representation of ζ_T (for simplicity, it is given for $n = K^0T^0$) :

$$\zeta_T = \sum_{k=1}^{K^0} \min_{j\in\{1,2\}} \sum_{i=1}^{p} \max\left(0, \sum_{l=1}^{T} \ln \frac{p\left(x_{l+(i-1)T+(k-1)T^0}; \theta_{\bar{j}}^0\right)}{p\left(x_{l+(i-1)T+(k-1)T^0}; \theta_j^0\right)}\right).$$

For $T^0 \neq pT$, let us apply (7.56). Under the conditions of this theorem, the number of runs increases: $K = K(T) \to \infty$, and by the multivariate central limit theorem the random vector $(l_1(T), l_1(T^0))$ has asymptotically normal probability distribution. The parameters (moments) of this asymptotic distribution are determined by Lemma 7.8. Then we come to the asymptotic expression :

$$Q_2(T) - \Phi\left(-\frac{\mathbf{E}\{l_1(T^0)\} - \mathbf{E}\{l_1(T)\}}{\sqrt{\mathbf{D}\{l_1(T^0)\} + \mathbf{D}\{l_1(T)\} - 2\mathbf{Cov}\{l_1(T), l_1(T^0)\}}}\right) \to 0,$$

leading to (7.75) after the substitution of values of the moments.

■

Note that the value $Q_2(T)$ in (7.74), (7.75) represents the probability of error ($\tilde{T} \neq T^0$) for the situations where only two hypotheses are tested : run length is equal either to T^0 or to T.

Theorem 7.14 *If the decision rule $\hat{T}$ based on the homogeneity statistic (7.65) is used under the conditions of Theorem 7.13 and the classes are equiprobable :*

$$q_i = \frac{1}{n}\sum_{t=1}^{n} \delta_{d_t^0, i} \to \frac{1}{2} \ (i = 1, 2),$$

then the asymptotic estimate of error probability Q_2 satisfies the inequalities (7.74), where $Q_2(T)$ admits the approximation :

$$Q_2(T) \approx \frac{1}{2}\sum_{i=1}^{2}\begin{cases} \Phi\left(-U_{1i}\right) & \text{if } T^0 = pT, \\ \Phi\left(-U_{2i}\right) & \text{otherwise,} \end{cases} \tag{7.76}$$

where

$$U_{1i} = \frac{\gamma(i, \alpha, T) - \gamma(i, \alpha, T^0) - nT^{-3/2}\nu(T)}{\sqrt{nT^{-3/2}\mu(T)}},$$

$$U_{2i} = \sqrt{\frac{\pi n}{2(\pi - 1)}}\frac{1}{\sigma}\left(J - \sqrt{\frac{2}{\pi T}}\beta + \frac{2\nu(T^0)}{(T^0)^{3/2}} + \frac{2\left(\gamma(i, \alpha, T) - \gamma(i, \alpha, T^0)\right)}{n}\right).$$

Proof. According to (7.57), (7.60), (7.65), for a fixed value $s = i \in S$ the objective functions of the estimators $\tilde{T}$, $\hat{T}$ differ by the term $-\gamma(i, \alpha, T)$ only.

Let us clarify the asymptotic distribution of the statistic s defined in Theorem 7.10 :

$$s = \arg\max_{j \in S} \frac{1}{n} \sum_{t=1}^{n} f(x_t; \theta_j^0).$$

Under the conditions of this theorem,

$$\mathbf{E}\left\{\frac{1}{n}\sum_{t=1}^{n} f(x_t; \theta_2^0) - \frac{1}{n}\sum_{t=1}^{n} f(x_t; \theta_1^0)\right\} = (q_2 - q_1)J(2,1) \to 0.$$

Therefore by the Lindeberg central limit theorem the difference of the objective functions has asymptotically normal probability distribution with zero mean:

$$\mathcal{L}\left\{\frac{\sum_{t=1}^{n} f(x_t; \theta_2^0) - \sum_{t=1}^{n} f(x_t; \theta_1^0)}{\sqrt{n}\sigma}\right\} \to \mathcal{N}_1(0,1).$$

Hence the asymptotic distribution of the statistic s is uniform :

$$\mathbf{P}\{s = i\} \to \frac{1}{2} \; (i = 1, 2).$$

As a result, taking additional expectation with respect to this limit distribution similarly to (7.75), we obtain (7.76).

■

Let us analyze the performance of the classification procedures using the results given by Theorems 7.13, 7.14. For this purpose let us investigate the asymptotics of the function $Q_2(T)$.

First, let us consider the situation where the number T does not divide T^0 ($T^0 \neq pT$, $p = 2, 3, \ldots$). Under conditions of Theorems 7.13, 7.14, using the notations of Theorem 7.10, we have :

$$\nu_i(T) \to 0, \; \mu_i(T) \to 0, \; \frac{\nu_i(T)}{\sqrt{\mu_i(T)}} \to 0, \; \frac{\gamma(s, \alpha, T)}{n} \xrightarrow{a.s.} 0.$$

Then basing on (7.75), (7.76) we conclude

$$Q_2(T) = \mathcal{O}\left(\exp\left(-\frac{\pi}{4(\pi - 1)}n\right)\right) \to 0.$$

In this case, if the set A of admissible values of the run length (7.53) does not contain the divisors T for the true run length T^0 (i.e., $T^0 \neq pT$ for $p = 2, 3, \ldots$, and $T \in A$), then $Q_2 \to 0$, and both estimators $\tilde{T}$, $\hat{T}$ are consistent.

Second, let us consider now the situation where the set A contains a divisor T for T^0: $T^0 = pT$ ($p \in \{2, 3, \ldots\}$). In addition, let us assume that $n/(T_-)^{3/2} \to 0$. Then (7.75) implies for the estimator $\tilde{T}$:

•

$$Q_2(T) \to \Phi(0) = \frac{1}{2}.$$

By (7.76) we obtain an asymptotics for the estimator $\hat{T}$:

$$Q_2(T) \to \Phi\left(-(1-p)^{-3/4} g_{1-\alpha}\right) = \alpha + \frac{g_{1-\alpha}}{p^{3/4}} \varphi(g_{1-\alpha}) + \mathcal{O}(p^{-3/2}),$$

where $p = T^0/T \in \{2, 3, \ldots\}$. Comparing these two asymptotics of error probability in terms of (7.74) we may conclude the following :

1) the estimator $\tilde{T}$ loses consistency in the second situation : $Q_2 \geq 0.5$;

2) the error probability $Q_2(T)$ for the estimator $\hat{T}$ decreases if the multiplicity factor p increases and the significance level α decreases (or, equivalently, the quantile $g_{1-\alpha}$ increases);

3) the preassigned level of error probability Q_2 for $\hat{T}$ can be reached by a special choice of the value α (in particular, if $\alpha \to 0$, then $Q_2(T) \to 0$, $Q_2 \to 0$ under the considered asymptotics, and $\hat{T}$ becomes a consistent estimator of the true run length T^0).

Concerning applications of Theorems 7.13, 7.14, note in addition that if $\{J_k^0\}$ is a sequence of independent identically distributed Bernoulli random variables on $S = \{1, 2\}$, then (under the conditions of Theorem 7.13) σ^2, J, $\mu(T)$, $\nu(T)$, and β have limits (with probability one); it is recommended to use these limit values in (7.74) — (7.76).

In conclusion, let us illustrate an application of the classification procedures (with their performances) constructed above for the Gaussian model of data, common in practice.

Let $\mathcal{P}$ be a family of N-variate Gaussian probability density functions with a fixed nonsingular covariance matrix Σ ($|\Sigma| \neq 0$) :

$$p(x; \theta) = n_N(x \mid \theta, \Sigma).$$

Let us assume that $\{J_k^0\}$ in (7.55) is a sequence of independent identically distributed discrete random variables :

$$\mathbf{P}\{J_k^0 = l\} = \frac{1}{L},\ l \in S = \{1, 2, \ldots, L\},\ k = \overline{1, K^0}.$$

First of all, let us present analytical expressions for the performance characteristics Q_1, Q_2 given by Theorems 7.12 — 7.14. Let

$$\Delta_{il} = \left((\theta_i^0 - \theta_l^0)^T \Sigma^{-1} (\theta_i^0 - \theta_l^0)\right)^{1/2}$$

be the Mahalanobis distance between the classes Ω_i, Ω_l. Theorem 7.12 gives the two-sided estimate of the probability of classification error :

$$\frac{1}{K^0}\sum_{k=1}^{K^0}\max_{j\neq J_k^0}\Phi\left(-\sqrt{T_k}\frac{\Delta_{jJ_k^0}}{2}\right)\leq Q_1\leq\frac{1}{K^0}\sum_{k=1}^{K^0}\sum_{j\neq J_k^0}\Phi\left(-\sqrt{T_k}\frac{\Delta_{jJ_k^0}}{2}\right),$$

which may be transformed to the exact formula for the case of $L=2$ classes ($n=K^0T^0$) :

$$Q_1=\Phi\left(-\sqrt{T^0}\frac{\Delta_{12}}{2}\right). \tag{7.77}$$

Note that if the pointwise classification algorithm, given in the Section 7.3.2, is used (without taking into account the "run structure" of the sample X), then the error probability Q_1 for this decision rule assumes the value

$$\bar{Q}_1=\Phi\left(-\frac{\Delta_{12}}{2}\right). \tag{7.78}$$

It follows from the comparison of (7.77), (7.78) that exploiting the "run structure" of the sample X we obtain a significant gain in potential accuracy of classification. This gain is especially evident for the case of "close classes" (small Δ_{12}) and "long runs" (large T^0).

For the estimator $\hat{T}$ (that uses the homogeneity statistic (7.65)), the error probability Q_2 satisfies the two-sided inequality (7.74), where the value of $Q_2(T)$ is determined by Theorem 7.13 :

$$Q_2(T)\approx\begin{cases}\Phi\left(-U_1\right), \text{if } T^0=pT,\\ \Phi\left(-U_2\right), \text{otherwise},\end{cases} \tag{7.79}$$

where

$$U_1=\frac{\gamma(\alpha,T)-\gamma(\alpha,T^0)-nT^{-3/2}\nu(T)}{2\sqrt{nT^{-3/2}\nu(T)}},$$

$$U_2=\sqrt{\frac{\pi n}{2(\pi-1)}}\left(\frac{\Delta_{12}}{2}-\sqrt{\frac{2}{\pi T}}+\frac{2\nu(T^0)}{\Delta_{12}(T^0)^{3/2}}+\frac{2\left(\gamma(\alpha,T)-\gamma(\alpha,T^0)\right)}{n\Delta_{12}}\right).$$

Here

$$\nu(T)=\frac{4\varphi(\sqrt{T}\Delta_{12}/2)}{\Delta_{12}},$$

$$\gamma(\alpha,T)=\frac{n\nu(T)}{T^{3/2}}+\frac{2g_{1-\alpha}\sqrt{n\nu(T)}}{T^{3/4}}.$$

For the estimator $\hat{T}$, it is necessary to substitute $\gamma(\cdot)\equiv 0$ in (7.79).

The decision rules described in this subsection are implemented by computer program for the Gaussian family $\mathcal{P}$ and investigated by the Monte-Carlo method.

Let us present now some results of computer experiments. Figure 7.6 plots the statistics $l_1(T)$, $l'_1(T) = \lambda(T) - \gamma(\alpha, T)$ (see (7.56), (7.65)) for one of computer experiments with known $\{\theta_i^0\}$ and

$$L = 2,\ N = 2,\ n = 200,\ T^0 = 15,\ T_- = 4,\ T_+ = 32,\ \Delta_{12} = 0.85,\ \alpha = 0.01.$$

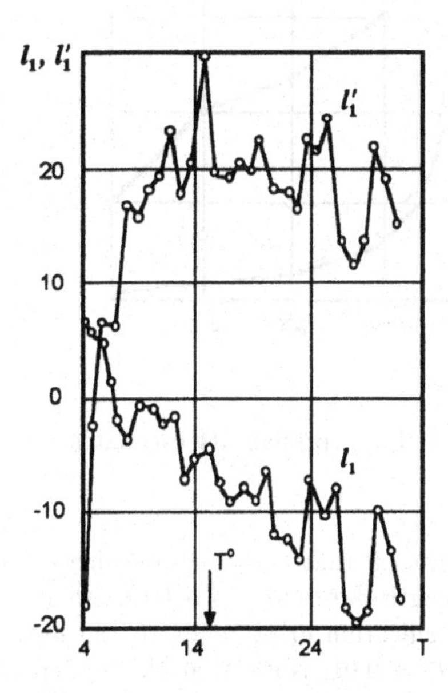

Figure 7.6: Plots of statistics for estimation of run length

The function $\gamma(s, \alpha, T)$ does not depend on s for Gaussian observations. These plots illustrate the multiplicity effect for the traditional ML–estimator $\tilde{T}$ and how the new estimator $\hat{T}$ based on the homogeneity statistic overcomes this effect. The multiplicity effect decreases the classification accuracy. For example, for sample size $n = 500$ the error rate is $\hat{Q}_1 = 0.03$ for decision rule $\hat{J}$, and $\tilde{Q}_1 = 0.23$ for the decision rule $\tilde{J}$. In addition, the error probability for the decision rule ignoring the "run structure" of the sample (according to (7.78)) is somewhat larger: $\bar{Q}_1 = 0.34$. At Figure 7.7 small circles indicate experimental values of $\hat{Q}_2$ for the estimator $\hat{T}$; here the solid lines display the 95%–confidence limits for the true value of Q_2, and the dashed line is the upper bound of Q_2 determined by the asymptotic expressions (7.74), (7.76). It is seen that the computer results are in good accordance with the analytically calculated upper bound for Q_2.

Figure 7.8 presents computer results of estimation of error probabilities Q_1, Q_2 (by error rates $\hat{Q}_1$, $\hat{Q}_2$) for the situation where the true parameter values $\{\theta_i^0\}$ were unknown, the decision rule (7.69), (7.71) was used, and

$$L = 2,\ N = 2,\ T^0 = 4,\ T_- = 2,\ T_+ = 5,\ \alpha = 0.2,\ \Delta = 2.35.$$

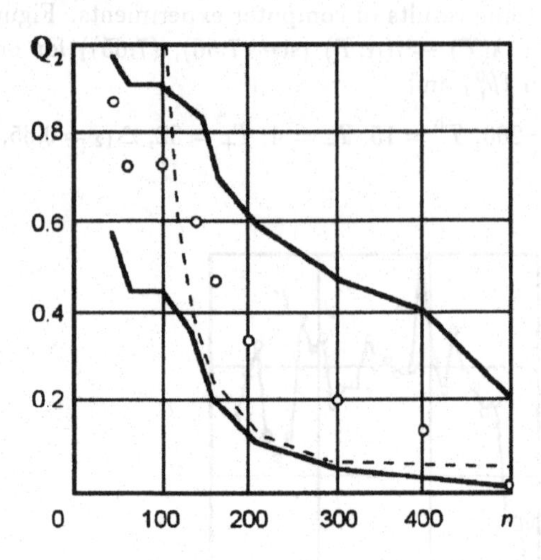

Figure 7.7: Error probability estimates

One can see from this Figure 7.8 that if the sample size n increases, then the error probability Q_2 (of run length determination) tends to zero sufficiently fast, and the probability Q_1 of classification error tends to the potentially attainable level determined by (7.77) : $Q_1^* = 0.01$. Note, that the decision rule ignoring the "run structure" of the sample has much larger error probability : $\bar{Q}_1 = 0.12$.

7.4 Cluster Analysis of Random Runs using Statistical Estimators of Interclass Distances

7.4.1 Mathematical Model

Let us consider once more the problem of cluster analysis of a sample X that has "run structure" described in Section 7.2 (its special case of "T^0-runs" was investigated in Section 7.3). As it was found in Sections 7.2, 7.3, this problem of cluster analysis becomes very difficult for the situations, common in practice, when the probability density functions are unknown *a priori*. In this section, we propose a new effective approach to cluster analysis in these situations.

Assume that a regular family $\mathcal{P}$ of N-variate probability density functions $p(x)$, $x \in R^N$, is defined on the observation space R^N, and $L \geq 2$ different unknown densities $p_1^0(\cdot), \ldots, p_L^0(\cdot) \in \mathcal{P}$ are fixed and describe random observations from the classes $\Omega_1, \ldots, \Omega_L$ respectively. An observed sample X of size n is composed of K^0 runs (homogeneous subsamples) of random lengths:

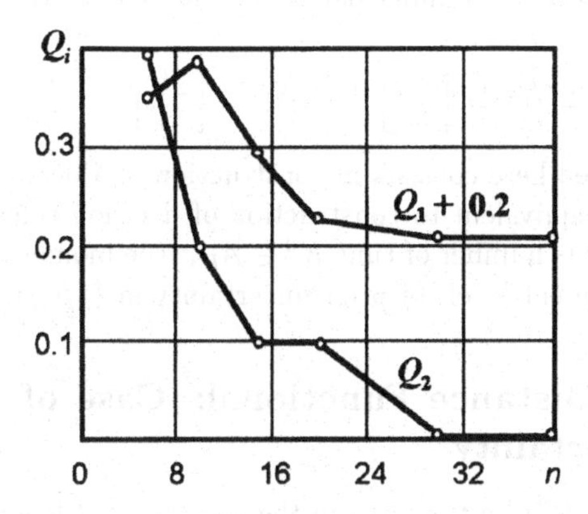

Figure 7.8: Computational results on estimation of error rate

$$X = (x_1, \ldots, x_n) = (X_1^0, \ldots, X_{K^0}^0),\ X_k^0 = (x_{T_{(k-1)}^0+1}, \ldots, x_{T_{(k)}^0}),$$

$$T_{(k)}^0 = T_{(k-1)}^0 + T_k^0\,(k = \overline{1, K^0},\ T_{(0)}^0 = 0,\ T_{(K^0)}^0 = n), \tag{7.80}$$

where X_k^0 is the k-th run of T_k^0 independent random observations from the same class $\Omega_{J_k^0}$ that have the same probability density function $p_{J_k^0}(\cdot) \in \mathcal{P}$. Here $J_k^0 \in S$ is an unknown true index of the class to which the run X_k^0 belongs; $T_{(k)}^0$ is an unknown true random moment of possible jump from the class $\Omega_{J_k^0}$ to $\Omega_{J_{k+1}^0}$ (the k-th "change point", if $J_k^0 \neq J_{k+1}^0$). Let us assume that the run lengths $T_1^0, \ldots, T_{K^0}^0$ are elements of a given finite set :

$$T_k^0 \in A_T = \{T_-, T_- + 1, \ldots, T_+\}\,(k = \overline{1, K^0}),$$

and that they can be considered as a realization of a random sequence with bivariate discrete probability distribution

$$q_2(\tau, \tau') = \mathbf{P}\{T_k^0 = \tau,\ T_{k+1}^0 = \tau'\},\ \tau, \tau' \in A_T\,(k = \overline{1, K^0}), \tag{7.81}$$

where $T_+(T_-)$ is the maximal (minimal) admissible value of run length. A special case of (7.81) is the case of independent identically distributed random variables $\{T_k^0\}$:

$$q_2(\tau, \tau') = q_1(\tau)q_1(\tau'),\ q_1(\tau) = \mathbf{P}\{T_k^0 = \tau\}, \tau, \tau' \in A_T\,(k = \overline{1, K^0}). \tag{7.82}$$

The true number of runs K^0 is assumed to be unknown :

$$K^0 \in A_K = \{K_-, K_- + 1, \ldots, K_+\},$$

where $K_+(K_-)$ is the maximal (minimal) admissible number of runs,

$$L \leq K_+ \leq \left\lfloor \frac{n}{T_-} \right\rfloor + 1,\ K_- \geq \left\lfloor \frac{n}{T_+} \right\rfloor .$$

The problem considered here consists in construction of a decision rule for classification of X, which is equivalent to construction of decision rules for estimation (from the sample X) of true number of runs $K^0 \in A_K$, true moments $\{T_k^0\}$ and true class indices $\{J_k^0\}$ for different levels of prior uncertainty in $\{p_k^0(\cdot)\}$.

7.4.2 Interclass Distance Functional: Case of Parametric Prior Uncertainty

Let X_k^0, X_l^0 $(1 \leq k < l \leq K^0)$ be two runs in the sample X of lengths T_k^0, T_l^0 from classes Ω_k, Ω_l respectively. Let us define the functional of interclass distance for the pair of classes Ω_k, Ω_l :

$$\rho_{kl}^0 = \rho(p_k^0, p_l^0) = f_2\left(\int_{R^N} f_1(p_k^0(x), p_l^0(x))\, dx \right). \tag{7.83}$$

Here $f_1(z_1, z_2) \geq 0$ is a twice continuously differentiable symmetric function in arguments $z_1, z_2 \geq 0$ such that

$$f_1(z_1, z_2) = 0 \leftrightarrow z_1 = z_2 ;$$

$f_2(y) \geq 0$ is a monotone increasing function in $y \geq 0$ such that

$$f_2(y) = 0 \leftrightarrow y = 0.$$

We shall distinguish three well-known special cases of the defined functional (7.83) :

- the L_2-distance is for

$$f_1(z_1, z_2) = \frac{(z_2 - z_1)^2}{2},\ f_2(y) = y ;$$

- the Bhattacharya distance (see, for example, (Fukunaga, 1972)) is for

$$f_1(z_1, z_2) = \frac{(\sqrt{z_2} - \sqrt{z_1})^2}{2},\ f_2(y) = -\ln(1 - y) ;$$

- the Kullback divergence (Kullback, 1967) is for

$$f_1(z_1, z_2) = (z_2 - z_1) \ln(z_2/z_1),\ f_2(y) = y.$$

For larger ρ^0_{kl} the difference of probability distributions for the k-th and l-th runs is larger. In practice, these probability distributions $p^0_k(\cdot)$, $p^0_l(\cdot)$ are usually unknown, therefore let us consider the problem of statistical estimation of the functional ρ^0_{kl} from the sample X. First, let us investigate the case of parametric uncertainty.

Let

$$\mathcal{P} = \{p(x;\theta),\ x \in R^N\ :\ \theta \in \Theta \subseteq R^m\}$$

be an m-parametric ($m < T_-$) family of N-variate probability density functions satisfying the Chibisov regularity conditions (Section 1.6), and let $\{\theta^0_1, \dots, \theta^0_L\} \subset \Theta$ be a subset of L different unknown parameter values: $p^0_i(x) = p(x;\theta^0_i)$, $i = \overline{1,L}$. Let us introduce the notations by (7.83) :

$$\rho_1(\theta_k, \theta_l) \;=\; \rho(p(\cdot;\theta_k), p(\cdot;\theta_l)),\ \theta_k, \theta_l \in \Theta\,;$$

∇^j_θ is the operator of calculation of the set of m^j partial derivatives of j-th order with respect to $\theta \in R^m$;

$$J(\theta) \;=\; -\int\limits_{R^N} p(x;\theta)\nabla^2_\theta \ln p(x;\theta)\,dx$$

is the positive definite Fisher information $(m \times m)$-matrix; $\hat{\theta}_k \in \Theta$ is the ML-estimator of the parameter θ^0_k calculated from the run of observations X^0_k;

$$\hat{\rho}_{kl} \;=\; \hat{\rho}_{kl}(T^0_{(k-1)}, T^0_{(k)}, T^0_{(l-1)}, T^0_{(l)}) \;=\; \rho_1(\hat{\theta}_k, \hat{\theta}_l) \tag{7.84}$$

is the consistent estimator of interclass distance ρ^0_{kl} by the runs of observations X^0_k, X^0_l. Let us define the hypothesis of homogeneity of the runs X^0_k, X^0_l

$$H_{0kl}\ :\ J^0_k = J^0_l.$$

Theorem 7.15 *If the family $\mathcal{P}$ is regular, the hypothesis H_{0kl} holds, and*

$$\nabla^2_{\theta_i,\theta_j}\rho_1(\theta_k,\theta_l)\big|_{\theta_k=\theta_l=\theta^0_k} = (-1)^{1+\delta_{ij}}B(\theta^0_k),\ i,j \in \{k,l\}, \tag{7.85}$$

where $B(\theta^0_k)$ is a positive definite symmetric matrix, then as $T_- \to \infty$ (asymptotics of increasing run length) the following stochastic expansion of the statistic (7.84) takes place :

$$\hat{\rho}_{kl} \;=\; \frac{1}{2}\left(\frac{1}{T^0_k} + \frac{1}{T^0_l}\right)\xi^T J^{-1/2}(\theta^0_k)B(\theta^0_k)J^{-1/2}(\theta^0_k)\xi + \mathcal{O}_{\mathbf{P}}(T_-^{3/2}), \tag{7.86}$$

where $\xi = (\xi_j) \in R^m$ is the Gaussian random vector with independent components having the standard normal distribution $\mathcal{N}_1(0,1)$, and $\mathcal{O}_{\mathbf{P}}(T_-^{3/2}) \xrightarrow{\mathbf{P}} 0$ is the remainder term (of order $T_-^{3/2}$) converging to zero in probability.

Proof. Let $\Delta\theta_k = \hat{\theta}_k - \theta_k^0$ denote the random deviation of the estimate. Taking into account (7.83), (7.85) and H_{0kl}, let us apply the Taylor formula of second order to the function (7.84) in the neighborhood of point (θ_k^0, θ_l^0):

$$\hat{\rho}_{kl} = \frac{1}{2}(\Delta\theta_k - \Delta\theta_l)^T B(\theta_k^0)(\Delta\theta_k - \Delta\theta_l) + \eta,$$

where $\eta = \mathcal{O}(|\Delta\theta_k|^3 + |\Delta\theta_l|^3)$ is the remainder term. From (Chibisov, 1973) we conclude that as $T_- \to \infty$, the random deviation $\Delta\theta_i$ has asymptotically normal distribution with zero mean and covariance matrix $(T_i^0)^{-1}J^{-1}(\theta_i^0)$, $i \in \{k, l\}$;

$$\eta = \mathcal{O}_{\mathbf{P}}(T_-^{3/2}) \xrightarrow{\mathbf{P}} 0.$$

The random runs X_k^0, X_l^0 are independent, therefore, the random vector

$$\xi = \left(\frac{1}{T_k^0} + \frac{1}{T_l^0}\right)^{-1/2} J^{1/2}(\theta_k^0)(\Delta\theta_k - \Delta\theta_l) \in R^m$$

has asymptotically normal distribution with zero mean and the identity covariance matrix. As a result, we come to the representation (7.86).

■

Corollary 7.11 *If*

$$B(\theta_k^0) = \frac{2}{c} J(\theta_k^0),$$

where c is a constant (c = 8 for the case of Bhattacharya distance; c = 1 for the case of Kullback distance), then the standardized estimator of interclass distance

$$r_{kl} = r_{kl}(T_{(k-1)}^0, T_{(k)}^0, T_{(l-1)}^0, T_{(l)}^0) = c\frac{T_k^0 T_l^0}{T_k^0 + T_l^0}\hat{\rho}_{kl} \tag{7.87}$$

has asymptotic χ^2-distribution with m degrees of freedom.

Proof. Substituting (7.86) and the expression of $B(\theta_k^0)$ into (7.87), the required fact becomes proved by the definition of χ^2-distribution.

■

Corollary 7.12 *Let $\gamma_{1-\alpha}(m)$ be the $(1-\alpha)$-quantile for the χ^2-distribution with m degrees of freedom. Then under conditions of Corollary 7.11 the asymptotic size of the test determined by the critical region $G = \{r_{kl} \geq \gamma_{1-\alpha}(m)\}$ of rejection of the hypothesis H_{0kl} coincides with the preassigned significance level $\alpha \in (0,1)$.*

Proof is conducted by an immediate calculation of the probability of type I error using the asymptotic distribution from Corollary 7.11.

■

7.4.3 Nonparametric Estimator of Interclass Distance

Assume that $\mathcal{P}$ is a family of triply continuously differentiable N-variate probability density functions $p(x)$, $x \in R^N$, bounded together with their derivatives, and

$$A_s(p) = \int_{R^N} p^s(x)\,dx < \infty, \; s = 1,2,3.$$

Let us construct the consistent nonparametric estimator of interclass L_2-distance (7.83) :

$$\hat{\rho}_{kl} = \hat{\rho}_{kl}(T^0_{(k-1)}, T^0_{(k)}, T^0_{(l-1)}, T^0_{(l)}) = \frac{1}{2}\int_{R^N} (\hat{p}_k(x) - \hat{p}_l(x))^2\,dx \geq 0. \qquad (7.88)$$

Here

$$\hat{p}_i(x) = \frac{1}{T^0_i} \sum_{t=T^0_{(i-1)}+1}^{T^0_{(i)}} \frac{1}{|H_i|} K\left(H_i^{-1}(x - x_t)\right),\; x \in R^N,\; i \in \{k,l\} \qquad (7.89)$$

is the Rosenblatt—Parzen nonparametric estimator (Rosenblatt, 1956), (Parzen, 1962) of the probability density function $p_i(x)$ calculated from the run of observations X^0_i, where

$$K(y) = (2\pi)^{-N/2} \exp(-y^T y/2)$$

is the N-variate Gaussian kernel, $H_i = \operatorname{diag}\{h_{ij}\}$ is a diagonal $(N \times N)$-matrix, whose diagonal elements are called *smoothness parameters*:

$$h_{ij} = b_j(T^0_{(i)} - T^0_{(i-1)})^{-\beta},\; \beta > 0,\; b_j > 0\,(j = \overline{1,N}).$$

Substituting (7.89) into (7.88) and making some equivalent transformations, we come to an explicit form of the statistic $\hat{\rho}_{kl}$:

$$\hat{\rho}_{kl} = \frac{1}{2}(a_{kk} + a_{ll}) - a_{kl}, \qquad (7.90)$$

$$a_{kl} = \frac{1}{T^0_k T^0_l} \left|H_k^2 + H_l^2\right|^{-1/2} \times$$

$$\times \sum_{t=T^0_{(k-1)}+1}^{T^0_{(k)}} \sum_{t'=T^0_{(l-1)}+1}^{T^0_{(l)}} K\left((H_k^2 + H_l^2)^{-1/2}(x_t - x_{t'})\right).$$

Theorem 7.16 *If $p^0_k(\cdot), p^0_l(\cdot) \in \mathcal{P}$, the hypothesis H_{0kl} holds, and*

$$\frac{1}{N+4} < \beta < \frac{1}{N} \;,$$

then as $T_- \to \infty$ the following asymptotic expansions for the moments of the non-parametric statistic (7.90) take place:

$$\mathbf{E}\{\hat{\rho}_{kl}\} = (2\pi)^{-N/2} 2^{-1-N/2} B^{-1} \left((T_k^0)^{-1+N\beta} + (T_l^0)^{-1+N\beta} \right) + o(T_-^{-1+N\beta}), \quad (7.91)$$

$$\mathbf{D}\{\hat{\rho}_{kl}\} = (2\pi)^{-N/2} 2^{-N-1} B^{-1} A_2(p) \Big((T_k^0)^{-2+N\beta} + (T_l^0)^{-2+N\beta} +$$

$$+2(T_k^0 T_l^0)^{-1} \left(\frac{(T_k^0)^{-2\beta} + (T_l^0)^{-2\beta}}{2} \right)^{-N/2} \Big) + o(T_-^{-2+N\beta}),$$

where

$$B = \prod_{j=1}^{N} b_j \quad .$$

Proof. It follows from (7.90) that

$$\mathbf{E}\{\hat{\rho}_{kl}\} = \frac{1}{2}(\mathbf{E}\{a_{kk}\} + \mathbf{E}\{a_{ll}\}) - \mathbf{E}\{a_{kl}\},$$

$$\mathbf{D}\{\hat{\rho}_{kl}\} = \frac{1}{4}(\mathbf{D}\{a_{kk}\} + \mathbf{D}\{a_{ll}\}) + \mathbf{D}\{a_{kl}\} - \mathbf{Cov}\{a_{kk}, a_{kl}\} - \mathbf{Cov}\{a_{ll}, a_{kl}\}.$$

Denoting

$$K_{kl}(z) = |H_k^2 + H_l^2|^{-1/2} K((H_k^2 + H_l^2)^{-1/2} z), \; z \in R^N,$$

we have from (7.90) :

$$\mathbf{E}\{a_{kk}\} = (2\pi)^{-N/2} 2^{-N/2} |H_k|^{-1} (T_k^0)^{-1} + (1 - (T_k^0)^{-1}) \mathbf{E}\{K_{kk}(x_1 - x_2)\},$$

$$\mathbf{E}\{a_{kl}\} = \mathbf{E}\{K_{kl}(x_1 - x_2)\},$$

$$\mathbf{D}\{a_{kk}\} = 2(T_k^0)^{-3}(T_k^0 - 1) \big(\mathbf{D}\{K_{kk}(x_1 - x_2)\} +$$

$$+2(T_k^0 - 2)\mathbf{Cov}\{K_{kk}(x_1 - x_2), K_{kk}(x_1 - x_3)\}\big),$$

$$\mathbf{D}\{a_{kl}\} = (T_k^0 T_l^0)^{-1} \big(\mathbf{D}\{K_{kl}(x_1 - x_2)\} +$$

$$+(T_k^0 + T_l^0 - 2)\mathbf{Cov}\{K_{kl}(x_1 - x_2), K_{kl}(x_1 - x_3)\}\big),$$

$$\mathbf{E}\{a_{kk} a_{kl}\} = (T_k^0)^{-2}(T_k^0 - 1) \big(2\mathbf{E}\{K_{kk}(x_1 - x_2) K_{kl}(x_1 - x_3)\} +$$

$$+(T_k^0-2)\mathbf{E}\{K_{kk}(x_1-x_2)\}\mathbf{E}\{K_{kl}(x_1-x_2)\}),$$

where $x_1, x_2, x_3 \in R^N$ are independent random vectors with the same probability density function $p_k^0(x)$.

Using Epanechnikov's method to calculate the moments of the statistics $K_{kk}(\cdot)$, $K_{ll}(\cdot)$, $K_{kl}(\cdot)$ (Epanechnikov, 1969) and then collecting the main terms of the asymptotic expansions, we come to (7.91).

■

By analogy with (7.87), let us construct the standardized nonparametric estimator of interclass distance using (7.90) :

$$r_{kl} = r_{kl}(T^0_{(k-1)}, T^0_{(k)}, T^0_{(l-1)}, T^0_{(l)}) =$$

$$= \left(\hat{\rho}_{kl} - (2\pi)^{-N/2} 2^{-1-N/2} B^{-1}\left((T_k^0)^{-1+N\beta} + (T_l^0)^{-1+N\beta}\right)\right) \times$$

$$\times \left((2\pi)^{-N/2} 2^{-N-1} B^{-1} A_2(\tilde{p})\left((T_k^0)^{-2+N\beta} + (T_l^0)^{-2+N\beta} +\right.\right.$$

$$\left.\left. +2(T_k^0 T_l^0)^{-1}\left(\frac{(T_k^0)^{-2\beta} + (T_l^0)^{-2\beta}}{2}\right)^{-N/2}\right)\right)^{-1/2}, \qquad (7.92)$$

where $\tilde{p}(x)$ is the nonparametric estimator of probability density constructed similarly to (7.89) from the whole sample X under the assumption of its homogeneity.

7.4.4 Decision Rules

For arbitrary $K \in A_K$, $T_1, \ldots, T_K \in A_T$ and "change points" $T_{(k)} = T_{(k-1)} + T_k$ ($k = \overline{1, K}$, $T_{(0)} = 0$, $T_{(K)} = n$) similarly to (7.80) let us split the observed sample X into K runs (subsamples) $X_1, \ldots, X_K$. By (7.81), (7.87), (7.92) let us construct the statistics $\{r_{kl}\}$ and the functional :

$$R = R(K, \{T_{(k)}\}) = \frac{1}{K-1}\sum_{k=1}^{K-1} q_2(T_{(k)} - T_{(k-1)}, T_{(k+1)} - T_{(k)}) \times$$

$$\times r_{k,k+1}(T_{(k-1)}, T_{(k)}, T_{(k)}, T_{(k+1)}).$$

Here R is the weighted average of interclass distances for all $K-1$ "change points". Let us find the estimator $\hat{K}$ of K^0 and the estimates $\{\hat{T}_{(k)}\}$ of $\{T^0_{(k)}\}$ as the solution of the following maximization problem (Kharin *et al.*, 1991):

$$R(K, \{T_{(k)}\}) \longrightarrow \max_{\{T_{(k)}\}} \max_{K \in A_K} \quad . \qquad (7.93)$$

By (7.80) let us transform the objective function in (7.93) to the form :

$$R(K, \{T_{(k)}\}) = \frac{1}{K-1} \sum_{k=1}^{K-3} \Psi_k(T_{(k)}, T_{(k+1)}, T_{(k+2)}), \tag{7.94}$$

where

$$\Psi_1(T_{(1)}, T_{(2)}, T_{(3)}) = q_2(T_{(1)}, T_{(2)} - T_{(1)}) r_{12}(0, T_{(1)}, T_{(1)}, T_{(2)}) +$$

$$+ q_2(T_{(2)} - T_{(1)}, T_{(3)} - T_{(2)}) r_{23}(T_{(1)}, T_{(2)}, T_{(2)}, T_{(3)}),$$

$$\Psi_j(T_{(j)}, T_{(j+1)}, T_{(j+2)}) = q_2(T_{(j+1)} - T_{(j)}, T_{(j+2)} - T_{(j+1)}) \times$$

$$\times r_{j,j+1}(T_{(j)}, T_{(j+1)}, T_{(j+1)}, T_{(j+2)}), \; j = \overline{2, K-4};$$

$$\Psi_{K-3}(T_{(K-3)}, T_{(K-2)}, T_{(K-1)}) = q_2(T_{(K-2)} - T_{(K-3)}, T_{(K-1)} - T_{(K-2)}) \times$$

$$\times r_{K-2,K-1}(T_{(K-3)}, T_{(K-2)}, T_{(K-2)}, T_{(K-1)}) +$$

$$+ q_2(T_{(K-1)} - T_{(K-2)}, n - T_{(K-1)}) r_{K-1,K}(T_{(K-2)}, T_{(K-1)}, T_{(K-1)}, n).$$

The representation (7.94) allows to use the method of dynamic programming (as in Section 7.2) for maximization in (7.93) with respect to $T_{(1)}, \ldots, T_{(K-1)}$ under the constraints (at fixed $K \in A_K$)

$$T_{(k-1)} + T_- \leq T_{(k)} \leq \min\{T_{(k-1)} + T_+, n - (K-k)T_-\} \; (k = \overline{1, K-1}).$$

Maximization with respect to $K \in [K_-, K_+]$ is performed by examination of $K_+ - K_- + 1$ values of the objective function (7.94). The computational complexity of this optimization algorithm is $\mathcal{O}((T_+ - T_-)^2 (T_+ + T_-)(K_+^3 - K_-^3))$.

For the case of parametric prior uncertainty, the statistical significance of each "change point" $T_{(k)}$ can be tested by Corollary 7.12: $\hat{T}_{(k)}$ is an α-significant "change point", if

$$r_{k,k+1} \geq \gamma_{1-\alpha}(m) \quad .$$

After finding the estimates $\hat{K}$, $\{\hat{T}_{(k)}\}$ we can easy classify the runs of observations (estimate the class indices $\{J_k^0\}$) by the hierarchial cluster procedure (see, for example, (Afifi *et al.*, 1979)). It consists in step-by-step merging of the clusters; the number of steps is $\hat{K} - L$. At the first step, the initial number of classes is assumed to be $L' = \hat{K}$. Using the criterion

$$r_{kl}(\hat{T}_{(k-1)}, \hat{T}_{(k)}, \hat{T}_{(l-1)}, \hat{T}_{(l)}) \to \min_{1 \leq k < l \leq L'} \quad , \tag{7.95}$$

we find the numbers (k^*, l^*) of two closest clusters (runs) which are to be merged into one cluster with number k^*. After this merging the number of clusters decreases: $L' = \hat{K} - 1$. Then the process is repeated: two closest clusters are found and merged into one cluster with the smallest number, and so on, until the preassigned number of classes $L' = L$ remains. After $\hat{K} - L$ steps of this kind we obtain a classification of the sample X for L classes, which determines the required estimates $\{\hat{J}_k\}$.

7.4.5 Computer Experiments

The decision rules (7.93)–(7.95) were implemented by a computer program and investigated by the Monte-Carlo method. The parametric statistical estimator (7.84), (7.87) of the Bhattacharya interclass distance was used. The values of run length had the probability distribution (7.82) determined by Table 7.3; $T_- = 4$, $T_+ = 9$.

Table 7.3: Probability distribution of T_k^0

τ	4	5	6	7	8	9
$q_1(\tau)$	0.10	0.22	0.35	0.18	0.10	0.05

The sample X was simulated according to (7.80) and consisted of $K^0 = 6$ runs of two-dimensional ($N = 2$) observations from $L = 2$ equiprobable classes ($\pi_1 = \pi_2 = 0.5$). $\mathcal{P}$ was the family of bivariate Gaussian probability densities $p(x) = n_2(x|\theta, \Sigma)$ with parameter $\theta \in R^2$ (mean vector). Eight series of computer experiments ($M = 50$ random samples X in each series) were conducted for 8 different values of Mahalanobis distance :

$$\Delta = ((\theta_2^0 - \theta_1^0)^T \Sigma^{-1} (\theta_2^0 - \theta_1^0))^{1/2} \in \{1.05, 1.35, 1.68, 2.07, 2.56, 3.29, 4, 4.65\}.$$

Figure 7.9 presents investigated dependence of classification error probability r_d on Mahalanobis distance Δ. Here the circles present the point estimates $\hat{r}_d$ (classification error rates), the dashed lines present the 95%-confidence limits, and the solid lines present the lower and upper bounds found in Sections 7.1, 7.3 :

$$r_- = \sum_{\tau=T_-}^{T_+} q_1(\tau) \Phi\left(-\sqrt{\tau}\frac{\Delta}{2}\right),$$

$$r_+ = \Phi\left(-\frac{\Delta}{2}\right) + \frac{1}{n}\left(\frac{N - 1 + \Delta^2/4}{\sqrt{2\pi}\Delta} e^{-\Delta^2/8} + \frac{\Delta^2}{16} e^{-\Delta^2/4}\right).$$

Table 7.4 presents the frequency distribution ν_j of random deviations $j = \hat{K} - K^0$ calculated from the results of 400 experiments. It characterizes the performance of the estimator $\hat{K}$.

Figure 7.10 presents the investigated mean deviations of $\{T_k^0\}$ and $\{\hat{T}_k\}$:

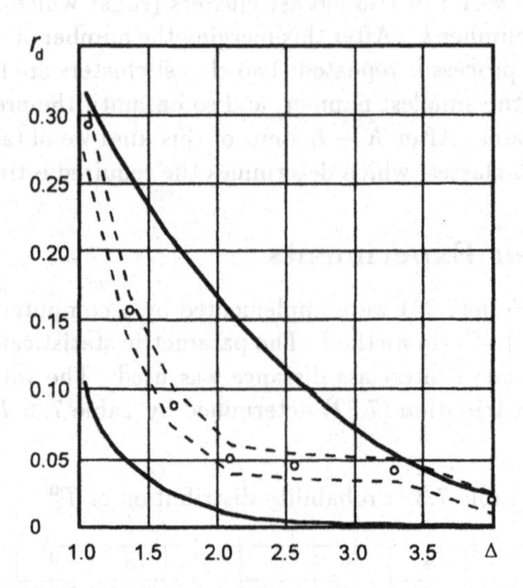

Figure 7.9: Error probability vs. interclass distance

Table 7.4: Frequency distribution of $j = \hat{K} - K^0$

j	-2	-1	0	1
ν_j	20	50	329	1

$$r_T = \mathbf{E}\left\{\min_{\{i_k\}} \sum_{k=1}^{K^0} \left|T_{i_k}^0 - \hat{T}_k\right|\right\} .$$

Here, circles present sample estimates of r_T obtained by computer experiment, and dashed lines present the 95%-confidence limits.

Computer experiments revealed a sufficiently high efficiency of the clustering procedure proposed in this section.

7.5 Case of Markov Dependence of Class Indices

7.5.1 Mathematical Model

In Sections 7.1–7.4 methods and algorithms were developed for the situation where the class indices $\{d_t^0\}$ are dependent deterministically: the class indices are invariable inside any run and may change only at the transitions between runs. This section

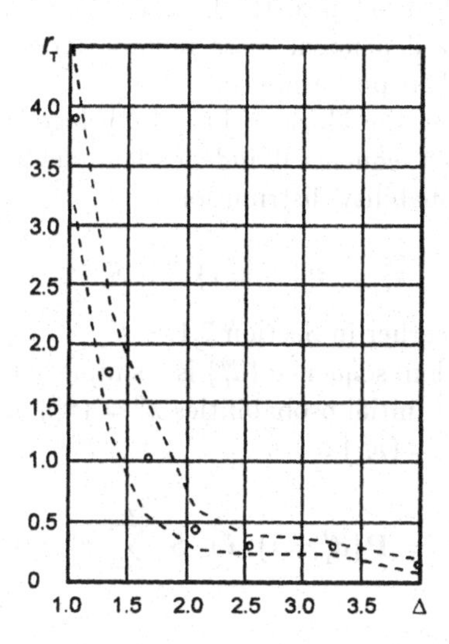

Figure 7.10: Mean deviations of $\{T_k^0\}$ and $\{\hat{T}_k\}$

is devoted to the situation where the indices of the classes $\{d_t^0\}$ are dependent stochastically: the random sequence $\{d_t^0\}$ is a finite Markov chain with discrete time and with state space $S = \{1, 2, \ldots, L\}$.

Let

$$\mathcal{P} = \{p(x;\theta) : x \in R^N, \theta \in \Theta \subseteq R^m\}$$

be any regular parametric family of N-variate probability density functions, and let $\{\theta_1^0, \ldots, \theta_L^0\} \subset \Theta$ be a subset of L different points in the parameter space Θ. Random observations from L classes are registered in the observation space R^N; Ω_i is the i-th class of observations with probability density function $p(x;\theta_i^0)$, where θ_i^0 is the true (possibly unknown) value of the parameter θ $(i \in S)$. Further, let $d_1^0, d_2^0, \ldots, d_n^0, \ldots \in S$ be a homogeneous K-order Markov chain with state space S, with initial probability distribution

$$\pi^0 = (\pi^0_{i_1,\ldots,i_K}), \ \pi^0_{i_1,\ldots,i_K} = \mathbf{P}\{d_1^0 = i_1, \ldots, d_K^0 = i_K\}$$

and with the set of transition probabilities :

$$P^0 = (p^0_{i_1,\ldots,i_K,j}), p^0_{i_1,\ldots,i_K,j} = \mathbf{P}\{d_{t+1}^0 = j \mid d_t^0 = i_K, \ldots, d_{t-K+1} = i_1\},$$

where $i_1, \ldots, i_k, j \in S$. The true values of the probabilities π^0, P^0 may be unknown. An observed sample $X = (x_1, \ldots, x_n)$ of size n is induced by $\{d_t^0\}$: x_t is a random observation from the class $\Omega_{d_t^0}$ $(d_t^0 \in S, t = \overline{1,n})$; for fixed $d_1^0, \ldots, d_n^0$, the random vectors $x_1, \ldots x_n$ are conditionally independent.

In this section we shall develop methods and algorithms for cluster analysis of a sample X with Markov dependent class indices under different levels of prior uncertainty and analyze their performance.

Note that if $K = 0$, then the Markov chain $\{d_t^0\}$ degenerates into a scheme of n independent trials: $\{d_t^0\}$ is a sequence of independent identically distributed random variables with discrete probability distribution :

$$\pi_i^0 = \mathbf{P}\{d_t^0 = i\}, \; i \in S \quad .$$

This case was considered earlier in Section 7.1.

If $K = 1$, then the random sequence $\{d_t^0\}$ is a simple homogeneous Markov chain specified by an L-vector of initial probabilities $\pi^0 = (\pi_i^0)$ and an $(L \times L)$-matrix of transition probabilities $P^0 = (p_{ij}^0)$:

$$\pi_i^0 = \mathbf{P}\{d_1^0 = i\}, \; i \in S; \; \sum_{i=1}^{L} \pi_i^0 = 1;$$

$$p_{ij}^0 = \mathbf{P}\{d_{t+1}^0 = j \mid d_t^0 = i\}, \; i, j \in S; \; \sum_{j=1}^{L} p_{ij}^0 = 1.$$

Depending on the explicit form of P^0, some important for practice special cases of the considered Markov model arise. For example, if P^0 is a matrix with dominant main diagonal :

$$p_{ii}^0 > 1 - \varepsilon_i, \; p_{ij}^0 < \varepsilon_i \, (j \neq i), \; 0 < \varepsilon_i < 0.5 \, (i \in S),$$

then the sequence of class indices $\{d_t^0\}$ has the "run structure", discussed in the Sections 7.2, 7.3.

Note in addition, that the presented model of data can be applied to speech recognition (Levinson *et al.*, 1983).

7.5.2 Optimal Decision Rules

Consider a situation where the probability characteristics π^0, P^0, $\{\theta_i^0\}$ of the model of a sample X are completely known *a priori*. Let $\hat{D} = (\hat{d}_1, \ldots, \hat{d}_n) \in S^n$ be a statistical decision rule for estimating D^0 :

$$\hat{D} = D(X), \; \hat{d}_t = d_t(X), \; t = \overline{1, n}, \; X \in R^{nN},$$

where $D(X)$ is any statistic: $R^{nN} \to S^n$.

Assume that a bounded loss function is given:

$$w = w(D', D''), \; D', D'' \in S^n;$$

here $w \geq 0$ is the loss value for the situation where the true classification vector is $D^0 = D'$, and the made decision is $\hat{D} = D''$. The following $(0-1)$-*loss function* is useful in practice :

$$w = w(D', D'') = \mathbf{1}(\|D' - D''\| - \delta), \tag{7.96}$$

where $\delta \geq 0$, and

$$\|D'' - D'\| = \sum_{t=1}^{n} (1 - \delta_{d'_t, d''_t})$$

is the number of disagreements in the sequences D', D''. The type (7.96) of loss function means that our losses are zero until the deviation $\|D^0 - \hat{D}\|$ does not exceed the given critical level δ.

We shall characterize the performance of the decision rule $\hat{D} = D(X)$ by the risk functional

$$r = r(D(\cdot)) = \mathbf{E}\{w(D^0, D(X))\} \geq 0 \quad . \tag{7.97}$$

For the case of loss function (7.96) the risk functional assumes the form :

$$r = r(D(\cdot)) = \mathbf{P}\{\|\hat{D} - D^0\| > \delta\} \quad . \tag{7.98}$$

In particular, if $\delta = 0$, then

$$r = r(D(\cdot)) = \mathbf{P}\{\hat{D} \neq D^0\}$$

is the probability of making at least one error when classifying the random sample X.

Theorem 7.17 *If the true parameters π^0, P^0, $\{\theta_i^0\}$ of the observed sample X are* a priori *known and $n > K$, then the optimal decision rule that minimizes the classification risk, is determined by the expressions :*

$$\hat{D} = D_0(X) = \arg \min_{D \in S^n} W_D(X), \tag{7.99}$$

$$W_D(X) = \sum_{J=(j_1,\dots j_n) \in S^n} w(J, D)\mathbf{P}(X, J),$$

$$\mathbf{P}(X, J) = \left(\prod_{t=1}^{n} p(x_t; \theta_{j_t}^0) \right) \left(\pi_{j_1,\dots,j_K}^0 \prod_{t=1}^{n-K} p_{j_t, j_{t+1}, \dots, j_{t+K}}^0 \right) .$$

Proof. According to the mathematical model of observed data (this model is described in Section 7.5.1) and to the definition of risk (7.97) we have :

$$r = r(D(\cdot)) = \int_{R^{nN}} W_{D(X)}(X)\, dX \quad .$$

Minimizing this functional, we find the optimal decision rule in the form (7.99). ■

Let us denote

$$\Lambda_1(X; D, \{\theta_i^0\}) = \sum_{t=1}^{n} \ln p(x_t; \theta_{d_t}^0),$$

$$\Lambda_2(D, \pi^0, P^0) = \ln \pi_{d_1,\ldots,d_K}^0 + \sum_{t=1}^{n-K} \ln p_{d_t,\ldots,d_{t+K}}^0, \qquad (7.100)$$

$$\Lambda(X; D, \{\theta_i^0\}, \pi^0, P^0) = \Lambda_1(X; D, \{\theta_i^0\}) + \Lambda_2(D, \pi^0, P^0),$$

$$V_\delta(D) = \{J : J \in S^n, \| J - D \| \leq \delta\}, \quad v_\delta = |V_\delta(D)| = \sum_{i=0}^{\delta} (L-1)^i C_n^i,$$

where v_δ denotes the cardinality of the set $V_\delta(D)$.

Corollary 7.13 *For the case of (0-1)-loss function (7.96), the optimal decision rule that minimizes the risk (7.98) has the form*

$$\hat{D} = D_0(X) = \arg \max_{D \in S^n} \frac{1}{v_\delta} \sum_{J \in V_\delta(D)} e^{\Lambda(X; J, \{\theta_i^0\}, \pi^0, P^0)}. \qquad (7.101)$$

Proof. Using the notations (7.99), (7.100), we have :

$$\mathbf{P}(X, J) = e^{\Lambda(X; J, \{\theta_i^0\}, \pi^0, P^0)}.$$

Substituting (7.96) into (7.100) and applying the previous expression, we come to the decision rule (7.101).

■

Corollary 7.14 *The optimal decision rule minimizing the probability of classification error* $r = \mathbf{P}\{\hat{D} \neq D^0\}$ *has the form*

$$\hat{D} = D_0(X) = \arg \max_{D \in S^n} \Lambda(X; D, \{\theta_i^0\}, \pi^0, P^0), \qquad (7.102)$$

where the objective function admits the following equivalent representation :

$$\Lambda(X; D, \{\theta_i^0\}, \pi^0, P^0) = \sum_{t=1}^{n-K} f_t(d_t, \ldots, d_{t+K}),$$

$$f_t(d_t, \ldots, d_{t+K}) = \delta_{t1} \left(\ln \pi_{d_1,\ldots,d_K}^0 + \sum_{l=1}^{K} \ln p(x_l; \theta_{d_l}^0) \right) +$$

$$+ \ln p_{d_t,\ldots,d_{t+K}}^0 + \ln p(x_{t+K}; \theta_{d_{t+K}}^0).$$

Proof. Assuming $\delta = 0$ in (7.101), taking into consideration that the set $V_0(D)$ in (7.100) is the singleton $\{D\}$ and making some equivalent transformations, we come to (7.102).

■

To evaluate the classification vector $\hat{D}$, it is necessary to solve the maximization problem in (7.102) :

$$\Lambda(X; D, \{\theta_i^0\}, \pi^0, P^0) \to \max_{D \in S^n} .$$

The special form (7.102) of the objective function allows to use the efficient method of dynamic programming (see, e.g., (Gabasov *et al.*, 1975)) to solve this maximization problem. Let us describe the procedure for computation of the classification vector $\hat{D} = (\hat{d}_1, \ldots, \hat{d}_n) \in S^n$.

At the first stage of this procedure we determine the sequence of $n - K$ Bellman functions $B_{K+1}(i_1, \ldots, i_K), \ldots, B_n(i_1, \ldots, i_K)$. Each of them is a function in K discrete variables $i = (i_1, \ldots, i_K) \in S^K$. The Bellman functions are computed recursively for each $i \in S^K$:

$$B_{K+1}(i_1, \ldots, i_K) = \max_{j \in S} f_1(j, i_1, \ldots, i_K);$$

for $k = K+1, K+2, \ldots, n-1$:

$$B_{k+1}(i_1, \ldots, i_K) = \max_{j \in S} \left(f_{k+1-K}(j, i_1, \ldots, i_K) + B_k(j, i_1, \ldots, i_{K-1})\right).$$

At the second stage the classification vector $\hat{D}$ is computed recursively (in inverse time):

$$(\hat{d}_{n-K+1}, \ldots, \hat{d}_n) = \arg \max_{(i_1, \ldots, i_K) \in S^K} B_n(i_1, \ldots, i_K);$$

for $k = 0, 1, \ldots, n-K-2$:

$$\hat{d}_{n-K-k} = \max_{j \in S} \Big(f_{n-k-K}(j, \hat{d}_{n-K-k+1}, \ldots, \hat{d}_{n-k}) +$$
$$+ B_{n-k-1}(j, \hat{d}_{n-K-k+1}, \ldots, \hat{d}_{n-k-1})\Big); \qquad (7.103)$$

$$\hat{d}_1 = \max_{j \in S} f_1(j, \hat{d}_2, \ldots, \hat{d}_{K+1}).$$

Computer implementation of the decision rule (7.103) needs memory of size $\mathcal{O}((n-K)L^K)$ (mainly for tabulating the Bellman functions), and the computation time is $\mathcal{O}((n-K+1)L^{K+1})$. So, the decision rule (7.103) is computationally feasible for the samples of large size n.

On the contrary, the decision rule (7.101) has exponential computational complexity and in fact is computationally prohibitive for large n. For this reason, let us construct a suboptimal decision rule. By the Jensen inequality we find the lower bound of the objective function in (7.101) :

$$\ln\left(\frac{1}{v_\delta}\sum_{J\in V_\delta(D)} e^{\Lambda(X;J,\{\theta_i^0\},\pi^0,P^0)}\right) \geq$$

$$\geq \frac{1}{v_\delta}\sum_{J\in V_\delta(D)} \Lambda(X;J,\{\theta_i^0\},\pi^0,P^0) \quad .$$

Instead of the true objective function, let us maximize this lower bound. As a result, we obtain the following suboptimal decision rule :

$$\hat{D} = D_*(X) = \arg\max_{D\in S^n} \frac{1}{v_\delta}\sum_{J\in V_\delta(D)} \Lambda(X;J,\{\theta_i^0\},\pi^0,P^0), \tag{7.104}$$

which has an obvious interpretation. For the suboptimal decision we take the point $\hat{D} \in S^n$, in whose δ-neighborhood $V_\delta(\hat{D})$ the arithmetic mean of $\Lambda(\cdot)$ is maximal.

The objective function in (7.104) can be equivalently transformed to a form similar to (7.102) and convenient for application of the dynamic programming. For example, if $K=1, L=2, \delta=1$, then the suboptimal decision rule may be represented in an equivalent form :

$$\hat{D} \;=\; D_*(X) \;=\; \arg\max_{D\in S^n} \sum_{t=1}^{n-1} g_t(d_t, d_{t+1}),$$

$$g_t(i,j) = (1-\frac{2}{n+1})f_t(i,j) + \frac{1}{n+1}(f_t(\bar{i},j)+f_t(i,\bar{j})),$$

where $\bar{i} = 3-i$, and the functions $\{f_t(\cdot)\}$ are defined in (7.102). This decision rule may be implemented by dynamic programming (7.103), where the functions $\{f_t(\cdot)\}$ are replaced by $\{g_t(\cdot)\}$.

7.5.3 Unknown Probability Characteristics π^0, P^0

Let the true matrix $P^0 = (p^0_{i_1,\ldots,i_k,j})$ of transition probabilities and the true matrix $\pi^0 = (\pi^0_{i_1,\ldots,i_K})$ of initial probability distribution of the Markov chain $\{d_t^0\}$ be *a priori* unknown. Let further $P = (p_{i_1,\ldots,i_k,j})$ be a $(K+1)$–dimensional stochastic matrix of L-th order subject to the constraints

$$0 \leq p_{i_1,\ldots,i_K,j} \leq 1, \sum_{j=1}^{L} p_{i_1,\ldots,i_K,j} = 1, \; (i_1,\ldots,i_K,j\in S). \tag{7.105}$$

Still further, let $\pi = (\pi_{i_1,\ldots,i_k})$ be a K–variate probability distribution on S^K :

$$0 \leq \pi_{i_1,\ldots,i_K} \leq 1, \quad \sum_{i_1,\ldots,i_K\in S} \pi_{i_1,\ldots,i_K} = 1. \tag{7.106}$$

To perform cluster analysis of a sample X with unknown matrix P^0, let us apply r_+-optimal decision rules that minimize the upper bound of the risk density defined in the Section 2.1. Assume the notations :

$$\nu_{i_1,\dots,i_K,j} = \nu_{i_1,\dots,i_K,j}(D) = \sum_{t=1}^{n-K} \delta_{d_{t+K},j} \prod_{k=1}^{K} \delta_{d_{t+k-1},i_k};$$

$$\nu_{i_1,\dots,i_K\bullet} = \nu_{i_1,\dots,i_K\bullet}(D) = \sum_{j=1}^{L} \nu_{i_1,\dots,i_K,j}(D)\,(i_1,\dots,i_K \in S);$$

$$\hat{p}_{i_1,\dots,i_K,j} = \nu_{i_1,\dots,i_K,j}(D)/\nu_{i_1,\dots,i_K\bullet}(D) \text{ if } \nu_{i_1,\dots,i_K\bullet}(D) \neq 0.$$

The variable $\nu_{i_1,\dots,i_K,j}(D)$ is the absolute frequency of the combination $(i_1,\dots,i_K,j)$ in the sequence of class indices $D = (d_1,\dots,d_n)$. The variable $\nu_{i_1,\dots,i_K\bullet}$ denotes the absolute frequency of the combination $(i_1,\dots,i_K)$ in the sequence $(d_1,\dots,d_{n-1})$. These frequencies satisfy the norming condition :

$$\sum_{i_1,\dots,i_K \in S} \nu_{i_1,\dots,i_K\bullet} = n - K.$$

The variable $\hat{p}_{i_1,\dots,i_K,j}$ is the relative frequency of appearance of the class index j immediately after the combination $(i_1,\dots,i_K)$ in the sequence D; the norming condition takes place :

$$\sum_{j=1}^{L} \hat{p}_{i_1,\dots,i_K,j} = 1\,(i_1,\dots,i_K \in S).$$

Theorem 7.18 *If the probability characteristics π^0, P^0 of the Markov chain $\{d_t^0\}$ are unknown, then the decision rule maximizing the upper bound of probability of error-free recognition of D^0 is determined by the expressions :*

$$\hat{D} = D_1(X) = \arg\max_{D \in S^n} \Lambda(X;\, D, \{\theta_i^0\}) \quad ,$$

$$\Lambda(X; D, \{\theta_i^0\}) = \Lambda_1(X; D, \{\theta_i^0\}) + \Lambda_2(D), \tag{7.107}$$

$$\Lambda_2(D) = \sum_{i_1,\dots,i_K \in S} \nu_{i_1,\dots,i_K\bullet} \sum_{j=1}^{L} \hat{p}_{i_1,\dots,i_K,j} \ln \hat{p}_{i_1,\dots,i_K,j},$$

where the statistic $\Lambda_1(\cdot)$ is defined by (7.100).

Proof. Using the joint probability distribution of the sample X and the classification vector D^0 (determined by (7.99)), we find an explicit form of error-free recognition of D^0 by the decision rule $\hat{D} = D(X)$:

$$Q(D(\cdot); \{\theta_i^0\}, \pi^0, P^0) = \int_{R^{nN}} \mathbf{P}(X, D(X))\,dX \quad .$$

As in Section 2.1, let us construct an upper bound for this functional with unknown π^0, P^0 :

$$Q(D(\cdot); \{\theta_i^0\}, \pi^0, P^0) \le \int_{R^{nN}} \max_{\pi,P} e^{\Lambda_1(X;D,\{\theta_i^0\})+\Lambda_2(D,\pi,P)}\, dX \quad ,$$

where the expressions (7.100) are taken into consideration and the maximization is under constraints (7.105), (7.106). The monotonicity of exp(·) implies:

$$\max_{\pi,P} \exp\left(\Lambda_1(X; D, \{\theta_i^0\}) + \Lambda_2(D, \pi, P)\right) =$$

$$= \exp\left(\Lambda_1(X; D, \{\theta_i^0\}) + \max_{\pi,P} \Lambda_2(D, \pi, P)\right).$$

By virtue of (7.100) we come to the maximization problem (for fixed value $D = (d_1, \ldots, d_n)$) :

$$\Lambda_2(D, \pi, P) = \ln \pi_{d_1,\ldots,d_K} + \sum_{t=1}^{n-K} \ln p_{d_t,\ldots,d_{t+K}} \to \max_{\pi,P}.$$

Under the constraints (7.106) we have :

$$\max_{\pi} \Lambda_2(D, \pi, P) = \sum_{t=1}^{n-K} \ln p_{d_t,\ldots,d_{t+K}} = \Lambda_{2+}(D, P).$$

This maximum is attained at

$$\pi^*_{j_1,\ldots,j_K} = \delta_{j_1,d_1} \ldots \delta_{j_K,d_K} \quad .$$

By the definition of $\nu_{i_1,\ldots,i_K,j}$, let us represent the objective function $\Lambda_{2+}(D, P)$ in the equivalent form :

$$\Lambda_{2+}(D, P) = \sum_{i_1,\ldots,i_K,j \in S} \nu_{i_1,\ldots,i_K,j} \ln p_{i_1,\ldots,i_K,j}.$$

It is easy to maximize this function with respect to P under the constraints (7.105) by the method of indefinite Lagrange multipliers. The maximal value is attained at $P^* = (\hat{p}_{i_1,\ldots,i_K,j})$ and equals to

$$\max_{P} \Lambda_{2+}(D, P) = \sum_{i_1,\ldots,i_K,j \in S} \nu_{i_1,\ldots,i_K,j} \ln \hat{p}_{i_1,\ldots,i_K,j} = \Lambda_{2+}(D).$$

As a result, the upper bound for Q assumes the form :

$$Q(D(\cdot); \pi^0, P^0, \{\theta_i^0\}) \le \int_{R^{nN}} \exp\left(\Lambda(X; D, \{\theta_i^0\})\right) dX,$$

where $\Lambda(\cdot)$ is defined by (7.107). Maximizing this upper bound with respect to $D(\cdot)$, we obtain the decision rule $\hat{D} = D_1(X)$ defined by (7.107).

■

To calculate the classification vector $\hat{D}$ by (7.107), it is necessary to maximize the objective function $\Lambda(X; D, \{\theta_i^0\})$ in (7.107) with respect to $D \in S^n$. This is a problem of nonlinear integer programming with n discrete variables $d_t \in S = \{1, \ldots, L\}$ $(t = \overline{1,n})$. Using some properties of the objective function $\Lambda(\cdot)$, let us formulate a special method for solving this problem.

Note that if the submatrices $P_j^0 = (p_{i_1,\ldots,i_K,j}^0 : i_1, \ldots, i_K \in S)$ of P^0 are close to each other for all $j \in S$, then the Markov chain $\{d_t^0\}$ is in fact close to the scheme of independent trials. This limiting model was considered in Section 7.1. Weak Markov properties (if P^0 is unknown) do not give any significant gain in accuracy of cluster anlaysis (in comparison with the methods given in Section 7.1). The case of high practical interest arises when the Markov properties of $\{d_t^0\}$ are strongly manifested: the submatrices P_1^0, ..., P_L^0 differ significantly. Let us assume that for each combination of K class indices $(i_1, \ldots, i_K) \in S^K$ there exists a class index $J^* = J^*(i_1, \ldots, i_K) \in S$ such that the following inequalities are true :

$$p_{i_1,\ldots,i_K,J^*}^0 \geq 1 - \varepsilon_{i_1,\ldots,i_K}; p_{i_1,\ldots,i_K,j}^0 \leq \varepsilon_{i_1,\ldots,i_K} \ (j \neq J^*), \tag{7.108}$$

where $\varepsilon_{i_1,\ldots,i_K} \in [0, 0.5)$ is a given (or assumed) critical value of transition probability. The conditions (7.108) have an obvious interpretation: among all L^{K+1} different combinations of $K + 1$ neighboring class indices

$$\{(i_1, \ldots, i_K, j) : i_1, \ldots, i_K, j \in S\}$$

in $\{d_t^0\}$ the combinations $(i_1, \ldots, i_K, J^*(i_1, \ldots, i_K))$ are present with dominant frequency. In other words, the appearance of the class index J^* immediately after the combination $(i_1, \ldots, i_K)$ is more probable than its non-appearance.

Let us formulate now two methods of finding an approximate solution of the maximization problem (7.107) under conditions (7.108). The first method exploits an upper bound for $\Lambda_2(\cdot)$.

Lemma 7.9 *Assume that $(p_1, \ldots, p_L)$ is a discrete probability distribution, $p_1 = \max_i p_i$ is the maximal elementary probability, and*

$$H_L(p_1, \ldots, p_L) = -\sum_{i=1}^{L} p_i \ln p_i$$

is the entropy functional for this probability distribution. Then entropy has the following lower bound :

$$H_L(p_1, \ldots, p_L) \geq H_2(p_1, 1 - p_1) = -(p_1 \ln p_1 + (1 - p_1) \ln(1 - p_1)).$$

Proof. Let us apply an obvious inequality :

$$H_L(p_1, \ldots, p_L) \geq -p_1 \ln p_1 - \max_{\substack{p_2, \ldots, p_L \\ p_2 + \cdots + p_L = 1}} \sum_{i=2}^{L} p_i \ln p_i.$$

Let us calculate the second term on the right side of this inequality :

$$\max \sum_{i=2}^{L} p_i \ln p_i = (1-p_1) \max \left(\sum_{i=2}^{L} \frac{p_i}{1-p_1} \ln \frac{p_i}{1-p_1} + \frac{p_i}{1-p_1} \ln(1-p_1) \right) =$$

$$= (1-p_1)\ln(1-p_1) + (1-p_1) \max_{q_1,\dots,q_{L-1}} \left(-H_{L-1}(q_1,\dots,q_{L-1})\right) =$$

$$= (1-p_1)\ln(1-p_1),$$

where $q_i = p_i/(1-p_1)$, $i = \overline{2,L}$. The nonnegativity property of entropy is used here. Substituting this maximal value into the initial inequality, we get the desired lower bound.

■

Corollary 7.15 *Under the conditions (7.108) the function $\Lambda_2(D)$ defined by (7.107) has the following upper bound :*

$$\Lambda_2(D) \leq - \sum_{i_1,\dots,i_K \in S} \nu_{i_1,\dots,i_K \bullet} H_2\left(\hat{p}_{i_1,\dots,i_K,J^*}, 1-\hat{p}_{i_1,\dots,i_K,J^*}\right).$$

Proof is based on the representation

$$\Lambda_2(D) = - \sum_{i_1,\dots,i_K \in S} \nu_{i_1,\dots,i_K \bullet} H_L\left(\hat{p}_{i_1,\dots,i_K,1}, \cdots, \hat{p}_{i_1,\dots,i_K,L}\right),$$

on the invariance property of entropy (with respect to permutations of the elementary probabilities), and on Lemma 7.9.

■

Lemma 7.10 *If $0 \leq \varepsilon < 0.5$, then the following piecewise linear approximation optimal in the mean square sense is valid :*

$$H_2(p, 1-p) \approx \begin{cases} a^* p, & \text{if } 0 \leq p \leq \varepsilon, \\ a^*(1-p), & \text{if } 1-\varepsilon \leq p \leq 1, \end{cases}$$

where the approximation coefficient is

$$a^* = a^*(\varepsilon) = \frac{2+\varepsilon}{4\varepsilon^2} + \frac{(1-\varepsilon)^2(1+2\varepsilon)}{2\varepsilon^3} \ln(1-\varepsilon) - \ln\varepsilon > 0.$$

Proof. Let us approximate the function $H_2(p, 1-p)$ on the intervals $[0,\varepsilon]$ and $[1-\varepsilon, 1]$. We shall find the optimal value of the approximation coefficient $a > 0$ by minimizing the mean square approximation error :

$$\int_0^{\varepsilon} \left(H_2(p,1-p) - ap\right)^2 dp \to \min_{a>0} .$$

Solving this minimization problem, we obtain the required approximation.

■

If we replace the function $\Lambda_2(D)$ in the decision rule (7.107) by its upper bound given by Corollary 7.15 and by Lemma 7.10, we obtain the following decision rule :

$$\hat{D} = D_2(X) = \arg \max_{D \in S^n} \Lambda_+(X; D, \{\theta_i^0\}),$$

$$\Lambda_+(X; D, \{\theta_i^0\}) = \sum_{t=1}^{n} \ln p(x_t; \theta_{d_t}^0) + \sum_{i_1,\dots,i_K \in S} a^*(\varepsilon_{i_1,\dots,i_K}) \times$$

$$\times (\nu_{i_1,\dots,i_K,J^*} - \nu_{i_1,\dots,i_K \bullet}) = \sum_{t=1}^{n} \ln p(x_t; \theta_{d_t}^0) +$$

$$+ \sum_{t=1}^{n-K} \sum_{i_1,\dots,i_K \in S} a^*(\varepsilon_{i_1,\dots,i_K}) \left(\delta_{d_{t+K}, J^*(i_1,\dots,i_K)} - 1\right) \prod_{k=1}^{K} \delta_{d_{t+k-1}, i_k}.$$

In particular, if

$$\varepsilon_{i_1,\dots,i_K} = \varepsilon \in [0, 0.5) \, (i_1, \dots, i_k \in S)$$

in (7.108), then this decision rule assumes the form :

$$\hat{D} = D_2(X) = \arg \max_{D \in S^n} \sum_{t=1}^{n-K} g_t(d_t, \dots, d_{t+K}), \tag{7.109}$$

$$g_t(d_t, \dots, d_{t+K}) = \delta_{t,1} \sum_{\tau=1}^{n} \ln p(x_\tau; \theta_{d_\tau}^0) +$$

$$+ \ln p(x_{t+K}; \theta_{d_{t+K}}^0) + a^*(\varepsilon) \delta_{d_{t+K}, J^*(d_t,\dots,d_{t+K-1})}.$$

The constructed decision rule (7.109) is similar to the optimal decision rule (7.102). The decision rule (7.109) formally can be derived from (7.102) by substituting $\{g_t(\cdot)\}$ for $\{f_t(\cdot)\}$. Therefore the decision rule (7.109) can be implemented by dynamic programming similar to (7.103).

The second method of finding an approximate solution of the maximization problem (7.107) under the conditions (7.108) is based on the local approximation of the function $\psi(p) = p \ln p$ from the expression for $\Lambda_2(D)$.

Lemma 7.11 *If $0 \le \varepsilon \le 1/e \le \delta \le 1$, then the following piecewise linear approximation optimal in mean square sense is valid :*

$$\psi(p) = p \ln p \approx \begin{cases} -b^* p, & \text{if } 0 \le p \le \varepsilon, \\ -c^*(1-p), & \text{if } \delta < p \le 1, \end{cases}$$

$$b^* = \frac{1}{3} + \ln \frac{1}{\varepsilon} \quad ,$$

$$c^* = \frac{1}{(1-\delta)^3}\left(\frac{3}{4}(1-\delta^2) - \frac{1}{3}(1-\delta^3) + \left(\frac{3}{2}-\delta\right)\delta^2 \ln\delta\right).$$

Proof is performed similarly to the proof of Lemma 7.10. The optimality criteria for the approximation coefficients are :

$$J_1(b) = \int_0^{\varepsilon} (\psi(p) + bp)^2\, dp \to \min_{b>0} \quad ,$$

$$J_2(c) = \int_{\delta}^{1} (\psi(p) + c(1-p))^2\, dp \to \min_{c>0} \quad .$$

The integrals $J_1(b)$, $J_2(c)$ are calculated by (Gradstein *et al.*, 1991), and minimization is performed by the classical method.

■

Corollary 7.16 *If $\varepsilon = \delta = 1/e$, then the optimal values of the approximation coefficients are*

$$b^* = \frac{4}{3}, \quad c^* = \frac{5e^3 - 27e + 16}{12(e-1)^3} \quad .$$

Using Lemma 7.11, let us construct an approximation of the function $\Lambda_2(D)$ from (7.107) :

$$\Lambda_2(D) = -A^*(n-K) + A^* \sum_{t=1}^{n-K} \delta_{d_{t+K}, J^*(d_t,\ldots,d_{t+K-1})},\ A^* = b^* + c^*.$$

Substituting this expression into (7.107), we obtain the decision rule of type (7.109), in which the coefficient a^* is replaced by $A^* = b^* + c^*$. In particular, under the condition of Corollary 7.16, we have

$$A^* = \frac{4}{3} + \frac{5e^3 - 27e + 16}{12(e-1)^3} \approx 2.040 \quad .$$

7.5.4 Parametric Prior Uncertainty Case

In this subsection we shall investigate the problem of cluster analysis of sample X generated by a Markov chain of class indices $\{d_t^0\}$ for the situations where the true parameter values $\{\theta_i^0\}$ of probability disitributions $\{p(x;\theta_i^0)\}$ are unknown. At the beginning, we shall assume that the probability characteristics π^0, P^0 of the Markov chain are given.

Theorem 7.19 *If the true parameter values $\{\theta_i^0\}$ are unknown, then the decision rule that maximizes the upper bound of probability of error-free classification assumes the form :*

$$\hat{D} = D_3(X) = \arg\max_{D \in S^n} \Lambda(X; D, \pi^0, P^0), \tag{7.110}$$

where

$$\Lambda(X; D, \pi^0, P^0) = \max_{\{\theta_i \in \Theta\}} \Lambda_1(X; D, \{\theta_i\}) + \Lambda_2(D, \pi^0, P^0),$$

and the statistics $\Lambda_1(\cdot)$, $\Lambda_2(\cdot)$ are defined in (7.100).

Proof is performed in the same way as the proof of Theorem 7.18, but an upper bound of the functional $Q(D(\cdot); \{\theta_i^0\}, \pi^0, P^0)$ is found by maximization with respect to $\{\theta_i^0\}$.

■

As it is seen from (7.110), (7.100), for the construction of decision rule $D_3(\cdot)$ it is necessary to solve a problem of nonlinear programming with Lm continuous variables $\{\theta_i\}$ and n discrete variables $\{d_t\}$. Even if the maximum of $\Lambda_1(\cdot)$ w.r.t. $\{\theta_i\}$ can be found in analytical form, to maximize $\Lambda(\cdot)$ w.r.t. $D \in S^n$, it is necessary to examine L^n values of the objective function. Therefore the decision rule (7.110) has exponential computer complexity and computationally prohibitive for large n. Using some asymptotic properties of the statistic $\Lambda_1(\cdot)$, we shall construct a family of suboptimal decision rules that may be implemented by dynamic programming.

Let us denote :

$$h(X; \theta) = \frac{1}{n} \sum_{t=1}^{n} \ln p(x_t; \theta), \; \theta \in \Theta; \tag{7.111}$$

$$\bar{\theta} = \arg\max_{\theta \in \Theta^*} h(X; \theta),$$

where Θ^* is the closure of the set Θ;

$$H(\theta; \theta_i^0) = \mathbf{E}_{\theta_i^0}\{-\ln p(x_t; \theta)\} = -\int_{R^N} p(x; \theta_i^0) \ln p(x; \theta) dx$$

$$(\theta, \theta_i^0 \in \Theta, \; i \in S);$$

$$G(\theta; \theta_i^0) = \mathbf{D}_{\theta_i^0}\{\ln p(x_t; \theta)\} = \int_{R^N} p(x; \theta_i^0) \left(\ln p(x; \theta) - H(\theta; \theta_i^0)\right)^2 dx > 0.$$

According to (7.111), the statistic $h(X; \theta)$ can be considered as the logarithmic likelihood function constructed under the homogeneity hypothesis for the observed sample X (i.e., $d_1^0 = d_2^0 = \ldots = d_n^0$); in this case, the statistic $\bar{\theta}$ can be considered as the corresponding statistical estimator. The function $H(\theta_i^0; \theta_i^0)$ in (7.111) defines entropy of the probability distribution $p(\cdot; \theta_i^0)$.

Theorem 7.20 *Let following conditions hold :*

A) the family $\mathcal{P}$ of probability densities satisfies the Chibisov regularity conditions (Section 1.6) and the variance $\mathbf{D}_{\theta_i^0}\{\ln p(x;\theta)\}$ $(i \in S)$ is finite;

B) the Markov chain $\{d_t^0\}$ is ergodic;

C) for any $i, j \in S$, $i \neq j$, the deviation is $|\theta_i^0 - \theta_j^0| = \mathcal{O}(\tau)$, as $\tau \to 0$.

Then there exists a point $\theta^* \in \Theta$ such that $|\theta_i^0 - \theta^*| = \mathcal{O}(\tau)$ $(i \in S)$, and the following almost sure convergence takes place as $n \to \infty$, $\tau \to 0$:

$$\bar{\theta} \xrightarrow{a.s.} \theta^*.$$

Proof. Let us investigate the asymptotic behavior of the statistic $h(X;\theta)$ defined by (7.111). By condition A) and the conditional independence of the observations $\{x_t\}$, the convergence of the series takes place :

$$\sum_{t=1}^{\infty} t^{-2}\mathbf{D}_{\theta_{d_t^0}^0}\{\ln p(x_t;\theta)\} \leq \max_{i\in S} G(\theta;\theta_i^0)\sum_{t=1}^{\infty}\frac{1}{t^2} < \infty.$$

Therefore $h(X;\theta)$ satisfies the strong law of large numbers. Using the notations (7.111), we find :

$$\mathbf{E}\{h(X;\theta) \mid D^0\} = -\sum_{i=1}^{L}\frac{\nu_i^0}{n}H(\theta;\theta_i^0),$$

where

$$\nu_i^0 = \sum_{t=1}^{n}\delta_{d_t^0,i}$$

is the frequency of class Ω_i in the sequence D^0. By condition B) and the ergodic theorem (see, e.g., (Koroljuk *et al.*, 1994)),

$$\nu_i^0/n \xrightarrow{a.s.} \pi_i^*,$$

where $(\pi_1^*,\ldots,\pi_L^*)$ is the stationary univariate probability distribution of $\{d_t^0\}$. Therefore,

$$h(X;\theta) \xrightarrow{a.s.} -\sum_{i=1}^{L}\pi_i^* H(\theta;\theta_i^0).$$

Further, following the proof of strong consistency of the ML–estimator (see, e.g., (Borovkov, 1984)), we conclude :

$$\bar{\theta} \xrightarrow{a.s.} \theta^*,\ \theta^* = \arg\min_{\theta\in\Theta^*}\sum_{i=1}^{L}\pi_i H(\theta;\theta_i^0).$$

By condition A), the Jensen inequality and the notations (7.111), we have :

$$\theta_i^0 = \arg\min_{\theta\in\Theta^*} H(\theta;\theta_i^0),\ i \in S.$$

Comparing θ^* and θ_i^0 under the condition C), we conclude that $|\theta^* - \theta_i^0| = \mathcal{O}(\tau)$, $i \in S$.

■

Note that the condition C) is typical in applications : interclass distances determined by the deviations $\{|\theta_i^0 - \theta_j^0| : i \neq j\}$ are sufficiently small, and high accuracy of classification can be achieved only exploiting the Markov dependence of class indices.

Let us construct now a quadratic approximation (w.r.t. the variables $\{\theta_i\}$) of the function $\Lambda_1(\cdot)$ entering the decision rule (7.110). According to Theorem 7.20, it is sufficient to construct this approximation in the τ-neighborhood of the point $\bar{\theta}$: $|\theta_i - \bar{\theta}| = \mathcal{O}(\tau)$, $i \in S$. Let us apply the Taylor formula to $\Lambda_1(\cdot)$:

$$\Lambda_1(X; D, \{\theta_i\}) = \sum_{i=1}^{L} \sum_{t=1,d_t=i}^{n} \ln p(x_t;\theta_i) = \sum_{t=1}^{n} \ln p(x_t;\bar{\theta}) + \tag{7.112}$$

$$+ \sum_{i=1}^{L} \nu_i(D) \left(\beta_i^T(X)(\theta_i - \bar{\theta}) - \frac{1}{2}(\theta_i - \bar{\theta})^T \gamma_i(X)(\theta_i - \bar{\theta}) \right) + \mathcal{O}(\tau^3),$$

where

$$\beta_i = \beta_i(X) = \frac{1}{\nu_i(D)} \sum_{t=1,\,d_t=i}^{n} \nabla_{\bar{\theta}} \ln p(x_t;\bar{\theta})$$

is an m-vector,

$$\gamma_i = \gamma_i(X) = -\frac{1}{\nu_i(D)} \sum_{t=1,d_t=i}^{n} \nabla_{\bar{\theta}}^2 \ln p(x_t;\bar{\theta})$$

is an $(m \times m)$–matrix, and

$$\nu_i(D) = \sum_{t=1}^{n} \delta_{d_t,i}\ (i \in S)$$

is the frequency of class Ω_i in the sequence D.

Under conditions of Theorem 7.20 by the strong law of large numbers, we have

$$\gamma_i \xrightarrow{a.s.} J(\theta_i^0) + \mathcal{O}(\tau)\mathbf{1}_{m\times m}\ (i \in S)\quad ,$$

as $n, \nu_i(D) \to \infty$, where

$$J(\theta_i^0) = \mathbf{E}_{\theta_i^0}\{-\nabla_{\theta_i^0}^2 \ln p(x_t;\theta_i^0)\}$$

is the Fisher information matrix for the parameter θ_i^0. Similarly, for $d_t^0 = i$ as $n \to \infty$,

$$\gamma = -\frac{1}{n}\sum_{t=1}^{n} \nabla_{\bar{\theta}}^2 \ln p(x_t;\bar{\theta}) \xrightarrow{a.s.} J(\theta_i^0) + \mathcal{O}(\tau)\mathbf{1}_{m\times m} \quad . \tag{7.113}$$

As we use only the main term of the asymptotic expansion of γ_i in (7.112), we substitute the matrix γ defined by (7.113) for γ_i in (7.112). Maximizing the obtained expression for Λ_1 w.r.t. $\{\theta_i\}$ at fixed $D \in S^n$, we find :

$$\max_{\{\theta_i\}} \Lambda_1\left(X; D, \{\theta_i\}\right) = \sum_{t=1}^{n} \ln p(x_t;\bar{\theta}) + \frac{1}{2}\sum_{t=1}^{n} \frac{B(t,t)}{\nu_{d_t}(D)} + \tag{7.114}$$

$$+\sum_{t'=1}^{n-1}\sum_{t''=t'+1}^{n} \frac{B(t',t'')}{\nu_{d_{t'}}(D)}\delta_{d_{t'},d_{t''}} + \mathcal{O}(\tau^3),$$

and this maximal value is attained at the point

$$\theta_i^* = \bar{\theta} + \gamma^{-1}\beta_i\,(i \in S).$$

The function $B(t', t'')$ in (7.114) is defined by the expression :

$$B(t',t'') = \left(\nabla_{\bar{\theta}} \ln p(x_{t'};\bar{\theta})\right)^T \gamma^{-1} \nabla_{\bar{\theta}} \ln p(x_{t''};\bar{\theta}).$$

Since $\mathcal{P}$ is regular, the Fisher information matrix $J(\theta_i^0)$ is positive definite, therefore, according to (7.113) the matrix γ also becomes positive definite as $n \to \infty$.

To simplify the maximization in (7.110), (7.114) w.r.t. D, let us introduce a constraint for D :

$$\sum_{i=1}^{L}\left(\frac{\nu_i(D)}{n} - \pi_i^*\right)^2 = \varepsilon_n^2, \tag{7.115}$$

expressing the above-mentioned ergodic property of $\{d_t^0\}$. Here $\varepsilon_n > 0$ is a decreasing number sequence such that $\varepsilon_n \to 0$. We shall take the constraint (7.115) into account in (7.110), (7.114) by the method of penalty function with the penalty factor $\mu_n \to \infty$. As a result, we obtain the following decision rule (from (7.110), (7.114), (7.115)) :

$$\hat{D} = D_3'(X) = \arg\max_{D\in S^n} \Lambda'(X; D, \pi^0, P^0), \tag{7.116}$$

$$\Lambda'(X; D, \pi^0, P^0) = \max_{\{\theta_i\}} \Lambda_1(X; D, \{\theta_i\}) + \Lambda_2(D, \pi^0, P^0) +$$

$$+\mu_n\left(\varepsilon_n^2 - \sum_{i=1}^{L}\left(\frac{\nu_i(D)}{n} - \pi_i^*\right)^2\right).$$

Applying (7.114) to (7.116) and neglecting the remainder $\mathcal{O}(\tau^3)$, we transform $\Lambda'(\cdot)$ to the form :

$$\Lambda'(X; D, \pi^0, P^0) = \left\{\mu_n\left(\varepsilon_n^2 - \frac{1}{n} - \sum_{i=1}^{L}(\pi_i^*)^2\right) + \sum_{t=1}^{n} \ln p(x_t;\bar{\theta})\right\} +$$

$$+\Lambda_2(D,\pi^0,P^0)+\frac{1}{n}\sum_{t=1}^{n}\left(\frac{B(t,t)}{2\pi^*_{d_t}}+2\mu_n\pi^*_{d_t}\right)+$$

$$+\frac{1}{n}\sum_{t'=1}^{n-1}\sum_{t''=t'+1}^{n}\left(\frac{B(t',t'')}{\pi^*_{d_{t'}}}-\frac{2\mu_n}{n}\right)\delta_{d_{t'},d_{t''}} .$$

Here the terms in $\{\cdot\}$ do not depend on D and therefore can be omitted for maximization in (7.116); the function $\Lambda_2(\cdot)$ is defined by (7.100). Keeping these facts in mind, let us define a family of statistics :

$$\Lambda_{(T)}(X;D,\pi^0,P^0)=\ln\pi^0_{d_1,\dots,d_K}+\sum_{t=1}^{n-K}\ln p^0_{d_t,\dots,d_{t+K}}+ \tag{7.117}$$

$$+\frac{1}{n}\sum_{t=1}^{n}\left(\frac{B(t,t)}{2\pi^*_{d_t}}+2\mu_n\pi^*_{d_t}\right)+$$

$$+\frac{n-1}{T\left(n-\frac{T+1}{2}\right)}\sum_{t'=1}^{n-1}\sum_{t''=t'+1}^{\min\{n,t'+T\}}\left(\frac{B(t',t'')}{2\pi^*_{d_{t'}}}-\frac{\mu_n}{n}\right)\delta_{d_{t'},d_{t''}} ,$$

where $T\in\{1,\dots,n\}$ is the parameter of this family. By the statistics (7.117) let us define a family of decision rules :

$$\hat{D}=D_{(T)}(X)=\arg\max_{D\in S^n}\Lambda_{(T)}(X;D,\pi^0,P^0). \tag{7.118}$$

Comparing the decision rule (7.116) with decision rules (7.117) and (7.118), we note that if the parameter T is equal to n, then the decision rule $D_{(n)}(X)\equiv D_3'(X)$. For $1\le T\le n-1$, one must consider the decision rule (7.118) as an approximation of the asymptotically optimal decision rule (7.116). For maximization w.r.t. $D\in S^n$ in (7.117), (7.118), the dynamic programming is suitable. The number of arguments of the Bellman functions (and the complexity of computation of $\hat{D}$) is determined by the value $K^*=\max\{K,T\}$, where K is the order of the Markov chain $\{d_t^0\}$ and T is the approximation parameter. Let us consider the case of simple Markov chain ($K=1$), which is common in applied pattern recognition problems. In this case, $K^*=T$, and the statistic $\Lambda_{(T)}$ can be represented in the form convenient for dynamic programming :

$$\Lambda_{(T)}(X;D,\pi^0,P^0)=\sum_{t=1}^{n-T}g_t(d_t,\dots,d_{t+T}), \tag{7.119}$$

where the functions $\{g_t(\cdot)\}$ are determined by the expressions :

$$g_1(d_1,\dots,d_{T+1})=\ln\pi^0_{d_1}+\sum_{\tau=1}^{T}\ln p^0_{d_\tau,d_{\tau+1}}+$$

$$+\frac{1}{n}\sum_{\tau=1}^{T+1}\left(\frac{B(\tau,\tau)}{2\pi^*_{d_\tau}}+2\mu_n\pi^*_{d_\tau}\right)+\frac{n-1}{T\left(n-\frac{T+1}{2}\right)}\sum_{\tau=2}^{T+1}\left(\frac{B(1,\tau)}{2\pi^*_{d_1}}-\frac{\mu_n}{n}\right)\delta_{d_1,d_\tau};$$

$$g_t(d_t,\ldots,d_{t+T})=\ln p^0_{d_{t+T-1},d_{t+T}}+\frac{1}{n}\left(\frac{B(t+T,t+T)}{2\pi^*_{d_{t+T}}}+2\mu_n\pi^*_{d_{t+T}}\right)+$$

$$+\frac{n-1}{T\left(n-\frac{T+1}{2}\right)}\sum_{\tau=t+1}^{t+T}\left(\frac{B(t,\tau)}{2\pi^*_{d_t}}-\frac{\mu_n}{n}\right)\delta_{d_t,d_\tau}\quad \text{if } 2\le t\le n-T-1;$$

$$g_{n-T}(d_{n-T},\ldots,d_n)=\ln p^0_{d_{n-1},d_n}+\frac{1}{n}\left(\frac{B(n,n)}{2\pi^*_{d_n}}+2\mu_n\pi^*_{d_n}\right)+$$

$$+\frac{n-1}{T\left(n-\frac{T+1}{2}\right)}\sum_{t=n-T}^{n-1}\sum_{\tau=t+1}^{n}\left(\frac{B(t,\tau)}{2\pi^*_{d_t}}-\frac{\mu_n}{n}\right)\delta_{d_t,d_\tau}.$$

The constructed decision rule (7.118), (7.119) can be implemented by the recurrent formulae (7.103), in which K is replaced by T, and the functions $\{f_t(\cdot)\}$ are replaced by $\{g_t(\cdot)\}$.

Let us turn now our attention to the problem of choice of the parameters in (7.119) : the approximation parameter $T\in\{1,\ldots,n\}$ and the penalty factor μ_n, $\mu_n\to\infty$. Multiple computer experiments show that a sufficient accuracy of classification can be achieved if $T\ge T^0$, where T^0 is the mean value of run length in D^0. In particular, for a simple stationary Markov chain ($K=1$, $\pi^0_i=\pi^*_i$, $i\in S$),

$$T\ge T^0,\quad T^0=\left\lfloor\sum_{i=1}^{L}\pi^*_i/(1-p^0_{ii})\right\rfloor.$$

After the analysis of the asymptotic behavior of the statistics $\{B(t,\tau)\}$ and the results of computer experiments, we propose the formula for μ_n :

$$\mu_n=\frac{nL}{4}\left(\Phi^{-1}(1-\alpha)\right)^2,$$

where α is the maximal probability of classification error for two classes Ω_i, Ω_j ($i,j\in S$, $i\ne j$) from a single observation $x_t\in R^N$; $\Phi^{-1}(1-\alpha)$ is the $(1-\alpha)$-quantile for the standard normal distribution.

Note in conclusion that if along with $\{\theta^0_i\}$ the probability characteristics π^0, P^0 of $\{d^0_t\}$ are unknown (as in Section 7.5.3), then for the construction of a decision rule maximizing (w.r.t. π^0, P^0, $\{\theta^0_i\}$) the probability of error–free classification, it is sufficient to replace the function $\Lambda_2(D,\pi^0,P^0)$ in (7.116) by the statistic $\Lambda_2(D)$ defined in Section 7.5.3. Such decision rule can be implemented by dynamic programming similarly to (7.118), (7.119).

7.5.5 Case of Gaussian Observations

Let us illustrate the construction of cluster analysis procedures by the methods proposed in Sections 7.5.2 — 7.5.4 under traditional in statistical pattern recognition assumption about Gaussian probability distribution of observations :

$$p(x;\theta_i^0) = n_N(x \mid \theta_i^0, \Sigma_i^0),\ x \in R^N,$$

where $\theta_i^0 = (\theta_{ij}^0) \in R^N$ is the mean vector, and $\Sigma_i^0 = (\sigma_{ijk}^0)$ is the $(N \times N)$–covariance matrix for class Ω_i $(i \in S)$. Let a sequence of class indices $\{d_t^0\}$ be a simple stationary Markov chain $(K = 1)$ with matrix of transition probabilities

$$P^0 = (p_{ik}^0), p_{ii}^0 = p^0 \geq 1 - \varepsilon, 0 \leq p_{ik}^0 \leq \varepsilon \ll 1 (i \neq k, i, k \in S) \qquad (7.120)$$

and with uniform initial probability distribution

$$\pi_i^* = \pi_i^0 = 1/L\,(i \in S).$$

Note two properties of this Markov chain of class indices $\{d_t^0\}$:

1) all class indices appear in $\{d_t^0\}$ with the same frequency;

2) a class index does not change in one step with high probability $p^0 \geq 1 - \varepsilon$, i.e., the sequence $\{d_t^0\}$ has a clear "run structure".

Case I : $P^0, \{\theta_i^0\}$ are *a priori* known.

An optimal decision rule minimizing the probability of classification error according to (7.102), (7.103) has the form :

$$\hat{D} = D_0(X) = \arg\max_{D \in S^n} \sum_{t=1}^{n-1} f_{0t}(d_t, d_{t+1}),\ X \in R^{nN}, \qquad (7.121)$$

$$f_{0t}(i,j) = \ln p_{ij}^0 - \frac{1}{2}\Big(\delta_{t1}\left(\ln|\Sigma_i^0| + (x_1 - \theta_i^0)^T(\Sigma_i^0)^{-1}(x_1 - \theta_i^0)\right) +$$

$$+ \ln|\Sigma_j^0| + (x_{t+1} - \theta_j^0)^T(\Sigma_j^0)^{-1}(x_{t+1} - \theta_j^0)\Big).$$

Case II : P^0 is unknown.

In this case we use the decision rule (7.109) with the function $J^*(i_1) = i_1$:

$$\hat{D} = D_2(X) = \arg\max_{D \in S^n} \sum_{t=1}^{n-1} f_{2t}(d_t, d_{t+1}),$$

$$f_{2t}(i,j) = a^*\delta_{ij} - \frac{1}{2}\Big(\delta_{t1}\left(\ln|\Sigma_i^0| + (x_1 - \theta_i^0)^T(\Sigma_i^0)^{-1}(x_1 - \theta_i^0)\right) +$$

$$+ \ln|\Sigma_j^0| + (x_{t+1} - \theta_j^0)^T(\Sigma_j^0)^{-1}(x_{t+1} - \theta_j^0)\Big),$$

where the coefficient a^* is defined in the Section 7.5.3.

Case III : $\{\theta_i^0\}$ (and P^0) are unknown.

Let us assume that the covariance matrices are equal: $\Sigma_i^0 = \Sigma^0$ $(i \in S)$, and apply the results of Section 7.5.4. According to (7.111), (7.113), (7.114), we have :

$$\bar{\theta} = \bar{x} = \frac{1}{n} \sum_{t=1}^{n} x_t \in R^N,$$

$$\gamma = (\Sigma^0)^{-1}, \ \nabla_{\bar{\theta}} \ln p(x; \bar{\theta}) = (\Sigma^0)^{-1}(x - \bar{x}),$$

$$B(t', t'') = (x_{t'} - \bar{x})^T (\Sigma^0)^{-1} (x_{t''} - \bar{x}).$$

Then the decision rule is determined by (7.118), (7.119) with $\pi_i \equiv 1/L$.

If in the considered case the matrix P^0 is also unknown, then it is enough to replace the term $\ln p^0_{ij}$ by $a^* \delta_{ij}$ in (7.119).

7.5.6 Performance Analysis

Let us investigate performance characteristics of the decision rules for the Gaussian case described in Section 7.5.5.

First, let us present analytical results of performance evaluation for the decision rule (7.121) in the case I, where

$$K = 1,\ L = 2,\ P^0 = \begin{pmatrix} p^0 & 1 - p^0 \\ 1 - p^0 & p^0 \end{pmatrix},\ \pi^0 = \pi^* = \begin{pmatrix} 0.5 \\ 0.5 \end{pmatrix},\ \Sigma^0_1 = \Sigma^0_2 = \Sigma^0.$$

To evaluate the performance of the decision rule $\hat{D} = D_0(X)$, let us introduce the probabilities of classification errors $q_1(D^0, D)$, q_2. Let $D = (d_t) \in S^n$, $D^0 = (d^0_t) \in S^n$ be two different sequences of class indices ($D \neq D^0$). Then

$$q_1 = q_1(D^0, D) = \mathbf{P}\{\Lambda(X; D^0, \{\theta^0_i\}, \pi^0, P^0) < \Lambda(X; D, \{\theta^0_i\}, \pi^0, P^0) \mid D^0\}$$

is the probability of error for discrimination of two decisions $\{D^0, D\}$ by the optimal decision rule (7.102) (or, equivalently, (7.121));

$$q_2 = \mathbf{E}\{q_1(D^0, D)\}$$

is the unconditional probability of error. Let us assume the following model of "competitive" decision D :

$$d_t = (d^0_t + \xi_t) \bmod 2 \, (t = \overline{1, n}),$$

where $\{\xi_t\}$ is a sequence of Bernoulli random variables :

$$\mathbf{P}\{\xi_t = 1\} = \alpha,\ \mathbf{P}\{\xi_t = 0\} = 1 - \alpha.$$

Here the parameter $\alpha \in (0, 1)$ characterizes the expected frequency of the event $\{d^0_t \neq d_t\}$. By the results presented in (Kharin, 1986), we obtain asymptotic expressions of q_1, q_2 as $n \to \infty$:

$$q_1 \approx \Phi\left(-\sqrt{||D - D^0||}\left(\frac{\Delta}{2} + \frac{1}{\Delta}\ln\frac{p^0}{1-p^0}\frac{\chi(D) - \chi(D^0)}{||D - D^0||}\right)\right),$$

$$q_2 \approx \Phi\left(-\sqrt{\alpha n}\left(\frac{\Delta}{2} + \frac{4(1-\alpha)}{\Delta}\left(p^0 - \frac{1}{2}\right)\ln\frac{p^0}{1-p^0}\right)\right),$$

where

$$\chi(D) = \sum_{t=1}^{n-1}(1 - \delta_{d_t, d_{t+1}})$$

is the number of "change points" in the sequence D. Figure 7.11 plots the dependence of error probability q_2 on sample size n for Mahalanobis distance $\Delta = 1$, $\alpha = 0.1$, and for different values $p^0 = 0.5; 0.7; 0.8$. It is seen that using the Markov dependence of class indices ($p^0 > 0.5$) results in significant gain. For example, for $n = 20$, $p^0 = 0.8$ the error probability q_2 is 0.01, and it is 24 times smaller than for $p^0 = 0.5$ (Markov dependence is absent).

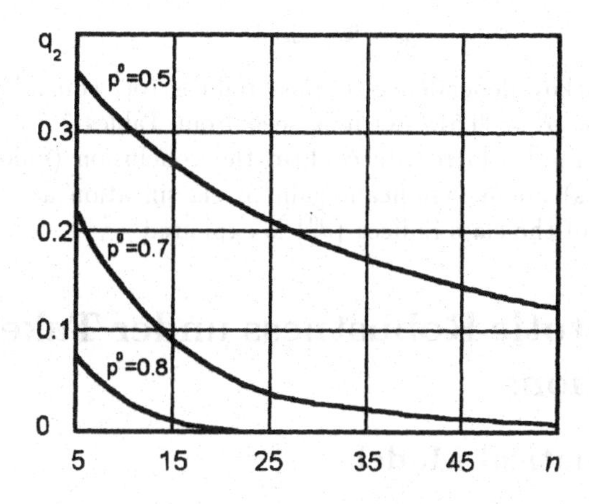

Figure 7.11: Error probability vs. sample size

Let us describe now the results of computer Monte-Carlo modeling. Tables 7.5 — 7.7 present the error rates $\hat{Q} = ||\hat{D} - D^0||/n$ for the Cases I, II (see Section 7.5.5), where

$$N = 2, n = 200, \varepsilon = 0.5, \theta_1^0 = \begin{pmatrix} -c \\ -c \end{pmatrix}, \theta_2^0 = \begin{pmatrix} c \\ c \end{pmatrix}, c > 0, \Sigma^0 = \mathbf{I}_2, \Delta = 2\sqrt{2}c.$$

At Figure 7.12 the solid lines plot the dependence of error rate $\hat{Q}$ on p^0 for the cases I, II and for $n = 1000$, $\varepsilon = 0.2$. The circles present the computer results for $\hat{Q}$. The dashed line indicates the level of error probability

Table 7.5: Error rate for $a = 0.6$

	p^0								
Case	0.600	0.675	0.700	0.750	0.825	0.875	0.925	0.950	0.975
I	0.198	0.162	0.183	0.141	0.143	0.109	0.071	0.056	0.044
II	0.332	0.302	0.288	0.336	0.200	0.160	0.211	0.076	0.144

Table 7.6: Error rate for $a = 0.37$

	p^0								
Case	0.600	0.675	0.700	0.750	0.825	0.875	0.925	0.950	0.975
I	0.312	0.284	0.265	0.244	0.250	0.210	0.168	0.104	0.113
II	0.396	0.353	0.354	0.446	0.296	0.246	0.298	0.112	0.262

$$r_0 = \Phi(-\Delta/2),$$

if we ignore the Markov dependence of class indices (or, equivalently, if we use the decision rules for $p^0 = 0.5$). As it is seen from Tables 7.5 — 7.7 and from Figure 7.12, the Monte-Carlo results confirm the conclusion (following from the asymptotical analysis) about significant gain in classification accuracy when the Markov dependence of the class indices $\{d_t^0\}$ is exploited.

7.6 Asymptotic Robustness under Tukey–Huber Distortions

7.6.1 Mathematical Model

In this subsection we shall consider the problem of cluster analysis of multivariate observations in the situation where their probability models are subjected to Tukey–Huber distortions, and parametric plug-in decision rules with minimum contrast estimators for unknown parameters are used.

Let us define a sample of n independent random observations $x_1, \ldots, x_n$ in R^N

Table 7.7: Error rate for $a = 0.18$

	p^0								
Case	0.600	0.675	0.700	0.750	0.825	0.875	0.925	0.950	0.975
I	0.404	0.366	0.413	0.404	0.373	0.358	0.337	0.242	0.262
II	0.464	0.433	0.464	0.427	0.389	0.385	0.381	0.246	0.356

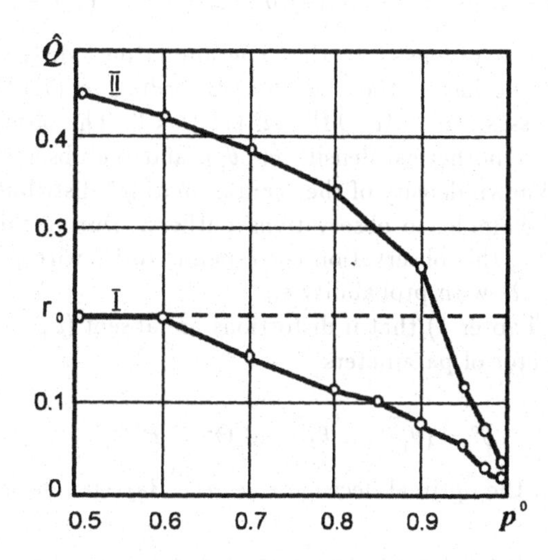

Figure 7.12: Error rate $\hat{Q}$ vs. elementary probability p^0

from $L \geq 2$ classes $\{\Omega_1, \ldots, \Omega_L\}$. Unlike the classical model of cluster analysis, an observation from Ω_i is a random N-vector with probability density function $p_i(x; \theta_i^o), x \in R^N$ that may be distorted:

$$p_i(\cdot : \theta_i^o) \in \mathcal{P}_i(\varepsilon_{+i}), \; i \in S = \{1, \ldots, L\}, \tag{7.122}$$

where $\mathcal{P}_i(\varepsilon_{+i})$ is the set of admissible probability densities for Ω_i; ε_{+i} is the level of distortions for Ω_i (if $\varepsilon_{+i} = 0$, then there are no distortions in Ω_i); $p_i(\cdot; \theta_i^o) \equiv q(\cdot; \theta_i^o)$ is the hypothetical parametric probability density function (in this case $\mathcal{P}_i(0)$ contains a single element $q(\cdot; \theta_i^o)$); $\theta_i^o \in \Theta \subseteq R^m$ is the true unknown value of the parameter for the i-th class. Let us introduce the notations: $d_t^o \in S$ is the unknown true index of the class to which the observation x_t belongs; $D^o = (d_1^o, \ldots, d_n^o)^T$ is the true classification vector of the sample $X = (x_1 \vdots \ldots \vdots x_n)^T$. *A priori*, $\{d_t^o\}$ are independent indentically distributed discrete random variables with distribution

$$\pi_i = \mathbf{P}\{d_t^o = i\}, \; i \in S, \; \sum_{i \in S} \pi_i = 1.$$

The loss matrix $W = (w_{ik})$ is given: $w_{ik} \geq 0$ is the loss at classifying the observation from Ω_i to $\Omega_k, i, k \in S$. The problem of cluster analysis consists in construction of a decision rule for classification of the sample X, i.e., in construction of the estimate $D = (d_1, \ldots, d_n)^T$ for D^o basing on X.

Let us consider the most common in practice model of type (7.122). It is the Tukey-Huber distortion model (see Section 2.2):

$$\mathcal{P}_i(\varepsilon_{+i}) = \{p_i(x; \theta_i^0), x \in R^N : p_i(x; \theta_i^0) =$$

$$= (1-\varepsilon_i)q(x;\theta_i^o) + \varepsilon_i h_i(x), 0 \le \varepsilon_i \le \varepsilon_{+i} < 1\}, i \in S, \tag{7.123}$$

where $h_i(\cdot)$ is an arbitrary density of the "contaminating" distribution; ε_i is the probability of contamination in the class Ω_i. According to (7.123), the class Ω_i consists of two subclasses: $\Omega_i = \Omega_i^o \cup \Omega_i^h$, $\Omega_i^o \cap \Omega_i^h = \emptyset$. The observation from Ω_i^o is determined by the hypothetical density $q(\cdot;\theta_i^o)$, and the observation from Ω_i^h is determined by an unknown density of the "contaminating" distribution $h_i(\cdot)$, which may correspond, for example, to observations–outliers. During the registration of an observation from Ω_i, this observation corresponds to Ω_i^o with probability $1-\varepsilon_i$ and it corresponds to Ω_i^h with probability ε_i.

It is known (see Chapter 1) that if distortions are absent ($\varepsilon_+ = \max_{i \in S} \varepsilon_{+i} = 0$) and the composite vector of parameters

$$\theta^o = (\theta_1^{oT} \vdots \dots \vdots \theta_L^{oT})^T \in \Theta^L \subseteq R^{Lm}$$

is given *a priori,* then the optimal decision rule (the Bayesian decision rule)

$$d = d(x;\theta^o) = \arg\min_{i \in S} \sum_{j \in S} \pi_j w_{ji} q(x;\theta_j^o), x \in R^N \tag{7.124}$$

delivers the minimal risk (expected losses):

$$r_o = R(\theta^o;\theta^o); \tag{7.125}$$

$$R(\theta;\theta^o) = \sum_{i,j \in S} \pi_i w_{ij} \int_{d(x;\theta)=j} q(x;\theta_i^o)dx \ge 0, \theta \in \Theta^L.$$

As θ^o is unknown, its statistical estimator $\hat{\theta}$ from the "contaminated" sample X is used. To solve the classification problem, the plug-in decision rule $d(\cdot;\hat{\theta})$ will be used, which is obtained by replacing θ^o by their estimator $\hat{\theta}$. This decision rule is characterized by the risk functional

$$r_\varepsilon = \mathbf{E}_{\theta^o}\{R(\hat{\theta};\theta^o)\}, \tag{7.126}$$

where $\mathbf{E}_{\theta^o}\{\cdot\}$ denotes the expectation w.r.t. the probability density

$$q_\pi^\varepsilon(x;\theta^o) = \sum_{i \in S} \pi_i p_i(x;\theta_i^o).$$

To quantify the robustness of the decision rule $d(\cdot;\hat{\theta})$, let us use the robustness factor (relative risk bias, if $r_o > 0$) defined in Section 2.3:

$$\kappa_\varepsilon = \frac{r_\varepsilon - r_o}{r_o}. \tag{7.127}$$

The smaller the robustness factor κ_ε, the more stable the decision rule $d(\cdot;\hat{\theta})$.

7.6.2 Asymptotic Expansions of Risk

Let $\hat{\theta}$ be a minimum contrast estimator (MCE) (Pfanzagl, 1969) determined by a contrast function (CF). Under the model (7.123) a random observation $x \in R^N$ has density

$$q_\pi^\varepsilon(x;\theta^o) = \sum_{i\in S} \pi_i p_i(x;\theta_i^o) = q_\pi^o(x;\theta^o) + \sum_{i\in S} \varepsilon_i \pi_i (h_i(x) - q(x;\theta_i^o)), \qquad (7.128)$$

where

$$q_\pi^o(x;\theta^o) = \sum_{i\in S} \pi_i q(x;\theta_i^o) \qquad (7.129)$$

is the hypothetical density (when $\varepsilon_+ = \max_i \varepsilon_{+i} = 0$) for which, according to (Chibisov, 1973), the contrast function $f(x;\theta)$ satisfies the following inequality:

$$b(\theta^o;\theta^o) < b(\theta;\theta^o), b(\theta;\theta^o) = \int_{R^N} f(x;\theta) q_\pi^o(x;\theta^o)dx, \theta^o \in \Theta^L, \theta \in \Theta^{*L}, \theta^o \neq \theta, \qquad (7.130)$$

where Θ^* denotes the closure of the set Θ. If $\varepsilon_+ = 0$, then we obtain the classical MC-estimator:

$$\tilde{\theta} = \arg\min_{\theta\in\Theta^{*L}} \tilde{L}_n(\theta), \tilde{L}_n(\theta) = n^{-1}\sum_{t=1}^{n} f(x_t;\theta). \qquad (7.131)$$

In particular, if $f(x;\theta) = -\ln q_\pi^o(x;\theta)$, then $\tilde{\theta}$ is the maximum likelihood estimator. If $\varepsilon_+ > 0$, then it is impossible to use the mixture (7.128) for the construction of the contrast function, because $\{h_i(\cdot)\}$ are not determined. Therefore, let us use the "truncation" principle from (Huber, 1981) and define the estimator $\hat{\theta}$:

$$\hat{\theta} = \arg\min_{\theta\in\Theta^{*L}} L_n(\theta), L_n(\theta) = n^{-1}\sum_{t=1}^{n} \psi(x_t;\theta), \qquad (7.132)$$

where

$$\psi(x;\theta) = f(x;\theta) - (f(x;\theta) - c)\mathbf{1}(f(x;\theta) - c), \qquad (7.133)$$

$$\mathbf{1}(z) = \{1, \text{ if } z > 0; 0, \text{ if } z \leq 0\}.$$

Clearly, if $c=+\infty$, then $\psi(\cdot) \equiv f(\cdot)$. Let us construct $\psi(x;\theta)$ as a contrast function for the distorted densities (7.128) by a special choice of the "truncation" constant $c \in R^1$.

Theorem 7.21 *The function $\psi(x;\theta)$ from (7.133) is a contrast function for the family (7.128), if the following regularity conditions are satisfied:*

C1) $f(x;\theta)$ is bounded from the below and differentiable w. r. t. $\theta \in \Theta^{*L}$;

C2) $f(x;\theta)$ is integrable w. r. t. the densities $\{h_i(x)\}$;

C3) the distortion level $\varepsilon_+ = \varepsilon_+(n)$ and the "truncation" constant $c = c(\varepsilon_+)$ satisfy the asymptotics (as $n \to +\infty$) :

$$\varepsilon_+(n) \to 0, \tag{7.134}$$

$$\sup_{\theta' \in \Theta, \theta \in \Theta^{*L}} \left| \int_{R^N} f(x;\theta)\mathbf{1}(f(x;\theta) - c)q(x;\theta')dx \right| = \mathcal{O}(\varepsilon_+). \tag{7.135}$$

Proof. The integrability of $\psi(x;\theta)$:

$$\int_{R^N} \psi(x;\theta)q^{\varepsilon}_{\pi}(x;\theta^o)dx < +\infty$$

is proved by the conditions C1, C2 and by the definition (7.133). The inequality, which is similar to (7.130) and defines the contrast function $\psi(\cdot)$ for the family (7.128), is verified by the regularity conditions C1, C3.

■

Let us construct an asymptotic expansion for $\hat{\theta}$ defined by (7.132). Let us introduce the notations: $M = mL$; $\mathbf{0}_M$ is an M-column-vector of zeros, ∇^k_θ is either the differential operator for the calculation of the vector (if $k = 1$) or of the matrix (if $k = 2$) of k-th order partial derivatives with respect to $\theta = (\theta_1, \ldots, \theta_M)^T$; $o_1(Z_n) \in R^K$ is a random sequence that depends on the random sequence $Z_n \in R^M$, in such a way that

$$\frac{| o_1(Z_n) |}{| Z_n |} \xrightarrow{a.s.} 0 \text{ as } n \to \infty;$$

$\mathbf{1}_M$ is the M-column-vector of ones; $\mathbf{1}_{M \times M}$ is the $(M \times M)$-matrix of ones.

Theorem 7.22 *Let the regularity conditions C1-C3 be satisfied and let the additional conditions C4-C6 take place:*

C4) the function $f(x;\theta)$, $x \in R^N$, is twice differentiable w. r. t. $\theta \in \Theta^{*L}$, so that the generalized functions:

$$\frac{\partial\psi(x;\theta)}{\partial\theta_i}, \frac{\partial\psi(x;\theta)}{\partial\theta_i} \cdot \frac{\partial\psi(x;\theta)}{\partial\theta_j},$$

$$\frac{\partial^2\psi(x;\theta)}{\partial\theta_i\partial\theta_j}, \frac{\partial\psi(x;\theta)}{\partial\theta_i} \cdot \frac{\partial\psi(x;\theta)}{\partial\theta_j} \cdot \frac{\partial\psi(x;\theta)}{\partial\theta_k}, x \in R^N,$$

are uniformly integrable in R^N w. r. t. $q(x;\theta')$, $\{h_l(x)\}$, $\theta \in \Theta^{*L}$, $\theta' \in \Theta$, $i,j,k = \overline{1,M}$; $l \in S$;

C5) integration and differentiation operations may be interchanged:

$$\nabla_\theta^k \int_{R^N} \psi(x;\theta)q(x;\theta')dx = \int_{R^N} \nabla_\theta^k \psi(x;\theta)q(x;\theta')dx;$$

$$\nabla_\theta^k \int_{R^N} \psi(x;\theta)h_l(x)dx = \int_{R^N} \nabla_\theta^k \psi(x;\theta)h_l(x)dx, l \in S,$$

$\theta \in \Theta^{*L}$, $\theta' \in \Theta$, $k = 1, 2$;

C6) for the functional $b(\theta;\theta^\circ)$ and its gradient vector $B(\theta;\theta^\circ) = \nabla_\theta b(\theta;\theta^\circ)$ we have:

$$B(\theta^\circ;\theta^\circ) = \mathbf{0}_M,$$

and the matrix of second order partial derivatives $a(\theta;\theta^\circ) = \nabla_\theta^2 b(\theta;\theta^\circ)$ is positive definite for $\theta = \theta^0$:

$$A(\theta^\circ) = a(\theta^\circ;\theta^\circ) \succ 0.$$

Then the following asymptotic expansion for the random deviation of the estimator (7.132) takes place as $n \to \infty$:

$$\triangle\,\theta = \hat{\theta} - \theta^\circ = -A^{-1}(\theta^\circ)\nabla_{\theta^\circ} L_n(\theta^\circ) + o_1(A^{-1}(\theta^\circ)\nabla_{\theta^\circ} L_n(\theta^\circ)). \qquad (7.136)$$

Proof. Under the conditions of Theorem 7.22, it follows from (7.132) that $\hat{\theta}$ is a root of the equation:

$$\nabla_\theta L_n(\theta) = \mathbf{0}_M. \qquad (7.137)$$

From the relation

$$\nabla_{\theta^\circ} L_n(\theta^\circ) \xrightarrow{a.s.} \mathbf{0}_M \text{ as } n \to +\infty,$$

which is based on the strong law of large numbers and on the following expression:

$$\mathbf{E}_{\theta^\circ}\{\nabla_{\theta^\circ} L_n(\theta^\circ)\} = B(\theta^\circ;\theta^\circ) + O(\varepsilon_+)\mathbf{1}_M = O(\varepsilon_+)\mathbf{1}_M,$$

we obtain by the known result from (Borovkov, 1984):

$$\hat{\theta} \xrightarrow{a.s.} \theta^\circ, n \to +\infty.$$

This fact allows us to apply the Taylor formula in the neighborhood of θ° :

$$\nabla_\theta L_n(\theta) = \nabla_{\theta^\circ} L_n(\theta^\circ) + \nabla_{\theta^\circ}^2 L_n(\theta^\circ)(\theta - \theta^\circ) + \mathbf{1}_M O(|\,\theta - \theta^\circ\,|^2) = \qquad (7.138)$$

$$= \nabla_{\theta^o} L_n(\theta^o) + A(\theta^o)(\theta - \theta^o) + (\nabla^2_{\theta^o} L_n(\theta^o) - \mathbf{E}_{\theta^o}\{\nabla^2_{\theta^o} L_n(\theta^o)\})(\theta - \theta^o) +$$

$$+ (\theta - \theta^o)\mathcal{O}(\varepsilon_+) + \mathbf{1}_M \mathcal{O}(|\,\theta - \theta^o\,|^2),$$

where the following expansion was used:

$$\mathbf{E}_{\theta^o}\{\nabla^2_{\theta^o} L_n(\theta^o)\} = A(\theta^o) + \mathcal{O}(\varepsilon_+)\mathbf{1}_{M\times M}.$$

Then the asymptotic expansion (7.136) is obtained by resolving (7.137), (7.138) with respect to the deviation $\Delta\theta$.

■

Let us define the matrix:

$$I_o(\theta^o) = A^{-1}(\theta^o) \int_{R^N} \nabla_{\theta^o} f(x;\theta^o)(\nabla_{\theta^o} f(x;\theta^o))^T q^o_\pi(x;\theta^o) dx A^{-1}(\theta^o)$$

by analogy with the inverse Fisher information matrix (where $\varepsilon_+ = 0$, $f(\cdot) \equiv -\ln q^o_\pi(\cdot)$).

Theorem 7.23 *Under the conditions of Theorem 7.22 the following asymptotic expansions are true:*

a) for the bias:

$$\mathbf{E}_{\theta^o}\{\Delta\theta\} = \beta(\theta^o) + \mathbf{1}_M o(\varepsilon_+),$$

where

$$\beta(\theta^o) = A^{-1}(\theta^o) \int_{R^N} \nabla_{\theta^o} f(x;\theta^o)(\mathbf{1}(f(x;\theta^o) - c) q^o_\pi(x;\theta^o) -$$

$$- \sum_{i\in S} \varepsilon_i \pi_i (h_i(x) - q(x;\theta^o_i))\mathbf{1}(c - f(x;\theta^o)))dx;$$

b) for the variance matrix:

$$\mathbf{E}_{\theta^o}\{\Delta\theta(\Delta\theta)^T\} = I_o(\theta^o)n^{-1} + \beta(\theta^o)\beta^T(\theta^o) + \mathbf{1}_{M\times M} o(\varepsilon^2_+ + n^{-1}).$$

Proof is based on (7.136), (7.128), (7.132), (7.133) and on the regularity conditions C1-C6.

■

Now let us costruct an asymptotic expansion for risk r_ε of the decision rule $d(\cdot;\hat{\theta})$ in the case of $L = 2$ classes. (But the results can be similarly constructed in the general case ($L \geq 2$)). The decision rule for $L = 2$ has the form:

$$d = d(x;\hat{\theta}) = \mathbf{1}(G(x;\hat{\theta})) + 1; \tag{7.139}$$

$$G(x;\theta^o) = a_2 q(x;\theta^o_2) - a_1 q(x;\theta^o_1),\ a_i = \pi_i(w_{i,3-i} - w_{ii}),$$

and the following representation for the risk (7.126) is valid (see Chapter 1):

$$r_\varepsilon = \pi_1 w_{11} + \pi_2 w_{21} - \mathbf{E}_{\theta^o}\{\int_{R^N} \mathbf{1}(G(x;\hat{\theta}))G(x;\theta^o)dx\}. \tag{7.140}$$

Theorem 7.24 *Suppose that under the conditions of Theorem 7.22 the density $q(x;\theta^*), \theta^* \in \Theta$ is differentiable with respect to $x \in R^N$ and the integrals:*

$$I_1 = \frac{1}{2}\int_\Gamma (\nabla_{\theta^o} G(x;\theta^o))^T I_o(\theta^o) \nabla_{\theta^o} G(x;\theta^o) \mid \nabla_x G(x;\theta^o) \mid^{-1} ds_{N-1},$$

$$I_2 = \frac{1}{2}\int_\Gamma ((\nabla_{\theta^o} G(x;\theta^o))^T \beta(\theta^o))^2 \mid \nabla_x G(x;\theta^o) \mid^{-1} ds_{N-1}$$

taken over the discriminant surface $\Gamma = \{x : G(x;\theta^o) = 0\}$ are finite. Then the risk r_ε admits the following asymptotic expansions (as $n \to +\infty$):

A1)under the asymptotics $\varepsilon_+ = o(n^{-1/2})$:

$$r_\varepsilon = r_o + I_1/n + o(n^{-1}); \tag{7.141}$$

A2)under the asymptotics $\varepsilon_+ = \mathcal{O}(n^{-1/2})$:

$$r_\varepsilon = r_o + I_2 + I_1/n + o(n^{-1}); \tag{7.142}$$

A3) under the asymptotics $\varepsilon_+/n^{-1/2} \to +\infty, \varepsilon_+ = \varepsilon_+(n) \to 0$:

$$r_\varepsilon = r_o + I_2 + o(\varepsilon_+^2). \tag{7.143}$$

Proof is based on the application of the Taylor formula in the neighborhood of θ^o with respect to $\triangle\theta = \hat{\theta} - \theta^o$ to the integral from (7.140) by Theorem 7.23 and by the generalized functions as in Chapter 3.

■

Corollary 7.17 *The risk $\tilde{r}_\varepsilon$ of the decision rule $d(\cdot;\tilde{\theta})$ that uses the classical estimator $\tilde{\theta}$ from (7.131) admits the following asymptotic expansions:*

under the asymptotics A1:

$$\tilde{r}_\varepsilon = r_o + I_1/n + o(n^{-1}); \tag{7.144}$$

under the asymptotics A2:

$$\tilde{r}_\varepsilon = r_o + \tilde{I}_2 + I_1/n + o(n^{-1}); \tag{7.145}$$

under the asymptotics A3:

$$\tilde{r}_\varepsilon = r_o + \tilde{I}_2 + o(\varepsilon_+^2), \tag{7.146}$$

where

$$\tilde{I}_2 = \frac{1}{2}\int_\Gamma ((\nabla_{\theta^o} G(x;\theta^o))^T \tilde{\beta}(\theta^o))^2 \mid \nabla_x G(x;\theta^o) \mid^{-1} ds_{N-1};$$

$$\tilde{\beta}(\theta^o) = A^{-1}(\theta^o) \sum_{i \in S} \varepsilon_i \pi_i \int_{R^N} \nabla_{\theta^o} f(x;\theta^o)(h_i(x) - q(x;\theta_i^o))dx.$$

Proof. If $c = +\infty$, then the condition (7.135) is satisfied, and the estimator $\hat{\theta}$, introduced in (7.132), turns into $\tilde{\theta}$, introduced by means of (7.131). The proof is concluded by substitution of $c = +\infty$ into (7.141)-(7.143).

■

From (7.141) and (7.144) it follows that under the asymptotics A1 it is senseless to use $\hat{\theta}$ instead of its classical version $\tilde{\theta}$, because the risk values $r_\varepsilon, \tilde{r}_\varepsilon$ (and hence the robustness factors κ_ε, $\tilde{\kappa}_\varepsilon$) coincide in main terms of their asymptotic expansions with the remainder $o(n^{-1})$, i.e.,

$$\tilde{r}_\varepsilon - r_\varepsilon = o(n^{-1});$$

$$\lim_{n\to+\infty} n\kappa_\varepsilon = \lim_{n\to+\infty} n\tilde{\kappa}_\varepsilon = I_1/r_o.$$

The results of Theorem 7.24 and its Corollary 7.17 allows to indicate the situations where under the asymptotics A2 (and A3) the use of the decision rule $d(\cdot;\hat{\theta})$ instead of its classical version $d(\cdot;\tilde{\theta})$ leads to significant gain in robustness factor (i.e., where $\kappa_\varepsilon < \tilde{\kappa}_\varepsilon$). In particular, if the following norm:

$$\| \nabla_{\theta^o} f(x;\theta^o) \| = \sqrt{(\nabla_{\theta^o} f(x;\theta^o))^T V(\theta^o) \nabla_{\theta^o} f(x;\theta^o)},$$

where

$$V(\theta^o) = A^{-1}(\theta^o) \int_\Gamma \nabla_{\theta^o} G(x;\theta^o)(\nabla_{\theta^o} G(x:\theta^o))^T \mid \nabla_x G(x;\theta^o) \mid^{-1} ds_{N-1} A^{-1}(\theta^o) \succ 0,$$

is unlimited in the region $U = \{x : f(x;\theta^o) \geq c\}$, then, obviously, there exist densities $\{h_i(\cdot)\}$ that are concentrated in U:

$$\int_U h_i(x)dx \to 1, \ \forall i \in S,$$

and for which $r_\varepsilon < \tilde{r}_\varepsilon$ ($\kappa_\varepsilon < \tilde{\kappa}_\varepsilon$). In this situation such densities $\{h_i(\cdot)\}$ may describe the observations-outliers.

7.6.3 Robustness under "Contaminated" Fisher Model

Let us illustrate the obtained results for the well known "contaminated" Fisher model (see the Section 5.1):

$$q(x;\theta_i^o) = n_N(x \mid \theta_i^o, \Sigma),$$

$$h_i(x) = n_N(x \mid \theta_i^+, \Sigma), \ i = 1,2 \ (L = 2),$$

where

$$n_N(x \mid \theta, \Sigma) = (2\pi)^{-\frac{N}{2}} |\Sigma|^{-\frac{1}{2}} \exp(-\frac{1}{2}(x-\theta)^T \Sigma^{-1}(x-\theta)), x \in R^N,$$

is the N-variate Gaussian density with mean vector θ and nonsingular covariance $(N \times N)$-matrix Σ ($|\Sigma| > 0$).

Let the classes Ω_1, Ω_2 be equiprobable ($\pi_1 = \pi_2 = 0.5$) and "equicontaminated" ($\varepsilon_1 = \varepsilon_2 = \varepsilon \le \varepsilon_+ < 1$); $w_{ij} = \{1, \text{ if } i \ne j;\ 0, \text{ if } i = j\}$ (in this case, the risk r_ε is in fact the classification error probability). Let us introduce the notations: $\varphi(z) = n_1(z \mid 0, 1)$, $\Phi(z)$ are the standard normal density and distribution function respectively; $\Psi(z) = \Phi(-z)/\varphi(z)$ is the Mills ratio;

$$\Delta = \sqrt{(\theta_1^o - \theta_2^o)^T \Sigma^{-1} (\theta_1^o - \theta_2^o)}$$

is the Mahalanobis interclass distance. Note that if $\varepsilon_+ = 0$ and $\theta^o = (\theta_1^{o^T} \vdots \theta_2^{o^T})^T$ is *a priori* known, then the error probability for the Bayesian (classical) decision rule is $r_o = \Phi(-\Delta/2)$ (see Chapter 1).

Let us investigate the robustness of the decision rule $d(\cdot; \tilde{\theta})$, where $\tilde{\theta}$ is the MC-estimator with contrast function

$$f(x; \theta^o) = -\ln \sum_{i=1}^{2} \exp(-\frac{1}{2}(x - \theta_i^o)^T \Sigma^{-1}(x - \theta_i^o)).$$

Under the asymptotics A3 ($\varepsilon_+ = n^{-0.5+\nu}, 0 < \nu \le 0.4$) and for the situation where

$$(\theta_i^+ - \theta_i^o)^T \Sigma^{-1} (\theta_i^+ - \theta_i^o) \le \delta^2 (i = 1, 2),$$

by Corollary 7.17 the maximal value of robustness factor (7.127) is obtained:

$$\tilde{\kappa}_\varepsilon \le \tilde{\kappa}_\varepsilon^+ + o(\varepsilon_+^2), \tag{7.147}$$

$$\tilde{\kappa}_\varepsilon^+ = \frac{\varepsilon_+^2 \delta^2 (4 + \Delta^2)}{\Delta \Psi(\Delta/2)}.$$

The smaller ε_+, the more stable the decision rule $d(\cdot; \tilde{\theta})$. From the condition $\tilde{\kappa}_\varepsilon^+ \le \gamma$ ($\gamma > 0$ is a preassigned value of the robustness factor) and the relations (7.147), we find the γ-admissible sample size:

$$n^* = n^*(\gamma) = \left\lfloor Z^{\frac{1}{2\nu - 1}} \right\rfloor + 1;$$

$$Z = \frac{\gamma \Delta \Psi(\Delta/2)}{\delta^2 (4 + \Delta^2)}.$$

The relations (7.147) help us to determine the "breakdown point" (see Section 2.3):

$$\varepsilon_+^* = \sqrt{\Delta(0.5 - \Phi(-\Delta/2))/(\delta^2\varphi(\Delta/2)(4+\Delta^2))};$$

if $\varepsilon_+ > \varepsilon_+^*$, then the error probability of the decision rule $d(\cdot;\tilde{\theta})$ can reach the "breakdown value" $\tilde{r}_\varepsilon^* = 0.5$, which corresponds to the equiprobable coin tossing.

If we use the estimator $\hat{\theta}$ with "truncated" contrast function (7.133), then the "truncation" constant is determined by the following relation, which is obtained from condition C3:

$$c = c(\varepsilon_+) = \frac{N}{2}(\Phi^{-1}(y))^2;$$

$$y = \frac{1 + (1-\varepsilon_+)^{1/N}}{2},$$

where $\Phi^{-1}(y)$ is the y-level quantile of the standard normal distribution.

Note in conclusion that multiple computer experiments with the decision rule (7.133), (7.139) confirmed its robustness to Tukey–Huber type of distortions.

Bibliography

[1] Abramovitch, M.S., Kharin, Yu.S., and Mashevskij, A.A., Application of robust discriminant analysis for diagnostique of cancer, *Proc. Internat. Conf. on Pattern Recognition and Image Analysis*, Belarussian Academy of Sciences, Minsk, pp.150-152, (1993) (in Russian).

[2] Abusev, R.A., and Lumelskij, Ya.P., *Statistical Group Classification,* Perm State University, Perm, (1987) (in Russian).

[3] Afifi, A.A., and Azen, S.P., *Statistical Analysis: A Computer Oriented Approach,* Academic Press, New York, (1979).

[4] Agrawala, A., Learning with a probabilistic teacher, *IEEE Trans. Information Theory,* IT-16 (1970), No. 3, pp. 373–380.

[5] Aitchinson, J., and Begg, C.B., Statistical diagnosis when basic cases are not classified with certainty, *Biometrika,* 63 (1976), pp. 1–12.

[6] Aitkin, M., and Tunnicliffe, W.G., Mixture models, outliers, and EM algorithm, *Technometrics,* 22 (1980), pp. 325-331.

[7] Aivazyan, S.A., *Classification of Multivariate Observations,* Statistika, Moscow, (1974) (in Russian).

[8] Aivazyan, S.A., To robustness problem of statistical inference, *Proc. 2nd Conf. on Multivariate Statistical Analysis and its Applications,* Academy of Sciences, Moscow, pp. 14–23, (1981) (in Russian).

[9] Aivazyan, S.S., Buchshtaber, B.M., Yenyukov, I.S., and Meshalkin, L.D., *Applied Statistics: Classification and Dimension Reduction,* Finansy i Statistika, Moscow, (1989) (in Russian).

[10] Aivazyan, S.A., Yenyukov, I.S., and Meshalkin, L.D., *Applied Statistics: Bases of Modeling and Initial Data Processing,* Finansy i Statistika, Moscow, (1983) (in Russian).

[11] Akahira, M., and Takeuchi, K., Asymptotic efficiency of statistical estimators: concepts and higher order asymptotic efficiency, *Lecture Notes in Statistics,* 7 (1981), Springer-Verlag, New York.

[12] Anderberg, M.R., *Cluster Analysis for Applications,* Academic Press, New York, (1973).

[13] Anderson, T.W., *The Statistical Analysis of Time Series,* John Wiley & Sons, New York, (1971).

[14] Anderson, T.W., *An Introduction to Multivariate Statistical Analysis,* John Wiley & Sons, New York, (1984).

[15] Anderson, T.W., and Bahadur, R.R., Classification into Two Multivariate Normal Distributions with Different Covariances, *Ann. Math. Stat.,* 33 (1962), pp. 422–431.

[16] Andrews, D.F., and Herzberg, A.M., *Data: A Collection of Problems from Many Fields for Student and Research Worker,* Springer-Verlag, New York, (1985).

[17] Artemjev, V.M., *Theory of Dynamic Systems with Random Changes of Structure,* High School Publ., Minsk, (1989) (in Russian).

[18] Asmikaga, T., and Chang, P.C., Robustness of Fisher linear discriminant function under two-component mixed normal models, *J. Amer. Statist. Assoc.,* 76 (1981), pp. 677–680.

[19] Balakrishnan, N., and Kocherlakota, S., Robustness to nonnormality of the linear discriminant function: mixtures of normal distributions, *Commun. Statist. - Theory Meth.,* 14 (1985), pp. 465–478.

[20] Balakrishnan, N., and Tiku, M.L., Robust classification procedures, In: *Classification and Related Methods of Data Analysis,* H.H. Bock (Ed.), North-Holland, Amsterdam, pp. 269–276, (1988).

[21] Barabash, Yu.L., *Problems of Statistical Pattern Recognition Theory,* Sovetskoye Radio, Moscow, (1967) (in Russian).

[22] Barnett, V., and Lewis, T., *Outliers in Statistical Data,* John Wiley & Sons, New York, (1978).

[23] Bellman, R., *Introduction to Matrix Analysis,* McGraw-Hill, New York, (1960).

[24] Berger, J.O., *Statistical Decision Theory and Bayesian Analysis,* Springer-Verlag, New York, (1985).

[25] Berger, J.O., Robust Bayesian analysis, *Journal of Statistical Planning and Inference,* 25 (1990), pp. 303–328.

[26] Berger, R.L., Gamma minimax robustness of Bayes rules, *Commun. Statist. - Theor. Meth.,* A8 (1979), pp. 543–560.

[27] Bertolino, F., On classification of observations structured into groups, *Appl. Stochast. Model. Data Anal.*, 4 (1988), pp. 240–251.

[28] Bock, H.H., Probabilistic aspects in cluster analysis, *Proc. 13th Conf. Gesellschaft für Klass. e. V.*, Springer-Verlag, New York, (1989), pp. 12–44.

[29] Bolshev, L.N., and Smirnov, N.V., *Tables of Mathematical Statistics,* Nauka, Moscow, (1983) (in Russian).

[30] Borovkov, A.A., *Mathematical Statistics,* Nauka, Moscow, (1984) (in Russian).

[31] Broemeling, L., and Son, M., The classification problem with autoregressive process, *Commun. Statist. – Theory Meth.,* 16 (1987), pp. 927–936.

[32] Broffit, B. *et al.*,, The effect of Huberizing and trimming on the quadratic discriminant function, *Commun. Statist. – Theory Meth.,* A9 (1980), pp. 13–25.

[33] Broffit, B. *et al.*,, Measurement errors - a local contamination problem in discriminant analysis, *Commun. Statist. – Theory Meth.*, B10 (1981), pp. 129–141.

[34] Campbell, N.A., Robust procedures in multivariate analysis: Robust covariance estimation, *Appl. Statist.,* 29 (1980), pp. 231–237.

[35] Chen, C.H., A review of statistical pattern recognition, In: *Pattern Recognition and Signal Processing* (Ed. Chen C.H.), Academic Press, New York, pp. 117–132, (1978).

[36] Chernoff, H., A measure of asymptotic efficiency for tests of a hypothesis based on the sum of observations, *Ann. Math. Statist.,* 23 (1962), pp. 493–497.

[37] Chernoff, H., The selection of effective attributes for deciding between hypothesis using linear discriminant functions, In: *Frontiers of Pattern Recognition* (Ed. S. Watanabe), Academic Press, New York, pp. 55–60, (1972).

[38] Chibisov, D.M., Asymptotic expansion for one class of estimators including ML-estimators, *Probability Theory and its Applications,* 18 (1973), No. 2, pp. 303–311.

[39] Chibisov, D.M., Asymptotic expansion for the distribution of statistic admitting a stochastic expantion, *Probability Theory and its Applications,* 25 (1980), No.4, pp. 745–756.

[40] Chittineni, C., Learning with imperfectly labeled patterns, *Pattern Recognition,* 12 (1980), pp. 281–291.

[41] Clark, W.R. and Lachenbruch, P.A., How nonnormality affects the quadratic discriminant function, *Commun. Statist. – Theory Meth.,* A8 (1979), pp. 1285–1301.

[42] Cox, D.R., and Hinkley, D.V., *Theoretical Statistics,* John Wiley & Sons : New York, (1974).

[43] Das Gupta, S., Theories and methods in classification: a review, In: *Discriminant Analysis and Prediction* (Ed. Cacoullos T.), Academic Press, New York, pp. 77–137, (1973).

[44] Das Gupta, S., Some problems in statistical pattern recognition, In: *Multivariate Analysis - IV* (Ed. Krishnaiah P.R.), North-Holland, New York, 457–471, (1977).

[45] Das Gupta, S., and Huang, D.Y., *Multiple Statistical Decision Theory. Recent Developments,* Springer-Verlag, Berlin, (1981).

[46] David, H.A., *Order Statistics,* John Wiley & Sons, New York, (1975).

[47] Deev, A.D., Representation of discriminant analysis statistics and asymptotic expansions for space dimensionalities compared with sample size, *Papers of Academy of Science of USSR,* 195 (1970), No.4, pp. 759–762 (translated in Soviet Math.).

[48] Devijver, P.A., and Kittler, J., *Pattern Recognition: A Statistical Approach,* Prentice Hall, London, (1982).

[49] Devore Jay, Reconstructing of noisy Markov chain, *J. Amer. Statist. Assoc.,* 68 (1973), No.342, pp. 394–398.

[50] Devroye, L.P., and Gyorfi, L., *Nonparametric Density Estimation: The* L_1 *View,* John Wiley & Sons, New York, (1985).

[51] Dorofeyuk, A.A., Algorithms of automatic classification, *Automatics and Remote Control,* 12 (1971), pp. 78–113.

[52] Dubes, R.S., and Jain, A.K., Random field models in image analysis, *J. Appl. Statist.,* 16 (1989), pp. 131–164.

[53] Duda, R.O., and Hart, P.E., *Pattern Classification and Scene Analysis,* John Wiley & Sons, New York, (1973).

[54] Dugue, D., *Traite de statistique theorique et appliquée,* Masson et C., Paris, (1958).

[55] Epanechnikov, V.A., Nonparametric estimator of multivariate probability density, *Probability Theory and its Applications,* 14 (1969), No.1, pp. 156–160 (translated in Soviet Math.).

[56] Ershov, A.A., Methods of robust parameter estimation (review), *Automation and Remote Control,* 8 (1978), pp. 66–100.

[57] Feller, W., *An introduction to Probability Theory and its Applications,* John Wiley & Sons, New York, (1971).

[58] Fix, E., and Hodges, J.L., *Nonparametric Discrimination,* Techn. Report, 11, US Air Force School, Texas, (1953).

[59] Fomin, V.N., *Mathematical Theory of Learning Recognition Systems,* Sankt-Petersburg State University, St.-Petersburg, (1976) (in Russian).

[60] Fomin, Ya.A., and Tarlovskij, G.R., *Statistical Theory of Pattern Recognition,* Radio and Svyaz, Moscow, (1986) (in Russian).

[61] Fu, K.S., *Sequential Methods in Pattern Recognition and Machine Learning,* Academic Press, New York, (1971).

[62] Fu, K.S., Recent developments in pattern recognition, *IEEE Trans. Computers,* C-20 (1980), No.10, pp. 845–854.

[63] Fukunaga, K., *Introduction to Statistical Pattern Recognition,* Academic Press, New York, (1972).

[64] Fukunaga, K., and Hayes, R., Effects of sample size in classifier design, *IEEE Trans. Pattern Anal. Machine Intell.,* PAMI-11 (1989), pp. 873–885.

[65] Fukunaga, K., and Hosteller, L.D., Optimization of k-nearest neighbour density estimates, *IEEE Trans. Information Theory,* IT-5 (1973), 320–326.

[66] Gabasov, R.F., and Kirillova, F.M., *Foundations of Dynamic Programming,* Belarussian State University, Minsk, (1975) (in Russian).

[67] Gelfand, I.M., and Shilov, G.E., *Generalized Functions and Operations on them,* Fizmatgiz, Moscow, (1959) (in Russian).

[68] Girko, V.L., *Multivariate Statistical Analysis,* High School, Kiev, (1988) (in Russian).

[69] Glick, N., Sample-based classification procedures derived from density estimators, *J. Amer. Statist. Assoc.,* 67 (1972), No.337, pp. 116–122.

[70] Gnanadesikan, R., *Methods for Satatistical Data Analysis of Multivariate Observations,* John Wiley & Sons, New York, (1977).

[71] Gnedenko, B.V., and Kolmogorov, A.N., *Limit Distributions for Sums of Independent Random Variables,* Addison-Wesley, New York, (1954).

[72] Golovkin, B.A., *Computer recognition and Linear Programming,* Sovetskoye Radio, Moscow, (1973).

[73] Gorelik, A.L., and Skripkin, V.A., *Classification Methods,* High School Publ., Moscow, (1984) (in Russian).

[74] Gorjan, I.S., Detection of statistically homogeneous image fragments, In: *Iconics. Computer Holography. Images Processing,* Nauka, Moscow, pp. 62–73, (1978) (in Russian).

[75] Gradstein, I.S., and Ryzhik, I.M., *Tables of integrals, Sums, Series and Products,* Nauka, Moscow, (1991) (in Russian).

[76] Greblicki, W., Learning to recognize patterns with a probabilistic teacher, *Pattern Recognition,* 12 (1980), No.1, pp. 159–180.

[77] Hamming, R.W., *Numerical Methods for Scientists and Engineers,* Mc-Graw-Hill, New York, (1962).

[78] Hampel, F.R., Ronchetti, E.M., Rousseeuw, P.J., and Stahel, W.A., *Robust Statistics: The Approach Based on Influence Functions,* John Wiley & Sons, New York, (1986).

[79] Hand, D.J., *Kernel Discriminant Analysis,* John Wiley & Sons, New York, (1982).

[80] Hartigan, J.A., *Bayes Theory,* (Springer Series in Statistics), Springer-Verlag, New York, (1983).

[81] Higgins, J., Some surface integral techniques in statistics, *American Statist.,* 29 (1975), No.1, pp. 43–46.

[82] Himmelblau, D.M., *Applied Nonlinear Programming,* Mc-Graw-Hill, New York, (1972).

[83] Huber, P.J., *Robust Statistics,* John Wiley & Sons, New York, (1981).

[84] Hufnagel, G., On estimating missing values in linear discriminant analysis, *Biometr. J.,* 30 (1988), pp. 69–75.

[85] Ibragimov, I.A., and Hasminskij, R.Z., *Asymptotic Theory of Estimation,* Springer-Verlag, New York, (1981).

[86] Igolkin, V.N., *Statistical Classification Based on Sample Distributions,* Sankt-Petersburg State Univ., St.-Petersburg, (1978) (in Russian).

[87] Kadane, J.B. (Ed.), *Robustness of Bayesian analysis,* North-Holland, New York, (1984).

[88] Kadane, J.B., and Chuang, D.T., Stable decision problems, *Ann. Statist.,* 6 (1978), pp. 1095–1110.

[89] Kailath, T., The divergence and Bhattacharya distance measures in signal selection, *IEEE Trans. Computers,* COM-15 (1967), 52–60.

[90] Kazakos, D., Statistical discrimination using inaccurate models, *IEEE Trans. Information Theory,* IT-28 (1982), No. 5, pp. 720–728.

[91] Kazakov, I.E., and Artemjev, V.M., *Optimization of Dynamic Systems with Random Structure,* Nauka, Moscow, (1980) (in Russian).

[92] Kendall, M.G., and Stuart, A., *The Advanced Theory of Statistics,* Vols.I and II, Griffin, London, (1958).

[93] Kharin, Yu.S., Adaptive construction of invariant features in pattern recognition problems, *Bulletin of USSR Academy of Sciences,* 5 (1977), pp. 87–97 (translated in Soviet Math.).

[94] Kharin, Yu.S., Adaptive classification and the criterion of risk density minimax, In: *Optimization of Dynamic Systems* (Ed. Medvedev G.A.), Belarussian State University, Minsk, pp. 125–130, (1978)(in Russian).

[95] Kharin, Yu.S., About robust decision rules in statistical classification problems, In: *Mathematical Statistics and its Applications* (Ed. Tarasenko F.P.), Tomsk State Univ., Tomsk, (1981) (in Russian).

[96] Kharin, Yu.S., About statistical classification accuracy at MC-estimators using, *Probability Theory and its Applications,* 26 (1981), No.4, pp. 866–867.

[97] Kharin, Yu.S., Robustness of decision rules in pattern recognition problems, *Automation and Remote Control,* 11 (1982), pp. 142–151.

[98] Kharin, Yu.S., Risk expansion for nonparametric classificator, In: *Adaptation and Learning in Systems of Control and Decision Making* (Ed. Medvedev A.V.), Nauka, Novosibirsk, 91–98, (1982) (in Russian).

[99] Kharin, Yu.S., Investigation and optimization of Rosenblatt-Parzen classificator by asymptotic expansions, *Automation and Remote Control,* 1 (1983), pp. 91–100.

[100] Kharin, Yu.S., Investigation of risk for statistical classificators using MC-estimators, *Probability Theory and its Applications,* 28 (1983), No.3, pp. 592–598.

[101] Kharin, Yu.S., About decision rule robustness under misclassifications presence in training samples, *Automation and Remote Control,* 11 (1983), pp. 100–110.

[102] Kharin, Yu.S., Asymptotic expansions for the risk of parametric and nonparametric decision functions, In: *Trans.IX Prague Conf. on Inform. Theory, Statist. Decision Funct., Random Process,* Academia, Prague, pp. 11–16, (1983).

[103] Kharin, Yu.S., Asymptotic robustness of Bayesian Decision rules in Statistical Decision Theory, In: *Robustness of Statistical Methods and Nonparametric Statistics* (Ed. Rasch D., Moti Lal Tiku), VEB Deutsche Verlag der Wissenschaften, Berlin, pp. 63–65, (1984).

[104] Kharin, Yu.S., Robustness investigation for the decision rules by risk asymptotic expansion method, In: *Proceedings of the Third Prague Symposium on Asymptotic Statistics* (Ed. Mandl P.), H.X.-Oxford, Elsevier, Amsterdam, pp. 309–317, (1984).

[105] Kharin, Yu.S., Classification of random series of unknown length, *Information Transmission Problems,* 21 (1985), No.4, pp. 64–75 (translated in Soviet Math.).

[106] Kharin, Yu.S., Risk asymptotic expansion for k-Nearest Neighbour classificator, *Notes in Statistics,* Nauka, 49 (1985), pp. 49–54 (in Russian).

[107] Kharin, Yu.S., Estimation and minimization of risk for k-Nearest Neighbour pattern recognition method, *Automation,* 4 (1986), pp. 99–103.

[108] Kharin, Yu.S., Detection of change points of the Markov type in random sequence of multivariate observations, In: *Detection of Changes in Random Processes,* Inc. Publ. Div., New York, (1986).

[109] Kharin, Yu.S., Robustness of discriminant analysis procedures, *Industrial Laboratory,* 10 (1990), pp. 69–72.

[110] Kharin, Yu.S., *Robustness in Statistical Pattern Recognition,* Universitetskoje, Minsk, (1992) (in Russian).

[111] Kharin, Yu.S., Robustness in discriminant analysis, *Lect. Notes in Statistics,* 109 (1996), pp. 225–234.

[112] Kharin, Yu.S., and Duchinskas, K.A., Asymptotic expansion of conditional risk moments for statistical classificators, In: *III Internat. Vilnius Conf. Probability Theory and Math. Statist.,* Vol.2, Vilnius, 224–225, (1981).

[113] Kharin, Yu.S., and Duchinskas, K.A., Regression experiments in evaluation of statistical classificator perfomance, In: *VII Conf. on Experiment Design and Automation,* MEI, Moscow, 56–57, (1983) (in Russian).

[114] Kharin, Yu.S., and Medvedev, A.G., Asymptotic robustness of discriminant procedures for dependent and nonhomogeneous observations, In: *Probability Theory and Mathematical Statistics. Proc. of the 5 Vilnius Conf.,* Utrecht, Netherlands, 602–610, (1991).

[115] Kharin, Yu.S., and Melnikova, E.N., Detection of multiple change points and classification of time series by statistical estimates of interclass distances, *Automation and Remote Control,* 12 (1991), pp. 76–84.

[116] Kharin, Yu.S., and Stepanova, M.D., *Computer Practicum in Mathematical Statistics,* Universitetskoje, Minsk, (1987) (in Russian).

[117] Kharin, Yu.S., and Zhuk, E.E., Asymptotic robustness in cluster-analysis for the case of Tukey-Huber distortions, In: *Information and Classification* (Ed. Opitz O., Lausen B., Klar R.), Springer-Verlag, New York, 31–39, (1993).

[118] Kharin, Yu.S., and Zhuk, E., Robustness in statistical pattern recognition under contaminations of training samples, In: *Proc. of the 12 IAPR Int. Conf. on Pattern Recogn.,* IEEE Computer Society Press, Washington, 504–507, (1994).

[119] Kharin, Yu.S. *et al.,*, *Robust Statistical Analysis (ROSTAN) Programs: User Manual,* Belarussian State University, Minsk, 256p. (1994).

[120] Kocherlakota, S., and Balakrishnan, N., The linear discriminant function: sampling from the truncated normal distribution, *Biometr. J.,* 29 (1987), pp. 131–139.

[121] Koroljuk, V.S. (Ed.), *Handbook on Probability Theory and Mathematical Statistics,* Nauka, Moscow, (1994) (in Russian).

[122] Kovalevskij, V.A., *Methods of Optimal Decisions in Images Recognition,* Nauka, Moscow, (1976) (in Russian).

[123] Kox, D.R., Computer analysis of electroencephalograms, blood pressure curves and electrocardiagrams, In: *Computer Pattern Recognition,* Mir, Moscow, (1974).

[124] Kracnenker, V.M., Robust methods of signal detection in noise presence (review), *Automation and Remote Control,* 5 (1980), pp. 65–88.

[125] Krishnaiah, P.R., and Kanal, L. (Ed.), *Handbook of Statistics, Vol.2: Classification, Pattern Recognition and Reduction Dimensionality,* North-Holl., Amsterdam, (1982).

[126] Krishnan, T., Efficiency of learning with imperfect supervision, *Pattern Recognition,* 21 (1988), pp. 183–188.

[127] Krzanowski, W.J., The performance of Fisher's linear discriminant function under non-optimal conditions, *Technometrics,* 19 (1977), pp. 191–200.

[128] Krzysko, M., Asymptotic distribution of the discriminant function, *Statist. Probability Letters,* 1 (1983), pp. 243–250.

[129] Krzysko, M., Unbiased estimators and statistical group classification problems, In: *Problems of Computer Data Analysis and Modeling* (Ed. Kharin Yu.S.), Belarussian State University, Minsk, (1991).

[130] Kullback, S., *Information Theory and Statistics,* John Wiley & Sons, New York, (1967).

[131] Kuznetzov, V.P., About robust rules of hypothesis testing, *Information Transmission Problems,* 18 (1982), No.1, pp. 51–63 (translated in Soviet Math.).

[132] Lachenbruch, P.A., Discriminant analysis when the initial samples are misclassified, *Technometrics,* 8 (1966), No.4, pp. 657–662.

[133] Lachenbruch, P.A., *Discriminant Analysis,* Hafner Press, New York, (1975).

[134] Lachenbruch, P.A., Note on initial misclassification effects on the quadratic discriminant function, *Technometrics,* 21 (1979), pp. 129–132.

[135] Launer, L., and Wilkinson, G.N. (Eds.), *Robustness in Statistics,* Academic Press, New York, (1979).

[136] Lawoko, C.R.O., and McLachlan, G.J., Asymptotic error rates of the W and Z statistics when the training observations are dependent, *Pattern Recognition,* 18 (1986), pp. 467–471.

[137] Lawoko, C.R.O., and McLachlan, G.J., Further results on discrimination with autocorrelated observations, *Pattern Recognition,* 21 (1988), pp. 69–72.

[138] Lbov, G.S., *Methods of Processing of Experimental Data,* Nauka, Novosibirsk, (1981) (in Russian).

[139] Le Cam, L., and Yang, G.L., *Asymptotics in Statistics. Some Basic Consepts,* Springer-Verlag, New York, (1990).

[140] Levinson, S.E., Rabiner, L.R., and Sondhi, M.M., An introduction to the application of the theory of probabilistic functions of a Markov process to automatic speech recognition, *The Bell System Technical Journal,* 62 (1983), No.4, pp. 1035–1074.

[141] Little, R.J.A., and Rubin, D.B., *Statistical Analysis with Missing Data,* John Wiley & Sons, New York, (1987).

[142] Loftsgarden, D.O., and Quesenberry, C.P., A nonparametric estimate of a multivariate density function, *Ann. Math. Statist.,* 36 (1965), pp. 1049–1051.

[143] Mack, Y.P., and Rosenblatt, M., Multivariate k-nearest neighbour density estimates, *Journal of Multivariate Analysis,* 9 (1975), pp. 1–15.

[144] Marron, J.S., Optimal rates of convergence to Bayes risk in nonparametric discrimination, *Ann. Statist.,* 11 (1983), pp. 1142–1155.

[145] Martynov, G.V., Computer program for probability distributions of quadratic forms, In: *Computer Methods of Mathematical Statistics: Algorithms and Programs,* Moscow State Univ., Moscow, 30–35, (1976) (in Russian).

[146] Mashevskij, A.A., *Diagnostics of Basic Forms of Cancer by Biochemical and Biophysical Blood Tests,* Doctoral dissertation, Minsk, (1994) (in Russian).

[147] *Mathematical Encyclopedia,* Sovetskaya Encyclopedia Publ., Moscow, (1984) (in Russian).

[148] McLachlan, G.J., Asymptotic results for discriminant analysis when the initial samples are misclassified, *Technometrics,* 14 (1972), pp. 415–422.

[149] McLachlan, G.J., Further results on the effect of intraclass correlation among training samples in discriminant analysis, *Pattern Recognition,* 7 (1976), pp. 273–275.

[150] McLachlan, G.J., *Discriminant Analysis and Statistical Pattern Recognition,* John Wiley & Sons, New York, (1992).

[151] McLachlan, G.J., and Basford, K.E., *Mixture Models: Inference and Applications to Clustering,* Marcel Dekker, New York, (1988).

[152] McLachlan, G.J., and Ganesalingam, S., Updating a discriminant function on the basis of unclassified data, *Commun. Statist. – Simula. Computa.,* 11 (1982) No. 6, pp. 753–767.

[153] Medvedev, G.A., Some problems of pattern recognition, In: *Selftraining Systems,* Nauka, Moscow, (1966) (in Russian).

[154] Meshalkin, L.D., Local methods of classification, In: *Statistical Methods of Classification, I,* Moscow State University, Moscow, (1969) (in Russian).

[155] Milenkij, A.V., *Classification of Signals under Uncertainty,* Sovetskoye Radio, Moscow, (1975) (in Russian).

[156] Mosteller, F., and Tukey, J.W., *Data Analysis and Regression,* Addison-Wesley, London, (1977).

[157] Mouder, J., *Operation Research. Vol. 2: Models and applications,* John Wiley & Sons, New York, (1981).

[158] Okamoto, M., An asymptotic expansion for the distribution of the linear discriminant function, *Ann. Math. Statist.,* 34 (1963), No.4, pp. 1286–1301.

[159] Omelchenko, V.A., Recognition of random signals by Spectrum, *Papers of Institutes: Radioelectronics,* 22 (1979), No.12, pp. 16–22.

[160] Orlov, A.I., *Robustness in Social-economic Models,* Nauka, Moscow, (1979) (in Russian).

[161] Orlov, A.I., Why you should not use iterative procedures for ML-estimators computation, *Industrial Laboratory,* 5 (1986), pp. 67–69.

[162] Orlov, A.I., Are observation often normally distributed, *Industrial Laboratory,* 7 (1991), pp. 64–66.

[163] Parzen, E., On the estimation of a probability density function and the mode, *Ann. Math. Statist.*, 40 (1962), pp. 1063–1076.

[164] Patrick, F.A., *Fundamentals of Pattern Recognition,* Prentice-Hall, N.J., (1972).

[165] Pau, L.F., Finite learning sample size problems in pattern recognition, In: *Pattern Recognition and Signal Processing* (Ed. Chen. C.H), Academic Press, New York, (1978).

[166] Pervozvanskij, A.A., and Gaizgori, V.G., *Decomposition, Aggregating and Approximate Optimization,* Nauka, Moscow, (1979) (in Russian).

[167] Pfanzagl, J., *Theory of Measurements,* Physica–Verlag, Wien, (1971).

[168] Pfanzagl, J., On the measurability and consistency of minimum contrast estimates, *Metrika,* 14 (1969), pp. 249–272.

[169] Pugachev, V.S., Statistical problems of pattern recognition theory, In: *Pattern Recognition,* Nauka, Moscow, (1967) (in Russian).

[170] Randless, R.H., Generalized linear and quadratic discriminant functions using robust estimates, *J. Amer. Statist. Assoc.,* 73 (1978), pp. 564–568.

[171] Rao, C.R., and Varadarajan, V.S., Discrimination of Gaussian processes, *Sankhya,* A.25 (1963), pp. 303–330.

[172] Raudys, Sh., Sample size finiteness in classification problems, In: *Statistical Control Problems,* 18 (1976), pp. 1–180.

[173] Raudys, Sh., and Pikelis, V., On dimensionality, sample size, classification error, and comlexity of classification algorithms in pattern recognition, *IEEE Trans. Pattern. Anal. Machine Intell.,* PAMI-2 (1980), pp. 242–252.

[174] Repin, V.G., and Tartakovskij, G.P., *Statistical Synthesis and Adaptation of Information Systems,* Sovetskoye Radio, Moscow, (1977) (in Russian).

[175] Rey, W.J.J., Robust statistical methods, In: *Lecture Notes in Math.,* 690 (1978), pp. 1–129.

[176] Rieder, H., *Robust Asymptotic Statistics,* Springer-Verlag. New York, (1994).

[177] Roussas, G.G., Extention to Markov processes of a result by A.Wald about the consistency of the maximum likelihood estimate, *Z. Wahrscheinlichkeitstheorie und verw. Geb.,* 4 (1965), pp. 69–73.

[178] Rosenblatt, M., Remarks on some nonparametric estimates of a density function, *Ann. of Math. Statist.,* 27 (1956), pp. 832–837.

[179] Sebestyen, G.S., *Decision-making Process in Pattern Recognition,* Macmillan, New York, (1962).

[180] Senin, A.G., *Recognition of Random Signals,* Nauka, Novosibirsk, (1974) (in Russian).

[181] Shanmugam, K., A parametric procedure for learning with an imperfect teacher, *IEEE Trans. Information Theory,* IT-18 (1972), No.3, pp. 300–302.

[182] Shilov, G.E., *Multivariate Mathematical Analysis,* Nauka, Moscow, (1972) (in Russian).

[183] Shurygin, A.M., The estimators of normal distribution parameters by exponential weighting: asymptotic theory, In: *Algorithms and Software of Applied Statistical Analysis,* Nauka, Moscow, (1980), pp. 241–259 (in Russian).

[184] Silvey, S.D., A note on maximum-likelihood in the case of dependent random variables, *J. Royal Statist. Soc. Ser. B.,* 23 (1961), pp. 444–452.

[185] Smoljak, S.A., and Titarenko, B.P., *Robust methods of estimation,* Statistika, Moscow, (1980) (in Russian).

[186] Snapp, R.R., and Venkatesh, S.S., Asymptotic predictions of the finite-sample risk of k-Nearest-Neighbour classifier, In: *Proc. of the 12th IAPR Int. Conf. on Pattern Recognition,* IEEE Computer Society Press, Washington, pp. 1–6, (1994).

[187] Subrahmaniam, K., and Chinganda, E.F., Robustness of the linear discriminant function to nonnormality: Edgeworth series distribution, *J. Statist. Planning Inference,* 2 (1978), pp. 79–91.

[188] Tiku, M.L., and Balakrishnan, N., Robust classification procedures based on the MML estimators, *Commun. Statist.-Theory Meth.,* 18 (1989), pp. 1047–1066.

[189] Tiku, M.L., Tan, W.Y., and Balakrishnan, N., *Robust Inference,* Marcel Dekker, New York, (1986).

[190] Tong, Y.L., *The Multivariate Normal Distribution* (Springer Series in Statistics), Springer-Verlag, New York, (1990).

[191] Troitskij, E.V., Asymptotic expansion of probability density function for adaptive classifing statistics, *Statistical Problems of Control,* 14 (1976), pp. 11–32.

[192] Tou, J.T., and Gonzalez, R.C., *Pattern Recognition Principles,* Addison-Wesley, London, (1974).

[193] Tukey, J.W., A survey of sampling from contaminated distributions, In: *Constributions to Probability and Statistic,* Stanford Univ. Press, Stanford, pp. 448–485, (1960).

[194] Van Ryzin, J., Nonparametric Bayesian decision procedures for pattern classification, In: *Trans. of 4th Prague Conf. on Inform. Theory,* Academic Press, New York, pp. 479–494, (1965).

[195] Van Ryzin, J. (Ed.), *Classification and Clustering,* Academic Press, New York, (1977).

[196] Vapnik, V.N., *Estimation of Dependencies based on Empirical Data,* Springer-Verlag, Berlin, (1982).

[197] Vapnik, V.N., and Chervonenkis, A.Ya., *Theory of Pattern Recognition,* Nauka, Moscow, (1974) (in Russian).

[198] Vapnik, V.N., and Stefanjuk, A.R., Nonparametric methods of probability density function estimation, *Automation and Remote Control,* 8 (1978), pp. 38–52.

[199] Vasenkova, E.I., About minimax robustness of discriminant functions under distortions described by mixtures of probability distributions, In: *Problems of Computer Data Analysis and Modeling* (Ed. Kharin Yu.S.), Belarussian State University, Minsk, (1991) (in Russian).

[200] Vasiljev, V.I., *Recognition Systems: Handbook,* Naukova Dumka, Kiev, (1983) (in Russian).

[201] Verhagen, C.J.D.M., Progress report on pattern recognition, In: *Reports on Progress in Physics,* 43 (1980), No.6, pp. 785–831.

[202] Vintsjuk, T.K., Words recognition by dynamic programming method, *Kibernetika,* (1968), No. 1, pp. 81–88 (in Russian).

[203] Wald, A., *Sequential Analysis,* John Wiley & Sons, New York, (1947).

[204] Zagoruiko, N.G., *Methods of Pattern Recognition and their Application,* Sovetskoye Radio, Moscow, (1972) (in Russian).

[205] Zhivogljadov, V.P., and Medvedev, A.V., *Nonparametric Algorithms of Adaptation,* Ilim, Bishkek, (1974) (in Russian).

[206] Zhurbenko, I.G., and Kozhevnikova, I.A., Detection of periodical components in sequences of pseudorandom numbers, *Kibernetika,* (1984) No. 4, 89–96.

[207] Zypkin, Ya.Z., *Fundamentals of Learning Systems Theory,* Nauka, Moscow, (1970) (in Russian).

[208] Zypkin, Ya.Z., *Fundamentals of Information Theory of Identification,* Nauka, Moscow, (1984) (in Russian).

Index

MAIN NOTATIONS AND ABBREVIATIONS

DR – decision rule

ADR – adaptive decision rule

BDR – Bayesian decision rule

RDR – robust decision rule

PDR – plug-in decision rule

MC-estimator – minimum contrast estimator

ML-estimator – maximum likelihood estimator

$\blacksquare$ – end of proof

$\{\cdots\}$ – set of parameter sets over all possible combinations of indices, e.g.,

$$\{(a_i, b_{ij})\} = \{(a_1, b_{11}), (a_1, b_{12}), \ldots, (a_2, b_{21}), (a_2, b_{22}), \ldots, \ldots\}$$

R^N – N-dimensional Euclidean space

$|A|$ – cardinality of A if A is a set

$|A|$ – Euclidean norm if A is a vector

$|A|$ – determinant if A is a matrix

$\lfloor z \rfloor$ – integer part of number z (floor function)

$[z]'$ – rounding to the nearest integer

$\mathrm{tr}(A)$ – trace of matrix A

A^T – transposed matrix A

$\mathrm{Arg\ min}_{x \in A} f(x) = \{x : f(x) = \min_{x \in A} f(x)\}$ – the set of values of the argument at which the function $f(\cdot)$ attains its minimum

$\arg\min_{x \in A} f(x)$ – element $x^* \in \mathrm{Arg\ min}_{x \in A} f(x)$ of minimal Euclidean norm

$\delta_{ij} = \begin{cases} 1 \text{ if } j=i \\ 0 \text{ if } j \neq i \end{cases}$ – Kronecker symbol

$\mathbf{1}(z) = \begin{cases} 1 \text{ if } z>0 \\ 0 \text{ if } z \leq 0 \end{cases}$ – unit step function

$\mathbf{I}_A(z) = \begin{cases} 1 \text{ if } z \in A \\ 0 \text{ if } z \notin A \end{cases}$ – indicator of the set A

$\mathbf{I}_N$ – identity $(N \times N)$-matrix

$\mathbf{O}_{M \times N}$ – $(M \times N)$-matrix of zeros

$\mathbf{O}_N$ – N-vector of zeros

$\mathbf{1}_N$ – N-vector of ones

$\mathbf{1}_{M\times N}$ – $(M \times N)$-matrix of ones

$\nabla_x f(x,y)$ – gradient, i.e., the column-vector of partial derivatives of $f(\cdot)$ with respect to $x \in R^N$

$\nabla_x^2 f(x,y)$ – the matrix of second partial derivatives of $f(\cdot)$ with respect to $x \in R^N$

$\mathcal{O}(\epsilon)$ – a function of ϵ such that if $\epsilon \to 0$ then $|\mathcal{O}(\epsilon)/\epsilon| \le$ const $< \infty$

$o(\epsilon)$ – a function of ϵ such that if $\epsilon \to 0$ then $o(\epsilon)/\epsilon \to 0$

L – number of classes

$\Omega = \{\Omega_1, \dots, \Omega_L\}$ – alphabet of classes (patterns)

$S = \{1, 2, \dots, L\}$ – decision space

S^n – Cartesian product of n copies of set S (n-th Cartesian power)

$\mathcal{L}\{\xi\}$ – probability distribution law of random variable ξ

$R[a,b]$ – uniform (rectangular) probability distribution of random variable on $[a,b]$

$\mathcal{N}_m(\mu, \Sigma)$ – normal (Gaussian) probability distribution law of random m-vector with mean μ and covariance matrix Σ

$n_m(x \mid \mu, \Sigma)$ – density of normal distribution $\mathcal{N}_m(\mu, \Sigma)$ at point x

$\text{mes}_N(A)$ – Lebesgue measure of set $A \subset R^N$

$\mathbf{P}\{\cdot\}$ – probability symbol

$\mathbf{D}\{\cdot\}$ – variance symbol

$\mathbf{E}\{\cdot\}$ – expectation symbol

$\mathbf{D}_\theta\{\cdot\}$ – variance under the distribution $\mathbf{P}_\theta$ with parameter θ

$\mathbf{E}_\theta\{\cdot\}$ – expectation under the distribution $\mathbf{P}_\theta$ with parameter θ

$\mathbf{Cov}\{\xi, \eta\}$ – covariance of random variables ξ, η

$\phi(x) = \frac{1}{\sqrt{2\pi}} e^{-x^2/2}$ – probability density function of standard distribution $N_1(0,1)$

$\Phi(x) = \frac{1}{\sqrt{2\pi}} \int_{-\infty}^{x} e^{-t^2/2} dt$ – distribution function for $\mathcal{N}_1(0,1)$

$\mathcal{L}\{\xi_n\} \to \mathcal{L}\{\xi\}$ – symbol of weak distribution convergence as $n \to \infty$

$\xrightarrow{P}$ – symbol of convergence in probability

$\xrightarrow{a.s.}$ – symbol of almost sure convergence (with $\mathbf{P} = 1$)

$\xrightarrow{m.s.}$ – symbol of mean square convergence

Let us agree that if a $(m \times n)$-matrix A_t has subscript t, then this subscript is attached to all elements of the matrix: $A_t = (a_{tij}), i = 1, \dots, m, j = 1, \dots, n$.